Fluid Mechanics and Its Applications

Founding Editor

René Moreau

Volume 127

The purpose of this series is to focus on subjects in which fluid mechanics plays a fundamental role. As well as the more traditional applications of aeronautics, hydraulics, heat and mass transfer etc., books will be published dealing with topics, which are currently in a state of rapid development, such as turbulence, suspensions and multiphase fluids, super and hypersonic flows and numerical modelling techniques. It is a widely held view that it is the interdisciplinary subjects that will receive intense scientific attention, bringing them to the forefront of technological advancement. Fluids have the ability to transport matter and its properties as well as transmit force, therefore fluid mechanics is a subject that is particulary open to cross fertilisation with other sciences and disciplines of engineering. The subject of fluid mechanics will be highly relevant in such domains as chemical, metallurgical, biological and ecological engineering. This series is particularly open to such new multidisciplinary domains. The median level of presentation is the first year graduate student. Some texts are monographs defining the current state of a field; others are accessible to final year undergraduates; but essentially the emphasis is on readability and clarity.

Springer and Professor Thess welcome book ideas from authors. Potential authors who wish to submit a book proposal should contact Dr. Mayra Castro, Senior Editor, Springer Heidelberg, e-mail: mayra.castro@springer.com

Indexed by SCOPUS, EBSCO Discovery Service, OCLC, ProQuest Summon, Google Scholar and SpringerLink

Huilin Lu · Guodong Liu · Qinghong Zhang ·
Xiaoxue Jiang · Boxue Pang · Wenjian Cai

Computational Transport Phenomena of Multiphase Systems and Fluidization

Formulation and Application of Kinetic Theory of Granular Flow

Springer

Huilin Lu
School of Energy Science and Engineering
Harbin Institute of Technology
Harbin, Heilongjiang, China

Guodong Liu
School of Energy Science and Engineering
Harbin Institute of Technology
Harbin, Heilongjiang, China

Qinghong Zhang
School of Transportation and Vehicle Engineering
Shandong University of Technology
Zibo, Shandong, China

Xiaoxue Jiang
School of Petroleum and Natural Gas Engineering
Changzhou University
Changzhou, Jiangsu, China

Boxue Pang
School of Resources and Civil Engineering
Northeastern University
Shenyang, Liaoning, China

Wenjian Cai
Institute of Aerospace Solid Propulsion Technology
China Aerospace Science and Technology Corporation
Xi'an, Shaanxi, China

ISSN 0926-5112 ISSN 2215-0056 (electronic)
Fluid Mechanics and Its Applications
ISBN 978-981-96-0697-9 ISBN 978-981-96-0698-6 (eBook)
https://doi.org/10.1007/978-981-96-0698-6

This Springer imprint is published by the registered company Springer Nature Singapore Pte Ltd.
The registered company address is: 152 Beach Road, #21-01/04 Gateway East, Singapore 189721, Singapore

Preface

Multiphase flow and fluidization occur in many operations in the chemical, petroleum and power generation industries. The majority of chemical engineering unit operations, such as coal and biomass gasification, combustion, and other related waste treatment techniques and resource recovery processes, involve multiphase flow and fluid-solid fluidized bed reactors. The fluid catalytic cracking technology commonly employed in the modem petroleum refineries is also based on fluidization principles. Hence, fluidized beds are ubiquitous in a variety of chemical and physical processing applications. Despite intense research efforts on fundamentals and applications, fluid-solid multiphase flows and fluidized beds still disclose previously unrevealed features and continue to pose challenging issues to researchers and process engineers. Poor solids processing performance caused by particles' impacts also threaten the economic viability of clean combustion of coal and biomass in fluidized bed reactors. Furthermore, the widespread application of transport phenomena of multiphase systems and fluidization process in industry demands improvements in efficiency as well as the development of predictive numerical simulations of industrial processes for design and scale-up purposes. The feasibility of application of computational models is certainly gaining significant traction. This book is intended to address some of these issues, while also providing fundamental multiphase transport and fluidization knowledge for graduate students in energy, chemical, mechanical, biological and environmental engineering, as well as professional scientists and engineers working in areas related to fluid-solid fluidization and multiphase flow systems.

Modelling multiphase flows and fluidization encompasses a wide spectrum of different length scales and a multitude of different computational approaches. For practical multiphase flows and fluid-solid fluidized beds, appropriate models for aptly handling the complexity associated with the state of the different phases are, in general, required. To model fluid–particle flows, both Eulerian–Eulerian and Eulerian–Lagrangian approaches are employed in the subject of multiphase flows and fluidized bed reactors. In the Eulerian–Eulerian two fluid methods with kinetic theory of granular flow (TFM-KTGF), both fluid phase and solid phase are considered as continuum phases. The KTGF, which is based on the oscillation of the particles, uses a granular temperature equation to determine the turbulent kinetic energy of particles, assumes a distribution function for instantaneous particle velocity, and defines a constitutive equation based on particle collision, interaction and fluctuation. With KTGF, granular transport equations for the closure laws for the stress tensor as well as the energy flux are derived. The Eulerian–Lagrangian approach utilizes a combination of computational fluid dynamics and the discrete element method (CFD-DEM) in order to accurately predict the particle–particle and particle–wall interactions. In CFD-DEM approach, each particle contact is calculated by variables such as the amount of overlap and duration, while the model has the advantage of representing multiple particle contacts. The various contact interaction models directly affect the computational cost. The contact interactions of particles can be analyzed in detail.

Because of the increasing importance of the use of computational modelling in multiphase flows and fluid-solid fluidized beds, the aim of this book is to try to present the basic techniques used in the treatment of multiphase flows and fluidization. For flow of fluid-particles suspension in multiphase systems and fluidized beds, the interstitial fluid affects in the KTGF enable more accurate predictions in fluidized beds and other multiphase flow systems. For heterogeneous flow with particle clusters, the criterion for the identification of clusters is, in general, required during computational simulations. The improvement of the prediction capability of TFM-KTGF and CFD-DEM approaches is a challenge due to the disagreement of predictions between them. Hence, this book is aimed at addressing these important issues. The unique emphasis of this book is on the sound formulation of the basic equations describing multiphase transport and fluidization. The clear underlying mathematical and physical bases of the interdisciplinary description of multiphase transport and fluidization are the main themes.

In Chap. 1, a survey of the elements of transport phenomena for fluid and solid phases is given (Qinghong, Guodong, Boxue). This theory serves as basis for the development of models for fluidized beds, chemical engineering processes and other systems involving multiphase flows. It also contains a summary of the basic kinetic theory of granular flow used in TFM-KTGF approach. This classical KTGF has been adopted in describing particulate flows in fluidized bed reactors and other multiphase flows. The theory is primarily used for dense bed reactors, but modified closures of this type have been employed for dilute and dense flows as well. The main purpose of introducing this kinetic theory is to improve the description of the particle transport and distribution in fluidized bed reactors and multiphase flows.

Chapter 2 provides an up-to-date description of the formulation and implementation of combined computational fluid dynamics and discrete element method (CFD-DEM) modeling (Xiaoxue, Qinghong, Guodong). Since DEM can track the motion behavior of each particle individually and CFD can qualitatively and quantitatively describe the fluid evolution process, CFD–DEM method has been realized from small scale to laboratory scale and even extended to large engineering scale. A comprehensive review of the application of coupled CFD–DEM methods is addressed. The detailed analysis is summarized mainly from the DEM modeling theory (particle–fluid interaction and particle-particle interaction) to CFD–DEM coupled mapping methods. The findings are summarized and discussed, and future developments and challenges are highlighted. This work will provide theoretical guidance for subsequent researchers using coupled CFD–DEM methods.

Chapter 3 contains a novel Euler fluid phase-Euler granular phase-Lagrangian discrete particle phase approach based on the kinetic theory of granular flow (KTGF) and discrete element method (DEM) model (Qinghong, Xiaoxue, Wenjian, Huilin). The model is known as a CFD-KTGF-DEM approach. The approach treats a combination of three momentum conservation equations of fluid phase, granular phase and discrete particles. The granular restitution coefficient and specularity coefficient used as an input in KTGF are computed according to the discrete particle restitution coefficient and wall-particle specularity coefficient using DEM. The coupling drag force and collisions are presented among fluid phase, granular phase and discrete particles. The CFD-KTGF-DEM approach makes up a continuum-discrete coupling method and gives a new tool to simulate quantitatively, multiphase flows and fluidized bed reactors from micro-scale interactions to the meso-scale interactions of turbulence and collisions. Parameters of fluid and granular phases and trajectories of discrete particles can be provided. The CFD-KTGF-DEM approach is free from empirical constants in computational simulations.

Chapter 4 provides an introduction to the concept of low-Stokes-number kinetic theory of granular flow (Guodong, Wenjian, Boxue, Xiaoxue). The low-Stokes-number KTGF is proposed according to the dilute and dense kinetic theory. Using Boltzmann equation, particle motion is described from Bhatnagar–Gross–Krook (BGK) operator in order to account for the interactions of interstitial fluid. The interactions of particles with fluid turbulence are explicitly described by a Fokker–Planck diffusion operator, and interparticle collisions are quantified by the Boltzmann operator. With the Chapman–Enskog perturbation method, solutions for the granular constitutive relations are derived, and a generalized constitutive model for particle motion in flows with dilute to dense solids concentrations is developed in terms of Stokes number. The low-Stokes-number KTGF model facilitates characterization of inelastic collisions of particles, particle-fluid interactions and interstitial fluid effect.

Chapter 5 deals with the effects of the presence of particle clusters in fluid-particle multiphase flows and numerical simulations (Xiaoxue, Wenjian, Qinghong, Boxue). One of the major numerical inaccuracies originates from the models used for the calculation of drag force due to the formation of particle clusters in the concentrated particulate phase of gas-particles fluidized beds. The key concepts and the fundamental derivations of different approaches for considering the presence of clusters are introduced. The criterion for the identification of heterogeneous flow with clusters and homogeneous flow without clusters is developed in terms of the minimum rate of energy dissipation (MRED) of the dilute phase and dense phase. The MRED-based drag model is proposed according to a set of transient equations of dilute phase and dense phase. The compromise in the competition of different dominant mechanisms, that is, mechanisms compromise among them to reach a relative overall extremum, is analyzed in gas-solid circulating fluidized bed risers.

Chapter 6 presents six case studies. Case 1 is the CFD modeling and simulation of active and passive regulations of particle clusters in circulating fluidized beds with reference to Dr. Xiaoxue's Ph.D. thesis. The occurrence of particle clusters is controlled and regulated through the manipulation of operating parameters and bed geometrical design in fluidized beds. Both active and passive methods are presented to show the distributions of cluster parameters in circulating fluidized beds. Case 2 presents the CFD-KTGF-DEM simulations in a gas jet fluidized bed and an internally circulating fluidized bed reactors and refers to Dr. Qinghong's Ph.D. thesis. The main feature is the granular collisional parameters obtain from the collisions of Lagrangian discrete particles using DEM. The turbulence effect of discrete particles is acquired from fluid-granular turbulence in gas-solid fluidized beds. Case 3 is the computational simulations of supercritical fluid fluidized beds and refers to Dr. Wenjian's Ph.D. thesis. Supercritical fluids may be alternatives to liquid solvents in chemical reactors. Hydrodynamics of both supercritical water and carbon dioxide fluid and organic supercritical methanol and ethanol fluids are presented in fluidized beds. Case 4 is the CFD modelling and simulation of drilling fluid flow and cuttings transport during well drilling and referred to Dr. Boxue Ph.D. thesis. Drilling fluid and cuttings mixture clearly exhibit non-Newtonian rheological characteristics in that their viscosities depend on the shear rate. Under the action of the drill string rotation and the carrying of the drilling fluid, the suspension, sedimentation, and transportation characteristics of cuttings are presented. Case 5 presents the CFD simulation of molten salt fluid-particle fluidized beds and refers to Dr. Wenjian's Ph.D. thesis. Molten salt fluidized bed reactor has the benefits of maximizing the yield of synthesis gas from gasification of biomass and other carbonaceous feed stock, ensuring effective transfer of heat to the reactants using solar energy. Fluidization of molten salt-particles mixture is presented in conventional fluidized bed and inverse fluidized bed reactors. For Case 6, Dr. Xiaoxue presented simulations of brownout phenomenon of dust particles induced by hovering helicopter from the sand ground. Simulations verify the advantages of TFM-KTGF approach to predict behavior of brownout, suggesting that modeling gas particles' flow might constitute the basis of a novel and interesting project to pursue mitigation of helicopter brownout.

The authors greatly appreciate several individuals who contributed in different ways to the completion of this book. I appreciate Prof. Dimitri Gidaspow at the Illinois Institute of Technology. I became engaged in kinetic theory from 1993 to 1997 in his group. His sudden demise is a great loss and source of sadness. He lives on in our hearts forever! I would like to acknowledge the original contributions of the following academicians/researchers: Prof. Hamid Arastoopour of Illinois Institute of Technology for the development of CFD-KTGF-DEM approach, Dr. Madhava Syamlal of National Energy Technology Laboratory for the development of dynamic cluster structure-dependent drag model in fluidized beds, Prof. Jesse Zhu of Western University for modeling of dynamic cluster structure-dependent drag in circulating fluidized bed risers, Prof. Xiaolong Yin of Eastern Institute of Technology for the development of minimum rate of energy dissipation-based drag model, and Dr. Yuqing Feng of CSIRO Mathematics for the modelling of CFD-KTGF-DEM approach and low-Stokes-number kinetic theory of granular flow in fluidized beds. I would also like to thank and greatly appreciate the National Science Foundation of China for their long-term financial support of this work (No. 51776059 and 91752115).

We express our sincere thanks to Mrs. Zhang Nan, Mr. Praveen Kumar Boominathan, and Springer for the grammatical editing of this book. We are also honored to further express our sincere thanks to all our colleagues who are not included in the list of authors. Their contributions are as important as those of the authors; these can be recognized by scanning the relevant references of each chapter.

Harbin, China Huilin Lu

Contents

Chapter 1
Introduction to Basic Equations and Kinetic Theory of Granular Flow

Abstract This chapter provides an up-to-date description of the formulation, implementation and applications of the Euler-Euler (E-E) two-fluid model (TFM) with mass and momentum exchange between the two phases. The detailed original kinetic theory of granular flow (KTGF) is presented as particles interact within the fluid phase and with the particles or walls. The hybrid models are presented, in which the frictional and collisional effects of particles are combined in a slightly different way, and stand out the granular kinetic theory in the dilute–dense regimes of fluid-particles multiphase flows.

A multiphase flow system consists of a number of single phase regions bounded by moving interfaces. In principle, a multiphase flow model could be formulated in terms of the local instantaneous variables pertaining to each phase and matching boundary conditions at the interfaces. The multiphase modeling framework represents a trade-off between accuracy and computational efforts for fluid-particles multiphase flow and fluidization processes. The multi-fluid models can be used analyzing chemical reactive systems on industrial scales considering both inherent physics and feasible computational efforts. These models have been used determining various quantitative parameters of reactor performance such as velocities, temperature, production rate, conversion and selectivity. The ability to establish such quantitative links between simulations of reactor performance and input and operating variables is essential in optimizing the operating conditions in design and scale-up of fluid–solid reactors.

For flows of fluid and particles mixture processes, the Euler-Euler two-fluid models (TFM), where a relatively large number of particles are considered determining a continuous phase in the control volume formulating the governing model equations, denote the mass and momentum conservation equations as formulated in the Eulerian frame of reference for both the dispersed and continuous phases describing the time-dependent motion. Conservation laws for mass, momentum and energy can be derived using the Reynolds transport theorem [3, 20, 39]. An advantage of the Euler-Euler TFM is that in principle it can be used to compute any multiphase flow regime, provided that an adequate closure relation for the interfacial coupling terms are provided. Besides, when these models are applied to

H. Lu et al., *Computational Transport Phenomena of Multiphase Systems and Fluidization*, Fluid Mechanics and Its Applications 127,
https://doi.org/10.1007/978-981-96-0698-6_1

describe dense dispersed flows, the Euler-Euler TFM approach is less computationally demanding in numerical simulations of the chemical reactors. As the flow in most operating chemical reactors is dense, the TFM combined with kinetic theory of granular flow (KTGF) have been found to represent a trade-off between accuracy and computational efforts for practical applications.

1.1 Basic Methods

For multiphase flow systems, the concept of the volume fraction for phase i, α_i, appears. For a single-phase system, $\alpha_i = 1$ [4, 6]. We need two concepts: conservation laws and constitutive equations. These conservation laws are obtained from the Reynolds transport theorem, in reference to Fig. 1.1 [3, 20, 39].

The point (x^0, y^0, z^0) represents the spatial coordinates of the mass at some fixed time t^0. Then, the spatial coordinates of the mass at any time t are given by:

$$x = x(t, x^0, y^0, z^0), \quad y = y(t, x^0, y^0, z^0)$$
$$\text{and} \quad z = z(t, x^0, y^0, z^0) \tag{1.1}$$

A property $\Im\,(t, \boldsymbol{x})$ per unit volume is defined, where t is time and $\boldsymbol{x}$ is the position vector such that

$$F(t) = \iiint\limits_{V(t)} \Im(t, \boldsymbol{x})dV \tag{1.2}$$

where $F(t)$ represents the system variable quantity that can change with time. The velocity of the phase i is $\boldsymbol{v}_i$. The balance made on $F(t)$ gives the Reynolds transport theorem

$$\frac{d}{dt}\iiint\limits_{V(t)} \Im(t, \boldsymbol{x})dV = \iiint\limits_{V(t)} \left(\frac{\partial \Im}{\partial t} + \nabla \cdot \Im \boldsymbol{v}_i\right)dV \tag{1.3}$$

In multiphase flow, the volume occupied by phase i cannot be occupied by any other phase at the same position in space at the same time. This distinction introduces the concept of the volume fraction α_i of phase i [20].

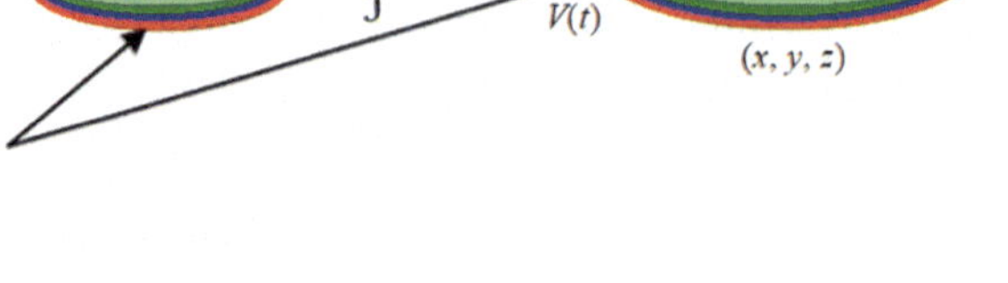

Fig. 1.1 Motion for a system of constant mass

$$V_i = \iiint\limits_{V(t)} \alpha_i dV \tag{1.4}$$

where V_i is the volume of phase i, and V is the element volume in the system. For multicomponent systems the sum of α_i is one, because it is assumed that molecules of component i occupy the same space at the same time.

$$\sum_{i=1}^{n} \alpha_i = 1 \tag{1.5}$$

In multiphase flow the volume fraction for phase i cannot be occupied by the remaining phases at the same position in space at the same time.

1.2 Conservations Laws

The Reynolds transport theorem will be used to derive the multiphase balances of mass, momentum and energy [3, 7].

1.2.1 Mass Balances

The mass of phase i in a volume element V is as follows:

$$m_i = \iiint\limits_{V(t)} \alpha_i \rho_i dV \tag{1.6}$$

where α_i is the volume fraction of phase i and ρ_i is defined as the density of phase i. This volume fraction is the only new variable in multiphase flow.

The balance on mass m_i moving with the phase i velocity $\boldsymbol{v}_i$ is [3, 20]

$$\frac{dm_i}{dt^i} = \frac{d}{dt^i} \iiint \alpha_i \rho_i dV = \iiint m_i' dV \tag{1.7}$$

where $m_i\prime$ is the rate of production of phase i. d/dt^i is the substantial derivative moving with phase i velocity.

The Reynolds transport theorem gives the continuity equation for phase i:

$$\frac{\partial(\alpha_i \rho_i)}{\partial t} + \nabla \cdot (\alpha_i \rho_i \boldsymbol{v}_i) = m_i' \tag{1.8}$$

Conservation of mass requires that

$$\sum_{i=1}^{n} m_i' = 0 \tag{1.9}$$

1.2.2 Momentum Balances

The rate of change of momentum of phase *i* moving with velocity v_i equals the forces acting on the system and can be mathematically expressed by the following [3, 26]:

$$\frac{d}{dt}\iiint\limits_{V(t)} \rho_i \boldsymbol{v}_i \alpha_i dV = \oiint\limits_{S(t)} \boldsymbol{T}_i da + \iiint\limits_{V(t)} \rho_i \boldsymbol{F}_i \alpha_i dV + \bar{p}_i dV + m_i' \boldsymbol{v}_i dV \tag{1.10}$$

The first term on the right-hand side represents the surface forces acting on the differential area *da*, the second term is the external forces, the third term denotes the interaction forces between phases, and the last term represents the force due to the phase change.

Figure 1.2 represents the stress tensor acting on surfaces of the volume of phase *i*. The Reynolds transport theorem is applied to the right-hand side of Eq. (1.10) and the divergence theorem to the first term of Eq. (1.10).

$$\oiint\limits_{S(t)} \boldsymbol{T}_i da = \iiint\limits_{V} \nabla \cdot T_i dV \tag{1.11}$$

The stress tensor $\boldsymbol{T}_i$ for phase *i* is composed of 9 components and can be written as follows.

$$\boldsymbol{T}_i = \begin{pmatrix} T_{ixx} & T_{ixy} & T_{ixz} \\ T_{iyx} & T_{iyy} & T_{iyz} \\ T_{izx} & T_{izy} & T_{izz} \end{pmatrix} \tag{1.12}$$

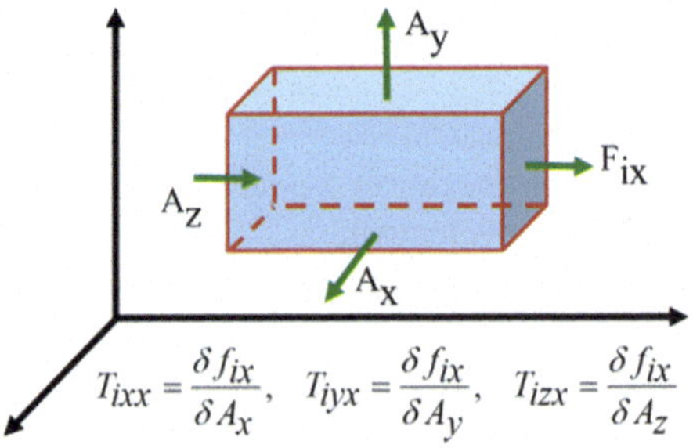

Fig. 1.2 Stress tensor acting on surfaces of volume of phase *i*

From Eq. (1.10), the momentum balance for phase i is [3]:

$$\frac{\partial}{\partial t}(\alpha_i \rho_i) + \nabla \cdot (\alpha_i \rho_i \boldsymbol{v}_i \boldsymbol{v}_i) = \nabla \cdot \boldsymbol{T}_i + \rho_i \alpha_i \boldsymbol{F}_i + \overline{p}_i + m_i' \boldsymbol{v}_i \tag{1.13}$$

Using the continuity Eq. (1.8), the momentum balance for phase i moving with phase velocity v_i becomes

$$\rho_i \alpha_i \frac{d\boldsymbol{v}_i}{dt^i} = \nabla \cdot \boldsymbol{T}_i + \rho_i \alpha_i \boldsymbol{F}_i + \overline{p}_i \tag{1.14}$$

$$\sum\nolimits_{\mathrm{i}}^{n} \overline{p}_i = 0 \tag{1.15}$$

The term on the left-hand side of Eq. (1.14) accounts for acceleration of phase i. The first term on the right-hand side is the gradient of stress due to the surface forces. The second term is the body force, and the last term is the interaction forces between phases, called the momentum supply.

The interaction term between phases is given by the expression [20]

$$\overline{p}_i = \sum\nolimits_j \beta_j (\boldsymbol{v}_j - \boldsymbol{v}_k) \tag{1.16}$$

where β_j represents the friction coefficient that accounts for the drag force between the phases. Note that moving with the velocity $\boldsymbol{v}_i$ there is no force due to the phase change, because the mass in the balance Eq. (1.10) moves with velocity $\boldsymbol{v}_{\mathrm{i}}$.

The simplest expression for the stress $\boldsymbol{T}_i$ is given by

$$\boldsymbol{T}_i = -P_i \boldsymbol{I} \tag{1.17}$$

where $\boldsymbol{I}$ is a unit tensor. Each phase i has its own pressure P_i similar to Euler's equation for single-phase flow.

1.2.2.1 Incompressible Viscous Flow

To meet the requirement of objectivity for each phase k [20]

$$\boldsymbol{T}_k = T_k(\nabla^s \boldsymbol{v}_k) \tag{1.18}$$

where $\nabla^s \boldsymbol{v}_k$ is the symmetrical gradient of velocity $\boldsymbol{v}_k$. Linearization of $\boldsymbol{T}_k$ gives

$$\boldsymbol{T}_k = A_k \boldsymbol{I} + B_k (\nabla^s \boldsymbol{v}_k) \tag{1.19}$$

where the deformation tensor of fluid k is expressed by

$$\nabla^s \boldsymbol{v}_k = \frac{1}{2}[\nabla \boldsymbol{v}_k + (\nabla \boldsymbol{v}_k)^T] \tag{1.20}$$

For incompressible fluids, A_k is chosen to be the negative of the pressure, and the derivative of the traction with respect to the symmetric gradient is the viscosity μ_k of fluid k. The traction for phase k is given by

$$\boldsymbol{T}_k = -P_k \boldsymbol{I} + 2\mu_k \nabla^s \boldsymbol{v}_k \tag{1.21}$$

1.2.2.2 Incompressible Navier–Stokes Equation

Substitution of stress tensors for each phase k into the momentum balance Eq. (1.14) gives the incompressible Navier–Stokes equation for each phase k:

$$\begin{aligned} &\frac{\partial}{\partial t}(\alpha_k \rho_k \boldsymbol{v}_k) + \nabla \cdot (\alpha_k \rho_k \boldsymbol{v}_k \boldsymbol{v}_k) = \\ &-\nabla p_k + \mu_k \nabla^2 \boldsymbol{v}_k + \sum\nolimits_j \beta_j (\boldsymbol{v}_j - \boldsymbol{v}_k) + \rho_k \alpha_k \boldsymbol{F}_k + m_k' \boldsymbol{v}_k \end{aligned} \tag{1.22}$$

The friction coefficients β_j can be expressed in terms of standard drag coefficients [20, 50]. In this model the shear viscosity μ_k of each phase k is an input into the model.

1.2.2.3 Compressible Viscous Flow

For compressible viscous flow there is another parameter, the bulk viscosity λ_k of phase k:

$$A_k = -P_k + \lambda_k \nabla \cdot \boldsymbol{v}_k \tag{1.23}$$

Then the stress for phase k is expressed by

$$\boldsymbol{T}_k = -P_k \boldsymbol{I} + 2\mu_k \nabla^s \boldsymbol{v}_k + \lambda_k \boldsymbol{I} \nabla \cdot \boldsymbol{v}_k \tag{1.24}$$

where, $2\mu_k + 3\lambda_k = 0$, similar to that for single phase flow.

The compressible Navier–Stokes equation for each phase k in substantial derivative form becomes as follows:

$$\begin{aligned} \alpha_k \rho_k \frac{d\boldsymbol{v}_k}{dt^k} &= \alpha_k \rho_k \boldsymbol{F}_k + \sum_j \beta_j (\boldsymbol{v}_j - \boldsymbol{v}_k) \\ &- \nabla P_k + \nabla \left(2\mu_k \nabla^s \boldsymbol{v}_k - \frac{2}{3} \mu_k \boldsymbol{I} \nabla \cdot \boldsymbol{v}_k \right) \end{aligned} \tag{1.25}$$

1.2.3 Energy Balances

The energy balance for an open system with phase change can be written for phase i as follows [21, 38]

$$\frac{dU_i}{dt^i} = \frac{dQ_i}{dt} - p_i \frac{dV_i}{dt^i} + D_{iss} + (U_i + \frac{p_i}{\rho_i}) \frac{dm_i}{dt^i} \tag{1.26}$$

where, U_i is the internal energy per unit mass. Q_i is the heat per unit mass flowing into the system and the only work done by the system is the mechanical work due to the volume changes $P_i\ dV_i$. There is inflow due to phase change, consisting of the energy inflow U_i and pressure P_i. From the internal energy per unit volume χ_i of phase i, the energy dissipation by the system is expressed

$$D_{iss} = \iiint\limits_{V(t)} \chi_i dV \tag{1.27}$$

The rate of heat transfer Q_i is related to the rate of heat transfer by conduction q_i through the following relations:

$$-\frac{dQ_i}{dt} = \oint\!\!\oint\limits_{A(t)} q_i \alpha_i da = \iiint\limits_{V(t)} (\nabla \cdot \alpha_i q_i) dV \tag{1.28}$$

The rate of heat transfer q_i is often expressed according to the enthalpy of phase i. In Eq. (1.26), the rate of heat transfer is also given in terms of heat transfer coefficient.

The Reynolds transport theorem applied to the energy balance Eq. (1.26) produces the energy equation for phase i in the form [3]

$$\begin{aligned} \frac{\partial(\alpha_i \rho_i U_i)}{\partial t} + \nabla \cdot (\alpha_i \rho_i \mathbf{v}_i U_i) &= -\nabla \cdot \alpha_i q_i - P_i \frac{\partial \alpha_i}{\partial t} \\ &\quad - P_i \nabla \cdot \alpha_i \mathbf{v}_i + h_i m_i' + \chi_\mathrm{i} \end{aligned} \tag{1.29}$$

where h_i is the enthalpy flowing into the system and q_i is the rate of heat transfer by conduction.

A set of averaged partial differential equations describing a fluid-particles two-phase flow needs extra equations to close the system. The closure laws are of three types: constitutive, transfer and topological laws. The constitutive laws relate physical properties using experimental data. The transfer laws are empirical equations that describe the interaction between the phases occurring at the interface. The topological laws are needed to describe the spatial distribution of a flow variable. The different types of closure laws are simply referred to as constitutive laws. The key to accurate modeling lies to a large extent in the closure laws used. The considerable number of models has been developed to allow a coarser mesh and a longer time step to be used in numerical simulations, but, it may still be difficult to find suitable closure laws for flow of fluid-particles mixture. For the particulate phase, there are two quite different ways of modeling such flow parameters as dynamic viscosity and pressure of particles. One is to make empirical models based on the particle properties and the local fluid volume fractions. These traditional models are relatively simple, and are thus easy to implement in a computer code. The other way to model these properties is to use the so-called kinetic theory of granular flow. Generally, these kinds of models are more complex and time-consuming to solve, but they are applicable for a wider range of problems. Nevertheless, the closure models become necessary for performance optimization, scale-up and design of industrial scale multiphase flows and fluidized bed reactors.

1.3 Conservation Equations of Fluid–Solid Two-Fluid Models

In a two-fluid framework, the system is considered as a mixture of two fluids, in which the dispersed phase and the continuous phase occupy the same space in a macroscopic sense but occupying different spaces in a microscopic sense. Different phases interpenetrate each other and have their own sizes, velocities, volume fractions and temperatures. With computational fluid dynamics (CFD), fluid phase and solids phase can be modeled by the Euler–Euler approach, which is a way of looking at the motion of fluid and particles from a continuum point of view. Two-fluid models (TFM) combined with kinetic theory of granular flow (KTGF) are applied to model flows of fluid-particles processes. The flow field is divided into a number of small grids to capture motions of both phases, provided that the grid size is larger than the particle size. The two continuous phases are modeled by applying laws of momentum and mass conservations. The TFM-KTGF models have been successfully utilized to obtain the flow behavior of various multiphase flows in laboratory, pilot and industrial scales.

The basic balances for the fluid and solid phases are based on mass and momentum conservation laws. The fluid phase and solid phase continuity equations are as follows [20, 26]:

$$\frac{\partial}{\partial t}(\alpha_f \rho_f) + \nabla \cdot (\alpha_f \rho_f \boldsymbol{u}_f) = 0 \tag{1.30}$$

$$\frac{\partial}{\partial t}(\alpha_s \rho_s) + \nabla \cdot (\alpha_s \rho_s \boldsymbol{u}_s) = 0 \tag{1.31}$$

where ρ is the density. α is the volume fraction. t is the time and $\boldsymbol{u}$ represents the velocity vectors of fluid phase f and solid phase s.

The momentum balance of fluid phase and solid phase is [20, 26]

$$\frac{\partial}{\partial t}(\alpha_f \rho_f \boldsymbol{u}_f) + \nabla \cdot (\alpha_f \rho_f \boldsymbol{u}_f \boldsymbol{u}_f) = -\alpha_f \nabla p + \nabla \cdot \boldsymbol{\tau}_f + \beta(\boldsymbol{u}_s - \boldsymbol{u}_f) + \alpha_f \rho_f \boldsymbol{g} \quad (1.32)$$

$$\frac{\partial}{\partial t}(\alpha_s \rho_s \boldsymbol{u}_s) + \nabla \cdot (\alpha_s \rho_s \boldsymbol{u}_s \boldsymbol{u}_s) = -\alpha_s \nabla p - \nabla p_s + \nabla \cdot \boldsymbol{\tau}_s - \beta(\boldsymbol{u}_s - \boldsymbol{u}_f) + \alpha_s \rho_s \boldsymbol{g} \quad (1.33)$$

where p_s is the solid phase pressure, and p is the fluid pressure. The terms on the left-hand side denote the rate of change of momentum of fluid phase and solid phase. The terms on the right-hand side of Eq. (1.32) represent the summation of fluid pressure force, viscous force, drag force at the phase interface due to two phase interactions and gravity force. The second term on the right-hand side of Eq. (1.33) denotes the solid pressure force. $\boldsymbol{g}$ is the gravity acceleration, and β is the interface momentum exchange coefficient. $\boldsymbol{\tau}_f$ and $\boldsymbol{\tau}_s$ are the fluid and solid phases stress tensors, and they are expressed as follows:

$$\boldsymbol{\tau}_f = \alpha_f \mu_{eff} [\nabla \boldsymbol{u}_f + (\nabla \boldsymbol{u}_f)^{\mathrm{T}}] - \frac{2}{3} \alpha_f \mu_{eff} (\nabla \cdot \boldsymbol{u}_f) \boldsymbol{I} \quad (1.34)$$

$$\boldsymbol{\tau}_s = \alpha_s \mu_s [\nabla \boldsymbol{u}_s + (\nabla \boldsymbol{u}_s)^{\mathrm{T}}] - \alpha_s \left(\mu_b - \frac{2}{3} \mu_s\right)(\nabla \cdot \boldsymbol{u}_s) \boldsymbol{I} \quad (1.35)$$

where μ_{eff} is the effective fluid viscosity. μ_s and μ_b are the granular shear viscosity and bulk viscosity, and they are expressed as the function of granular temperature based on the kinetic theory model [20].

The summation of volume fractions for fluid and solid phases is equal to one

$$\alpha_f + \alpha_s = 1.0 \quad (1.36)$$

For flow of fluid and particles mixture, the drag models are commonly based either on the pressure drop over a fluidized bed or on the drag coefficient on a single particle. Gidaspow drag model [20] is widely applied to model momentum exchange, and the Ergun equation [16] and the equation of [52] are used for the calculation of the interphase momentum exchange coefficient β [27]. For solid volume fractions greater than 0.2, the Ergun correlation is used, whereas for values less than or equal to 0.2, the Wen and Yu expression is employed. The transition at the concentration of 0.2 makes the drag law discontinuous in solid concentration though it is continuous in Reynolds number. Physically, the drag force is a continuous function of both solid concentration and Reynolds number, and therefore the continuous forms of the drag law may be needed to correctly simulate fluid–solid fluidized beds. To avoid discontinuity of the two correlations, a switch function φ is proposed to give a smooth transition from the dilute regime to the dense regime [25].

$$\beta = (1-\varphi)\left[150\frac{\alpha_s^2\mu_f}{\alpha_f(d_p)^2} + 1.75\frac{\rho_f\alpha_s\left|\boldsymbol{u}_f-\boldsymbol{u}_s\right|}{d_p}\right] + \varphi\frac{3}{4}C_D\frac{\rho_f\alpha_f\alpha_s\left|\boldsymbol{u}_f-\boldsymbol{u}_s\right|}{d_p}\alpha_f^{2.65} \tag{1.37}$$

$$\varphi = \frac{\arctan[150\times 1.75(\alpha_f-0.8)]}{\pi} + 0.5 \tag{1.38}$$

$$C_D = \begin{cases} \frac{24}{Re}(1+0.15Re^{0.687}) & Re < 1000 \\ 0.44 & Re \geq 1000 \end{cases} \tag{1.39}$$

The fluid phase effective viscosity μ_{eff} is calculated according to the fluid dynamic viscosity μ_f and fluid turbulent viscosity μ_{ft}. The fluid turbulent viscosity is determined from a two equation turbulence model describing the conservation of turbulence kinetic energy k_f and energy dissipation rate ε_f.

$$\mu_{ft} = C_\mu\rho_f\frac{k_f^2}{\varepsilon_f} \tag{1.40}$$

These indicate that the Boussinesq approximation is valid. This condition states that the characteristic time scale of the fluctuating motion must be much smaller than the time scale of the mean flow. The fluid transport equations of turbulent kinetic energy k_f and turbulent kinetic energy dissipation rate ε_f are expressed according to fluid-particle velocity covariance k_{fs} and the interfacial term Π_f [9, 47]:

$$\frac{\partial}{\partial t}(\alpha_f\rho_f k_f) + \nabla\cdot(\alpha_f\rho_f\boldsymbol{u}_f k_f) = \nabla\cdot(\alpha_f\frac{\mu_{ft}}{\sigma_k}\nabla k_f) + \alpha_f\boldsymbol{G}_k + \boldsymbol{\Pi}_f - \alpha_f\rho_f\varepsilon_f \tag{1.41}$$

$$\frac{\partial}{\partial t}(\alpha_f\rho_f\varepsilon_f) + \nabla\cdot(\alpha_f\rho_f\boldsymbol{u}_f\varepsilon_f) = \nabla\cdot\left(\alpha_f\frac{\mu_{ft}}{\sigma_\varepsilon}\nabla\varepsilon_f\right) + \alpha_f\frac{\varepsilon_f}{k_f}(C_1\boldsymbol{G}_k - C_2\rho_f\varepsilon_f) + C_3\left(\frac{\varepsilon_f}{k_f}\right)\boldsymbol{\Pi}_f \tag{1.42}$$

where C_1, C_2, C_3, $C\mu$, σ_k and σ_ε are empirical constants, and they are 1.44, 1.92, 1.22, 0.09, 1.0 and 1.3, respectively. Keep in mind that these constants have standard for single-phase flow turbulence values. It is reasonable to assume that these values should relate to the type of particles and the type of fluid flow. The third term in the right-hand side of k_f and ε_f equations represents the fluctuation energy interaction term. They represent the kinetic energy necessary to accelerate particles. The interaction fluctuation energy accounts for the transfer of fluctuation energy within the fluid phase and the solid phase. The production of kinetic energy due to fluctuation in drag and the generation of turbulent kinetic energy due to the mean velocity gradients are

$$\boldsymbol{\Pi}_f = \beta(k_{fs} - 2k_f) \tag{1.43}$$

$$\boldsymbol{G}_k = \mu_{ft}\boldsymbol{S}^2 \tag{1.44}$$

where $\boldsymbol{S}$ is the modulus of the mean rate-of-strain tensor, defined as

$$\boldsymbol{S} = \sqrt{2\boldsymbol{S}_{ij}\boldsymbol{S}_{ij}} \tag{1.45}$$

and

$$\boldsymbol{S}_{ij} = \frac{1}{2}\left(\frac{\partial \boldsymbol{u}_{fi}}{\partial \boldsymbol{x}_j} + \frac{\partial \boldsymbol{u}_{fj}}{\partial \boldsymbol{x}_i}\right) \tag{1.46}$$

The fluid-particle velocity covariance k_{fs} represents the efficiency of turbulence to entrain particles, and can be expressed according to the ratio between the Lagrangian integral time scale and the particle relaxation time and the granular temperature θ. A transport equation for the fluid-particle velocity correlation tensor can be derived from a procedure similar to the one used in single-phase flows. On the other hand, [36] has calculated this term for a dilute fluid–solid suspension at low particle Reynolds number where the collisions determine the velocity distribution function of particles. The algebraic expressions for the fluid-particle velocity correlation tensor are based on the kinetic theory applied to a fluid–solid suspension.

$$k_{fs} = \frac{d_p}{4\sqrt{\pi}} \frac{(\boldsymbol{u}_f - \boldsymbol{u}_s)^2}{T_0\sqrt{\theta}} \tag{1.47}$$

$$T_0 = \frac{4d_p\rho_s}{3C_D\rho_f\left|\boldsymbol{u}_f - \boldsymbol{u}_s\right|} \tag{1.48}$$

where T_0 represents the particle relaxation time. This expression is based on the kinetic theory applied to a monodisperse gas–solid suspension under the following assumptions: collisions are elastic and dilute suspensions. Thus, the hypothesis of molecular chaos is valid. For fluid–solid suspensions where these conditions are fulfilled, one does not need to solve a transport equation for the fluid-particle velocity correlation tensor.

An alternative method to account for particle–particle collisions is proposed where interactions of collisions determine the particle velocity distribution function [47]. A model for the fluctuating fluid–solid velocity correlation must be consistent with the modeling of single-phase turbulence when the particle response time tends to zero with respect to the eddy–particle interaction time.

$$k_{fs} = \frac{\eta_r}{1 + \left(1 + \frac{\alpha_s\rho_s}{\alpha_f\rho_f}\right)\eta_r}\left(2k_f + 3\frac{\alpha_s\rho_s}{\alpha_f\rho_f}\theta\right) \tag{1.49}$$

where η_r represents the ratio between the Lagrangian integral time scale and the particle relaxation time, and exhibits the efficiency of turbulence to entrain particles.

$$\eta_r = \frac{3}{2}\frac{k_f}{\varepsilon_f}\frac{\beta C_\mu}{\alpha_s \rho_s \sqrt{1 + C_\beta[3/(2k_f)]|\boldsymbol{u}_f - \boldsymbol{u}_s|}} \tag{1.50}$$

where C_β is the direction-dependent parameter. When applying the two equation k_f-ε_f model presents, one should be aware of the treatment of the interaction term Π_f in Eq. (1.43) in which the relative velocity effect is neglected. It represents the energy necessary to accelerate the particles, or the energy transferred from the particles to the fluid phase if the particles have high fluctuation velocities in a region where the fluid phase turbulent kinetic energy is low. These formulations assume the turbulent production in the wakes behind particles to be in local equilibrium with the viscous dissipation. More detailed models can be suggested by developing the model of transport equations for the fluid–solid velocity correlation.

1.4 Kinetic Theory of Granular Flow

The shear stresses of solid phase depend on particles shear viscosity and bulk viscosity that are modelled using the kinetic theory of granular flow (KTGF). The KTGF is based on the similarities between the flow of a granular material, a population of particles without an interstitial fluid, and the molecules of a gas. This treatment uses classical results from the kinetic theory of gases [10] in order to predict the form of the transport equations of momentum and energy, and the corresponding specific quantities of the mean flux of momentum and energy and the mean rate of energy dissipation, for a granular material [20, 30, 37, 45]. The standard references for the kinetic theory of granular flows consist of the books of [3, 21, 26, 29, 42]. In addition, several reviews of the kinetic theory of granular flow and papers presenting locally averaged equations of motion for a mixture of identical spherical particles might be suggested. Among these reports are [2, 8, 12, 15, 24, 28].

A granular material represents as a collection of a large number of particles in a vacuum. Granular flow might thus be referred to as a flow of a powder in a vacuum. The KTGF models neglect the effect of the interstitial fluid, i.e., the equations are derived considering vacuum conditions. If particles are closely packed or if they are much denser than the interstitial fluid, particles alone, and not the fluid or the fluid-particle interactions, will play the greatest role in momentum transport, in which case the interstitial fluid can be ignored in describing the flow behavior. Flows of granular material are generally taken to fall into this limiting category. On the other hand, for reactive systems like fluidized beds the interstitial fluid is normally of outmost importance because the interstitial fluid contains one or more of the reactants, hence the transport processes within this phase have to be described in an appropriate manner to determine the chemical process operation.

In a granular material there are primarily three mechanisms by which the bulk stresses are generated. These are (i) dry friction, (ii) transport of momentum by particle translation like in dilute gases, and (iii) momentum transport by particle interactions like in dense gases and liquids. It is usually assumed that although all three mechanisms co-exist in most flow regimes, one of them will normally play a predominant role.

For the rapid-flow regime, each particle moves freely instead of moving in many-particle blocks. This flow regime evolves naturally from the quasi-static regime if the motion becomes rapid enough to transfer enough energy to the particles in the vicinity of the local slip regions to break these particles free of their blocks. In this way the rapid flow regime might expand until every particle in the granular mass is moving independently. These flows are of great interest to chemical reactor engineers because the fluidized bed reactors are operated within this regime.

In order to close the conservation equations for the momentum, one needs to calculate the solid phase stress tensors according to the KTGF. These models compute the particulate viscosity and the gradient of the solid pressure as a function of the granular temperature [20, 37]. The granular temperature is a measure of the random particle kinetic energy per unit mass. The granular temperature is defined as the random kinetic energy of particles.

$$\theta = \frac{1}{3} < \boldsymbol{C}^2 >= \frac{1}{3n} \int_{-\propto}^{\propto} \int_{-\propto}^{\propto} \int_{-\propto}^{\propto} (C_x^2 + C_y^2 + C_z^2) dC_x dC_y dC_z \qquad (1.51)$$

where n is the number of particles per unit volume. $\boldsymbol{C}$ is the peculiar velocity or fluctuation velocity of particles, and represents the difference between the instantaneous and the hydrodynamic velocities. In two dimensions, we would have only two random velocities, and we would divide $<\boldsymbol{C}^2>$ by two. In one dimension, we have only one random velocity, and the granular temperature is then simply the variance of the instantaneous velocities.

In kinetic theory of granular flow, it is postulated that the particulate system is represented by a collection of identical, smooth, rigid spheres, adapting a Boltzmann type of equation. The microscopic balance describes the rate of change of the distribution function with respect to position and time. The frequency distribution of velocities of particles f is a function of position $\boldsymbol{r}$, and the instantaneous velocity $\boldsymbol{c}$, as well as time t:

$$f = f(\boldsymbol{r}, \boldsymbol{c}, t) \qquad (1.52)$$

The particulate phase is modelled as a population of identical, smooth and inelastic spheres. A conservation equation can be formulated in terms of the single particle velocity distribution function $f(\boldsymbol{r}, \boldsymbol{c}, t)$ to yield the Maxwell–Boltzmann equation [20, 37]:

$$\frac{\partial f}{\partial t} + \boldsymbol{c} \cdot \nabla f + \nabla \cdot (\boldsymbol{F} f) = \left(\frac{\partial f}{\partial t}\right)_{\text{collision}} \tag{1.53}$$

where the six coordinates, the position $\boldsymbol{r}$ and the instantaneous velocity $\boldsymbol{c}$, are sufficient to determine the location of a particle, because Newton's second law has six integration constants. In general, the external force $\boldsymbol{F}$ includes the gravity, the external pressure gradient, the steady drag force, the added mass force, and several lift forces.

Suppose that ψ and f are expressed as functions of ($\boldsymbol{r}$, $\boldsymbol{c}$, t), then a mathematical moment can be defined by the following ensemble average:

$$< \psi(\boldsymbol{r}, t) >= \frac{1}{n(\boldsymbol{r}, t)} \int \psi(\boldsymbol{r}, \boldsymbol{c}, t) f(\boldsymbol{r}, \boldsymbol{c}, t) d\boldsymbol{c} \tag{1.54}$$

where the moment $<\psi(\boldsymbol{r}, t)>$ denotes a macroscopic quantity.

For the single-particle distribution function $f(\boldsymbol{r}, \boldsymbol{c}, t)$, we take the Maxwellian velocity distribution function as the single-particle distribution [14, 30, 45]

$$f(\boldsymbol{r}, \boldsymbol{c}, t) = \frac{n}{(2\pi\theta)^{3/2}} \exp\left[-\frac{\boldsymbol{C}^2}{2\theta}\right] \tag{1.55}$$

where $\boldsymbol{C}$ represents the fluctuation velocity of particles. For the uniform particles in which density, mean velocity and granular temperature are assigned, there is only one possible mode of distribution of isotropic particle velocities.

In order to deduce explicit expressions for the stress tensor, it is necessary to determine the form of the complete pair distribution function $f^{(2)}$ at collision. It is assumed that the $f^{(2)}$ function can be expressed as a product of a configurational pair-correlation function and two single particle distribution functions. The Enskog assumption for the pair distribution function is expressed [10, 14, 37].

$$f^{(2)}(\boldsymbol{r}_1, \boldsymbol{c}_1; \boldsymbol{r}_2, \boldsymbol{c}_2, t) = g_0 f^{(1)}(\boldsymbol{r}_1, \boldsymbol{c}_1, t) f^{(2)}(\boldsymbol{r}_2, \boldsymbol{c}_2, t) \tag{1.56}$$

where g_0 is the radial distribution function at contact. Assuming that the single-particle velocity distribution is Maxwellian, we can express the complete pair distribution function using Eq. (1.55) as

$$f^{(2)} = g_0 \frac{n_1 n_2}{(2\pi\theta)^{3/2}} \exp\left[-\frac{\boldsymbol{C}_1^2 + \boldsymbol{C}_2^2}{2\theta}\right] \tag{1.57}$$

where n_1 and n_2 are the number of particle 1 and particle 2 per unit volume. $\boldsymbol{C}_1$ and $\boldsymbol{C}_2$ represent the fluctuation velocity of particle 1 and 2. Bear in mind that the distribution function is restricted at equilibrium in which there is no mean deformation so spatial homogeneity of the mean flow prevails. Furthermore, the single distribution function of each particle is approximated by the Maxwellian.

The moment method can be employed to derive a generalized equation of change for a mean particle property $<\psi>$. In particular, the generalized transport equation for $<\psi(\boldsymbol{r}, t)>$ is derived by multiplying Eq. (1.53) by a microscopic quantity $\psi(\boldsymbol{r}, \boldsymbol{c}, t)$ and integrating the resulting relation over the whole velocity space [3, 20].

$$\frac{\partial}{\partial t}(n < \psi >) + \nabla \cdot (n < \boldsymbol{c}\psi >) - n\left(\frac{\partial < \psi >}{\partial t} + < \boldsymbol{c} \cdot \nabla\psi > + < \boldsymbol{F} \cdot \psi >\right)$$
$$= \int \psi \left(\frac{\partial f}{\partial t}\right)_{\text{collision}} \quad (1.58)$$

The term on the right-hand side denotes the collisional rate of change of the macroscopic variable $<\psi>$.

$$\left(\frac{\partial f}{\partial t}\right)_{\text{collision}} = \iint (f^{(2)}\boldsymbol{c}'_{12} \cdot \boldsymbol{k} - f^{(2)}\boldsymbol{c}_{12} \cdot \boldsymbol{k}) d_p^2 d\boldsymbol{k} d\boldsymbol{c}_1 \quad (1.59)$$

where the primes indicate the quantities after particle interaction–collision. $\boldsymbol{k}$ is a unit vector from the center of the first particle to that of the second upon contact. Hence, we see that the Boltzmann equation is an integral–differential equation.

The transport equation for the fluctuating kinetic energy expressed in terms of the granular temperature is achieved by subtracting the equation for the macroscopic kinetic energy from the equation for the sum of the fluctuating and macroscopic energy forms. We let $\psi = (1/2)mc^2$, hence, Eq. (1.58) yields the equation for the solid phase fluctuating energy (granular temperature), and it is expressed as follows [20, 37]:

$$\frac{3}{2}\left[\frac{\partial(\alpha_s \rho_s \theta)}{\partial t} + \nabla \cdot (\alpha_s \rho_s \boldsymbol{u}_s \theta)\right] = (-\nabla p_s \boldsymbol{I} + \boldsymbol{\tau}_s) : \nabla \boldsymbol{u}_s$$
$$- \nabla \cdot (k_s \nabla\theta) - \gamma_s + \beta\left(\sqrt{2k_f}\sqrt{3\theta} - 2k_f\right) \quad (1.60)$$

where p_s is the solid pressure. k_s is the thermal conductivity of particles. The first term on the right-hand side of Eq. (1.60) represents the energy production due to the deformation work, the second term is the energy transfer, the third term is the collisional energy dissipation due to inelastic collisions, and the last term is the net rate of transfer of fluctuation energy between the fluid phase and the solid phase.

1.4.1 Granular Constitutive Equations

In order to solve solid phase stress, the detailed derivation for the constitutive equation of the particle pressure p_s, shear viscosity μ_s and bulk viscosity μ_b as a function of granular temperature can be found elsewhere [14, 37]. By use of the Chapman-Enskog approximate solution method [10], the kinetic solid pressure part can be given by [20, 37]:

$$p_{s,k} = \alpha_s \rho_s \theta \tag{1.61}$$

It shows a relation between the solid pressure and the granular temperature, similar to the ideal equation of state for gases between the gas pressure and the gas temperature [26].

The solid kinetic part of solid stress tensor represents the deviatoric pressure tensor, and is given in terms of restitution coefficient e of particles [20]:

$$\boldsymbol{\sigma}_{kin} = -\frac{2}{(1+e)g_0}\left[1+\frac{4}{5}(1+e)\alpha_s g_0\right]2\mu_{dil}\boldsymbol{S} \tag{1.62}$$

where the variable $\boldsymbol{S}$ is the solid phase rate of strain (deformation) tensor, and is given by:

$$\boldsymbol{S} = \frac{1}{2}\left(\nabla \boldsymbol{u}_s + \nabla^T \boldsymbol{u}_s\right) - \frac{1}{3}(\nabla \boldsymbol{u}_s : \boldsymbol{I})\boldsymbol{I} \tag{1.63}$$

Except for a conversion to granular temperature, the expression for viscosity was given by [10]. The first approximation gives [3, 26]

$$\mu_{dil} = \frac{30\sqrt{2}}{96}\rho_s \alpha_s \min\left|\frac{1}{6\sqrt{2}}\frac{d_p}{\alpha_s}, \frac{1}{6\sqrt{2}}\frac{d_p}{\alpha_{s,min}}\right|\sqrt{\pi\theta} \tag{1.64}$$

where d_p is the diameter of particles. This relation for μ_{dil} becomes zero at $\alpha_s < \alpha_{s,min}$. A typical value, $\alpha_{s,min} = 1 \times 10^{-5}$, is used to limit the dilute flow of fluid-particles suspension.

The radial distribution function, g_0, accounts for the increase of the probability of collisions when the particles become denser. In a dilute flow of particles, where particles chaos can be assumed, the value of radial distribution function g_0 is equal to unity, whereas g_0 tends to infinity when the particles are closely packed, so that motion is almost impossible. The radial distribution function is given by [3, 20, 37]

$$g_0 = \left[1 - \left(\frac{\alpha_s}{\alpha_{s,max}}\right)^{1/3}\right]^{-1} \tag{1.65}$$

where $\alpha_{s,max}$ is the maximum solid volume fraction at dense random packing. As the solid volume fraction approaches the maximum solid volume fraction, the probability of a particle being in contact with a neighbor approaches infinite.

By use of the Chapman-Enskog approximate solution method, the collisional pressure tensor of particles was derived from granular kinetic theory in the kinetic–collisional regime [3, 20, 37]. Detailed derivation for the collisional pressure tensor can be found elsewhere [14, 20, 26]. It consists of two parts, and can be written as:

$$\boldsymbol{p}_c = \boldsymbol{p}_{c1} + \boldsymbol{p}_{c2} \tag{1.66}$$

where the first part is given by:

$$\begin{aligned}\boldsymbol{p}_{c1} &= \frac{2}{5}\alpha_s^2\rho_s g_0(1+e)(2 <\boldsymbol{CC}> + \boldsymbol{C}^2\boldsymbol{I}) \\ &= 2\alpha_s^2\rho_s g_0(1+e)\theta\boldsymbol{I} - \frac{4\mu_{dil}}{(1+e)g_0}\left(1+\frac{4}{5}\alpha_s g_0(1+e)\right)^2\boldsymbol{S}\end{aligned} \tag{1.67}$$

and the second part is:

$$\boldsymbol{p}_{c2} = -\frac{4}{3\sqrt{\pi}}\alpha_s^2\rho_s d_p g_0(1+e)\sqrt{\theta}\left[\frac{6}{5}\boldsymbol{S} + (\nabla\cdot\boldsymbol{u}_s)\boldsymbol{I}\right] \tag{1.68}$$

It is recognized that the first term in the formula for $\boldsymbol{p}_{c1}$ contains the collisional pressure part, defined by:

$$p_{s,c} = 2\alpha_s^2\rho_s g_0(1+e)\theta \tag{1.69}$$

It is proportional to the granular temperature θ and coefficient of restitution e of particles. The effect of the particle collisional pressure-gradient term is to keep the particles apart so that the calculated solid volume fraction does not exceed the maximum solid volume fraction obtainable for a given sphericity and size distribution of the particles. This term represents the particle–particle interaction force which relates to particle coefficient of restitution. The restitution coefficient represents the loss of energy during collision and varies between zero and one. If it is equal to one, there is no energy loss during elastic collision; otherwise, energy is dissipated during inelastic collision.

The total solid pressure tensor contains both the kinetic and collisional contributions. Hence it follows that the total solid pressure p_s is defined as the sum of the kinetic pressure part $p_{s,k}$ and collisional pressure part $p_{s,c}$:

$$p_s = p_{s,k} + p_{s,c} = \alpha_s\rho_s\theta + 2\rho_s(1+e)\alpha_s^2 g_0\theta \tag{1.70}$$

This indicates that the solid pressure is proportional to solid volume fraction at low solid volume fractions, whereas it is proportional to the square of the solid volume fractions at high solid volume fractions.

The collisional deviatoric pressure tensor of particles is given by [20]:

$$\begin{aligned}\boldsymbol{\sigma}_{s,c} = &-\frac{4\mu_{dil}}{(1+e)g_0}\left[1+\frac{4}{5}\alpha_s g_0(1+e)\right]^2\boldsymbol{S} \\ &-\frac{4}{3\sqrt{\pi}}\alpha_s^2\rho_s d_p g_0(1+e)\sqrt{\theta}\left[\frac{6}{5}\boldsymbol{S}+(\nabla\cdot\boldsymbol{u}_s)\boldsymbol{I}\right]\end{aligned} \tag{1.71}$$

The granular shear viscosity μ_s is expressed as a function of granular temperature from Eq. (1.71) [20, 37].

$$\mu_s = \mu_{s,k} + \mu_{s,c} = \frac{2\mu_{dil}}{(1+e)\alpha_s g_0}\left[1 + \frac{4}{5}(1+e)\alpha_s g_0\right]^2 + \frac{4}{5}\alpha_s^2 \rho_s d_p g_0 (1+e)\sqrt{\frac{\theta}{\pi}} \tag{1.72}$$

The first term on the right-hand side denotes the kinetic contribution and dominates in the dilute flow regime. The second term represents the collisional contribution and dominates in the dense flow regime.

The normal component of the collisional deviatoric pressure tensor resembles the bulk viscosity term, and the solid phase bulk viscosity is denoted by [20, 37]

$$\mu_b = \frac{4}{3}\alpha_s^2 \rho_s d_p g_0 (1+e)\sqrt{\frac{\theta}{\pi}}. \tag{1.73}$$

1.4.2 Low-Knudsen-Number GTSH Model

The granular constitutive correlations of kinetic theory developed by [19], referred to as the low-Knudsen-number GTSH model [40], incorporate the effects of the interstitial fluid from three contributions: mean drag, which is associated with the mean particle velocity, thermal drag, which corresponds to the fluctuating particle velocity; and the neighbor effect, a stochastic source of fluctuating particle velocity due to the fluctuations of nearby particles transmitted through the interstitial fluid. Here, the Knudsen number is defined as the ratio of the particles mean free path to a length scale that is the distance over which gradients in the hydrodynamic variables occur. The mean free path is a function of the solid volume fraction, while the length scale associated with the gradients depends on the specific fluid–solid flow system. The low Knudsen number means the mean free path is small compared to the macroscopic scale. The hydrodynamic description is derived by looking for a normal solution to the Enskog equation by means of the Chapman–Enskog method adapted to inelastic collisions [10], assuming the distribution $f(\boldsymbol{r}, \boldsymbol{c}, t)$ has a time dependence. It is also assumed that the instantaneous impulse can be modelled using a Langevin equation, meaning that the change in particle momentum due to neighbour particle effects occurs on time scales much smaller than those associated with drag due to the slip and particle velocity fluctuations. The constitutive equations are proportional to the first-order spatial derivatives in the hydrodynamic fields. Using the Langevin model for the instantaneous fluid phase force on a particle, the granular shear viscosity is expressed according to the kinetic part of the shear viscosity μ_k [18].

$$\mu_s = \mu_k\left[1 + \frac{4}{5}(1+e)\alpha_s g_0\right]^2 + \frac{4}{5}\alpha_s^2 g_0 \rho_s (1+e)\left(1 - \frac{a_2}{16}\right) d_p \sqrt{\frac{\theta}{\pi}} \tag{1.74}$$

$$\mu_k = \frac{[1 - 0.4\alpha_s g_0(1+e)(1-3e)]\rho_s\alpha_s\theta}{\frac{96\alpha_s g_0}{20d_p}\sqrt{\frac{\theta}{\pi}}(1+e)(3-e)\left(1-\frac{7a_2}{16}\right) - 0.5\left(\xi_0 - \frac{\xi}{\theta} - 2\frac{\gamma}{m}\right)} \tag{1.75}$$

where m is the mass of a particle. The collisional cooling rate ξ and γ are associated with the fluctuating solid velocity and particle momentum change caused by neighbour particles. The collisional cooling rate is proportional to $(1 - e^2)$ due to the particle dissipative collision. The zeroth-order cooling rate is

$$\xi_0 = \frac{8}{d_p}\alpha_s g_0(1-e^2)\left(1+\frac{3a_2}{16}\right)\sqrt{\frac{\theta}{\pi}} \tag{1.76}$$

The cooling rate ξ describes how granular temperature generated by the neighbour effect, and is expressed as the function of granular temperature θ and slip velocity $(u_s - u_f)$ between the solid phase and the fluid phase.

$$\xi = \frac{d_0}{6\sqrt{\pi}}\left(\frac{3\pi\mu_f d_p}{m}\right)^2 \frac{[10\alpha_s(1-\alpha_s)^{-2} + (1-\alpha_s)^2(1+1.5\sqrt{\alpha_s})]^2}{g_0(1+3.5\sqrt{\alpha_s}+5.9\alpha_s)} \frac{\left|\boldsymbol{u}_s - \boldsymbol{u}_f\right|^2}{\sqrt{\theta}} \tag{1.77}$$

where d_0 is the dimensionality of the system (i.e., $d_0 = 2$ for disks and $d_0 = 3$ for sphere particles). γ denotes the viscous energy dissipation in a suspension with a Maxwellian particle velocity. It represents the thermal drag which leads to a dissipation of granular temperature, and is expressed by Reynolds number Re_t.

$$\gamma = 3\pi\mu_f d_p\left[1 + 3\sqrt{\frac{\alpha_s}{2}} + \frac{135\ln\alpha_s}{64\alpha_s^{-1}} + \frac{11.26\alpha_s}{(1-5.1\alpha_s+16.6\alpha_s^2-21.8\alpha_s^3)^{-1}} - \frac{\ln 0.01}{(\alpha_s g_0)^{-1}} + \frac{Re_t\sqrt{0.3\alpha_s}}{(1-\alpha_s)^{3.6}}\right] \tag{1.78}$$

This term is proportional to fluid kinetic viscosity μ_f. The thermal Reynolds number associates with the particle velocity fluctuation, and is expressed as the function of particle mass and granular temperature.

$$Re_t = \frac{\rho_f d_p}{\mu_f}\sqrt{\frac{\theta}{m}} \tag{1.79}$$

The solid phase bulk viscosity has a collisional contribution, and is expressed as the function of coefficient of restitution and coefficient a_2.

$$\mu_b = \frac{4}{3}\alpha_s^2 g_0\rho_s(1+e)(1-\frac{a_2}{16})d_p\sqrt{\frac{\theta}{\pi}} \tag{1.80}$$

The coefficient a_2 is proportional to the fourth cumulant of the isotropic velocity function of particles, and follows a decay function of restitution coefficient of particles. It can be shown as

$$a_2 = \frac{(0.5 - e^2)\sqrt{2\pi} g_0 (1 - e^2)}{\sqrt{2\pi} g_0 \left[\frac{3}{32}(69 + 10e^2)(1 - e^2) + 2(1 - e) - \frac{95}{16}(1 - e^2) \right] + \frac{25}{2} \frac{\pi d_p \xi}{\alpha_s \theta \sqrt{72\theta}}} \tag{1.81}$$

where the radial distribution function at contact is given according to the maximum solid volume fraction (i.e., $\alpha_{s,max} = 0.64356$).

$$g_0 = \frac{1 + 2.5\alpha_s + 4.5094\alpha_s^2 + 4.51544\alpha_s^3}{[1 - (\alpha_s/\alpha_{s,max})^3]^{0.67802}} \tag{1.82}$$

From the low-Knudsen-number GTSH model, the fluid phase effects appear explicitly via the appearance of parameter γ and ξ, and implicitly via the appearance of the kurtosis a_2. Both shear viscosity and bulk viscosity have been generalized to granular flow by taking into account fluid hydrodynamics through the slip velocity between the fluid and solid phases. It is clear that the presence of the fluid phase lowers the solid phase shear and bulk viscosities. GTSH model solves the kinetic equation for systems away from equilibrium, and captures not only the energy non-equipartition but also the flow behavior for a wide range of restitution coefficients.

For a more detailed derivation and explanation for GTSH constitutive relation expressions, see [19, 40].

1.4.3 MFIX KT Model

The effect of the fluid phase on the particle–particle collisions was studied to modify the solid phase constitutive correlations, referred to as MFIX KT model [5, 40, 51]. The interstitial fluid dissipates a part of the particle fluctuating energy (i.e., granular temperature), indicating that interstitial fluid effects on the kinetic–collisional solids stresses and pseudo-thermal energy flux. The momentum and energy of neighboring particles are imparted to a particle through the fluid medium. The fluid-particle interaction force is expressed using a momentum exchange coefficient β and the slip velocity between the fluid and solid phases. Accounting for the role of the fluid phase contribution on the interactions of collision of particles, the solid phase shear viscosity expression is expressed using the term μ_{dil} [1, 5, 23].

$$\mu_s = \frac{(1 + a_0)}{3} \frac{\mu_{dil}}{g_0 \eta (2 - \eta)} \left(1 + \frac{8}{5} \alpha_s \eta g_0 \right)$$

$$\left(1 + \frac{8}{5}\eta(3\eta - 2)\alpha_s g_0 + \frac{48}{15\pi}\eta g_0 \rho_s d_p \alpha_s^2 \sqrt{\pi\theta}\right) \tag{1.83}$$

where η is the coefficient, and it relates to restitution coefficient of particles.

$$\eta = \frac{1+e}{2} \tag{1.84}$$

To account for the presence of an interstitial fluid, it is necessary to consider that particles lose energy during collisions by displacing the fluid. The dilute solid viscosity incorporates the momentum exchange term to account for the effects of the interstitial fluid.

$$\mu_{dil} = \frac{5\alpha_s g_0 \theta}{96\alpha_s g_0 \theta + 10\beta d_p \sqrt{\pi\theta}} \rho_s d_p \sqrt{\pi\theta} \tag{1.85}$$

These indicate that the solid phase shear viscosity takes into account the influence of the fluid phase on the constitutive relations through the momentum exchange coefficient. Additionally, a factor $(1 + a_0)/3$ is added to the solid shear viscosity, which was introduced by [31], where a_0 is a constant of order unity. The value of a_0 must tend to unity when the solid volume fraction approaches zero. A typical value of a_0 is 1.6 [31].

For granular flows at low solid volume fractions, the solid phase stresses arise because of collisional and translational transfer of momentum at the viscous or rapidly shearing regime. The solid phase shear viscosity proposed by [51] is given by

$$\mu_s = \mu_{s,k} + \mu_{s,c} = \frac{d_p \rho_s \sqrt{\pi\theta}}{6(3-e)}\left[1 + \frac{2}{5}(1+e)(3e-1)\alpha_s g_0\right] + \frac{4}{5}\alpha_s^2 \rho_s d_p g_0 (1+e)\sqrt{\frac{\theta}{\pi}} \tag{1.86}$$

The solid phase bulk viscosity depends on solid volume fraction and granular temperature, and is not influenced by the coefficient of restitution of particles provided that the quantity $(1 - e)$ is small [1, 5].

$$\mu_b = \frac{8}{3\pi} g_0 \rho_s d_p \alpha_s^2 \sqrt{\pi\theta} \tag{1.87}$$

For a detailed explanation and derivation of the MFIX KT constitutive equations, see [5, 40, 51].

1.4.3.1 Cao-Ahmadi KT Model

A mass-weighted averaging technique was used to establish a thermomechanical formulation for turbulent multiphase flows [9], and is also referred to as AHMADI model [40]. The formulation includes separate transport equations for the fluctuation kinetic energies of the solid and fluid phases,it also includes the solid phase collisional stresses. The solid pressure is given as:

$$p_s = \alpha_s \rho_s \theta [1 + 4\alpha_s g_0 + 0.5(1 - e^2)] \tag{1.88}$$

The first and third parts on the right-hand side are proportional to solid volume fraction, and the second part is proportional to the square of the solids volume fraction.

The solid shear viscosity and bulk viscosity are reduced due to a thermodynamically consistent turbulence of fluid–solid flows. The solid shear viscosity and bulk viscosity are given by the following equations.

$$\mu_s = \frac{\alpha_s \mu_f}{[1 - (\alpha_s/\alpha_{s,m})]^{2.5\alpha_{s,m}}} + \frac{0.1045\left[(\alpha_s g_0)^{-1} + 3.2\alpha_s + 12.1824 g_0 \alpha_s^2\right] d_p \rho_s \sqrt{\theta}}{1 + \tau_{ts}(1 - \alpha_s/\alpha_{s,m})} \tag{1.89}$$

$$\mu_b = \frac{5}{3} \frac{0.1045\left(12.1827 g_0 \alpha_s^2\right) d_p \rho_s \sqrt{\theta}}{1 + \tau_{ts}(1 - \alpha_s/\alpha_{s,m})^3} \tag{1.90}$$

where $\alpha_{s,m}$ is 0.644. τ_{ts} represents a dimensionless time scale that is the ratio of the time scale of turbulent eddies to the particle relaxation time scale. The former time scale relates to fluid kinetic energy and dissipation, and the latter depends on momentum exchange coefficient.

$$\tau_{ts} = \left(\frac{3}{2} \frac{C_\mu k_f}{\varepsilon_f}\right)\left(\frac{\beta}{\alpha_s \rho_s}\right) \tag{1.91}$$

The first term on the right-hand side of Eq. (1.89) is similar to those suggested by [48], and is consistent with Einstein's equation for effective viscosity of dilute suspensions. For the solid phase, as a stress model, Cao-Ahmadi KT model is considered as a modification of the fluid phase two-equation turbulence $k_f - \varepsilon_f$ model.

A detailed derivation and explanation for constitutive expressions can be found in [9, 40].

1.4.4 Granular Conductivity and Energy Dissipations

A similar derivation can be made for the granular kinetic energy flux vectors [20]. The total granular energy fluctuating flux is the sum of the kinetic flux q_k and the collisional flux q_c, and expressed by granular conductivity k_s.

$$q_s = q_k + q_c = k_s \nabla\theta \tag{1.92}$$

The granular conductivity consists of the kinetic part due to the elastic particles derived from dilute kinetic theory of gases [10] and the collisional part due to the inelastic collision of particles [3, 20, 37]:

$$\begin{aligned} k_s = & \frac{150\rho_s d_p\sqrt{\pi\theta}}{384(1+e)g_0}\left[1 + \frac{6}{5}\alpha_s g_0(1+e)\right]^2 \\ & + 2\rho_s\alpha_s^2 d_p(1+e)g_0\sqrt{\frac{\theta}{\pi}} \end{aligned} \tag{1.93}$$

The first term on the right-hand side of the conductivity closure denotes the kinetic contribution and dominates in the dilute regime. The second term denotes the collisional contribution and dominates in the dense flow regime.

From the low-Knudsen-number GTSH model, the granular conductivity consists of a kinetic part and collisional contribution, and is given as the function of dilute conductivity k_{dil} [19].

$$k_s = k_{dil}\left[1 + \frac{6}{5}\alpha_s g_0(1+e)\right] + k_0\left[\frac{256}{25}\alpha_s^2 g_0(1+e)\left(1 + \frac{7a_2}{16}\right)\right] \tag{1.94}$$

$$k_{dil} = \frac{64\alpha_s}{5\sqrt{\pi}d_p}\frac{k_0\theta^{3/2}[1 + 2a_2 + 0.6\alpha_s g_0(1+e)^2(2e - 1 + a_2(1+e))]}{k_1 + \frac{\xi}{2\theta} - 2\xi_0 - \frac{2\theta}{m}\frac{\partial\gamma}{\partial\theta} + \frac{\partial\xi}{\partial\theta}} \tag{1.95}$$

We see that the dilute conductivity correlates to parameters γ and ξ and their derivatives with respect to the granular temperature. The coefficients of k_0 and k_1 are given by

$$k_0 = \frac{25}{128}\rho_s d_p\sqrt{\pi\theta} \tag{1.96}$$

$$k_1 = \frac{32\alpha_s g_0(1+e)\theta^{1.5}}{5\sqrt{\pi}d_p}\left[1 + \frac{33}{16}(1-e) + \frac{a_2(947 - 579e)}{256}\right] \tag{1.97}$$

From MFIX KT model, an expression of the solid phase conductivity is expressed using the dilute granular conductivity k_{dil}. To account for the interstitial fluid effects, similar to the solid viscosity, the momentum exchange term is incorporated that relates to fluid-particles momentum transfer coefficient [1, 5].

$$k_s = \frac{k_{dil}}{g_0}\left[(1+\frac{12}{5}\eta\alpha_s g_0)(1+\frac{12}{5}\eta^2(4\eta-3)\alpha_s g_0) + \frac{64(41-33\eta)}{25\pi}\eta^2\alpha_s^2 g_0^2\right] \tag{1.98}$$

$$k_{dil} = \frac{75(\alpha_s\rho_s)^2 g_0 d_p\theta\sqrt{\pi\theta}}{48\eta(41-33\eta)\alpha_s^2\rho_s g_0\theta + 90\beta d_p\sqrt{\pi\theta}} \tag{1.99}$$

When the shear viscosity model in Eq. (1.86), proposed by [51], is adopted, an alternative diffusion coefficient for granular energy used in MFIX KT model is given by [40]

$$k_s = \frac{15 d_p\rho_s\alpha_s\sqrt{\pi\theta}}{4(41-33\eta)}\left[1+\frac{12}{5}\eta^2(4\eta-3)\alpha_s g_0 + \frac{16(41-33\eta)}{15\pi}\eta\alpha_s g_0\right] \tag{1.100}$$

From Cao-Ahmadi KT model [9], an equation of the solid phase conductivity is give as follows:

$$k_s = 0.1306 d_p\rho_s(1+e^2)\left(\frac{1}{g_0}+4.8\alpha_s+12.1184 g_0\alpha_s^2\right)\sqrt{\theta} \tag{1.101}$$

Recall that the fluid phase effect appears in the solid shear viscosity in Eq. (1.83) and granular conductivity in Eq. (1.98) via the appearance of momentum transfer coefficient in the dilute shear viscosity μ_{dil} and conductivity k_{dil}. Hence, it is worthwhile to consider the effect of fluid hydrodynamic associates with drag due to the mean slip velocity and particle velocity fluctuations on solid phase closure relations of the KTGF.

The dissipation term in the granular temperature equation assumes that the interaction force between fluid and particles is due to drag only. Hence, the energy dissipation due to inelastic collision of particles is expressed as a function of granular temperature [3, 20, 37].

$$\gamma_s = 3(1-e^2)g_0\rho_s\alpha_s^2\theta\left(\frac{4}{d_p}\sqrt{\frac{\theta}{\pi}} - \nabla\cdot\boldsymbol{u}_s\right) \tag{1.102}$$

From Eqs. (1.72) and (1.102), both shear viscosity and energy dissipation relate to granular temperature. The former is proportional to $\theta^{1/2}$, while the latter is proportional to $\theta^{3/2}$. These indicate that the energy dissipation of collisions has a stronger dependence on granular temperature than does the shear stress. Therefore, when the solid volume fraction is decreased, the granular temperature increases to provide the dissipation required for the energy balance. Thus, the increase in granular temperature causes the increase of shear viscosity of particles.

When a pair of particles approach each other within a distance comparable to the mean free path of the gas molecules, the force resulting from the viscous stresses is appeared around the particle. This results in energy dissipation due to the viscous forces on the particles. The viscous dissipation of granular energy is the process by which the granular mechanical energy is converted into heat. The rate of dissipation of granular kinetic energy by viscous damping is modelled using a model proposed by [33]. The dissipation of granular energy via viscous interaction of the fluid phase is

$$J_{vis} = -54\frac{\alpha_s \mu_{eff} \theta R_{dis}}{d_p^2} \tag{1.103}$$

$$\begin{aligned} R_{dis} = {} & 1 + 3\sqrt{\frac{\alpha_s}{2}} + \frac{135}{64}\alpha_s \ln \alpha_s \\ & + \frac{11.26\alpha_s}{(1 - 5.1\alpha_s + 16.57\alpha_s^2 - 21.77\alpha_s^3)^{-1}} - \alpha_s g_0 \ln(0.01) \end{aligned} \tag{1.104}$$

where μ_{eff} is the effective fluid viscosity. From Eq. (1.102) and Eq. (1.103), these indicate that the granular kinetic energy is dissipated through inelastic collisions of particles and viscous damping.

In fluid–solid flows, interactions between particles and the interstitial fluid can generate granular temperature. The production of granular energy through slip between the fluid phase and the solid phase is [33]

$$J_{pro} = \frac{81\alpha_s \mu_{eff}^2 \left|\boldsymbol{u}_f - \boldsymbol{u}_s\right|^2}{g_0 d_p^3 \rho_f \sqrt{\pi\theta}} \frac{R_{pro}}{1 + 3.5\sqrt{\alpha_s} + 5.9\alpha_s} \tag{1.105}$$

$$R_{pro} = \begin{cases} \frac{1+3\sqrt{0.5\alpha_s}+2.109\alpha_s \ln \alpha_s + 17.14\alpha_s}{1+0.681\alpha_s - 8.48\alpha_s^2 + 8.16\alpha_s^3} & \alpha_s < 0.4 \\ 10\alpha_s/(1-\alpha_s)^3 + 0.7 & \alpha_s \geq 0.4 \end{cases} \tag{1.106}$$

From Eqs. (1.60) and (1.105), it can be seen that the rate of production of granular kinetic energy is through the granular shear and fluid–particle slip, respectively. Generally, both the rate of dissipation of viscous damping from Eq. (1.103) and the rate of production of fluid–particle slip from Eq. (1.105) are taken into account in Eq. (1.60) of TFM-KTGF model for simulations of fluid-particles multiphase flows.

A detailed derivation and discussion of the models for the production and dissipation of granular energy can be found in [33, 40].

1.4.5 Algebraic Granular Energy Equation

From the kinetic theory of granular flow, it can be seen that the solid phase kinetic energy of mean flow first degrades into the kinetic energy of random particle fluctuations, and then dissipates as heat because of inelastic collisions. The solid phase kinetic energy of fluctuations is accounted for in the kinetic theory using a granular temperature θ. From the orders of magnitude analysis, the terms accounting for momentum transfer due to particle translation (kinetic contribution) are discarded because they make the granular temperature unbounded in the dilute limit of gas volume fractions going to one. From Eq. (1.60), an algebraic expression for granular temperature was proposed by [51] on the basis of the energy equation of [37], by assuming that the granular energy is dissipated locally, neglecting the convection and diffusion contributions; and retaining only the generation and dissipation terms. The resulting algebraic granular energy equation is

$$\theta = \left\{ \frac{-K_1\alpha_s D_{ii} + \sqrt{K_1^2 D_{ii}^2 \alpha_s^2 + 4K_4\alpha_s[K_2 D_{ii}^2 + 2K_3(D_{ij}D_{ij})]}}{2\alpha_s K_4} \right\}^2 \tag{1.107}$$

where D_{ii} is the solid phase rate of strain tensor. K_i is the granular stress constants, and they are given by

$$K_1 = 2(1+e)\rho_s g_0 \tag{1.108}$$

$$K_2 = \frac{4d_p\rho_s(1+e)\alpha_s g_0}{3\sqrt{\pi}} - \frac{2}{3}K_3 \tag{1.109}$$

$$K_3 = \frac{d_p\rho_s}{2}\left\{ \frac{\sqrt{\pi}[1+0.4(1+e)(3e-1)\alpha_s g_0]}{3(3-e)} + \frac{8\alpha_s g_0(1+e)}{5\sqrt{\pi}} \right\} \tag{1.110}$$

$$K_4 = \frac{12(1-e^2)\rho_s g_0}{\sqrt{\pi}d_p} \tag{1.111}$$

These constants relate to solid volume fraction, particle diameter and coefficient of restitution. From simulations of gas–solid fluidized beds, it indicates that this simplification of algebraic granular energy equation does not lead to significantly different results, but it does reduce the computational effort of the simulations comparing to the applications of the granular temperature partial differential equation.

A detailed explanation of the algebraic granular energy equation can be found in two papers by [42, 51].

1.5 Hybrid Theories of Granular Flow

It is widely believed that the KTGF can apply to flows of particles interacting through binary and instantaneous collisions. Many industrially important fluid–solid systems often include slow and dense solid flows, which are dominated by sustained frictional contacts between particles. The interactions of particles can be thought of neither as permanent nor as instantaneous. The interactions of particles are a combination of the effect of short-term collisions and the sustained contact between particles. For simulations of these dense flows, in addition to the kinetic and collisional stresses, the model should account for the frictional stresses. The most common approach to account for the effect of frictional stresses is the kinetic-frictional model based on the addition of stresses from the two limiting regimes at a critical solid volume fraction α_{cr} [31, 44, 51].

$$\boldsymbol{\tau}_s = (\boldsymbol{\tau}_k + \boldsymbol{\tau}_c) + \boldsymbol{\tau}_{fr} \qquad \alpha_s \geq \alpha_{cr} \tag{1.112}$$

This approach is based on the assumptions that the solid stress comes from the kinetic and collisional stresses ($\tau_k + \tau_c$), and frictional contributions τ_{fr} in an additive manner, where the frictional contributions appear only at higher solid volume fractions (i.e., greater than α_{cr}). Although, this approach lacks a strong physical justification and the hypothetical assumption of the critical solid volume fraction remains without experimental proof, the theory has been shown to capture the qualitative features of slow dense solid flows [49]. The frictional stress model accounts for compressibility of dense granular flows according to solid phase friction pressure p_{fr} and viscosity μ_{fr}.

$$\boldsymbol{\tau}_{fr} = -p_{fr}\boldsymbol{I} + 2\mu_{fr}\boldsymbol{S} \tag{1.113}$$

From Eq. (1.112), the constitutive equation for the solids pressure p_s consists of three sources, namely, kinetic, collision, and friction, which could be written in an additive manner as

$$p_s = (p_{s,k} + p_{s,c}) + p_{fr} \qquad \alpha_s \geq \alpha_{cr} \tag{1.114}$$

where the critical solid volume fraction α_{cr} is probably taken to be 0.5.

The constitutive equation for the solids shear viscosity μ_s consists of kinetic part $\mu_{s,k}$, collisional part $\mu_{s,c}$ and frictional part μ_{fr}.

$$\mu_s = (\mu_{s,k} + \mu_{s,c}) + \mu_{fr} \qquad \alpha_s \geq \alpha_{cr}. \tag{1.115}$$

1.5.1 Schaeffer, Johnson-Jackson and Syamlal Frictional Models

When particles are close together and deformation of the assembly is slow, contacts are sustained for long times. These forces consist of normal reactions and the associated tangential forces due to friction. [46] derived the expression for the frictional viscosity by assuming the system to be perfectly rigid-plastic, incompressible, non-cohesive, Coulomb powder with a yield surface of von Mises type, and the eigenvectors of the strain rate and stress tensors are parallel, as

$$\mu_{fr} = \frac{p_{fr}\sqrt{2}\sin\varphi}{2\sqrt{\boldsymbol{S}:\boldsymbol{S}}} \tag{1.116}$$

where φ is the angle of internal friction of particles. The model equation expresses the shear stress in relation to the solid phase normal stress.

Johnson and Jackson [31] proposed a critical state solid frictional pressure that allows for a slight compressibility with very limited particle concentration change. The Johnson-Jackson correlation for frictional pressure can be written as [31, 42]

$$p_{fr} = Fr\frac{(\alpha_s - \alpha_{min})^q}{(\alpha_{max} - \alpha_s)^p} \qquad \alpha_{min} \le \alpha_s \le \alpha_{max} \tag{1.117}$$

where the empirical constants q and p are to be 2.0 and 3.0. Fr is a coefficient with different values reported in the literature from 0.05 to 5 [31]. The coefficient Fr is modified by means of a function of the volume fraction as $Fr = 0.1\alpha_s$ while limiting the solid volume fraction to values less than 0.629 to prevent divergence.

Syamlal et al. [51] proposed that the frictional stresses need to be considered only for regions where the solid volume fraction is higher than the maximum packing limit $\alpha_{s,max}$.

$$p_{fr} = 10^{25}(\alpha_s - \alpha_{max})^{10} \qquad \alpha_{min} \le \alpha_s \le \alpha_{max} \tag{1.118}$$

where α_{min} and α_{max} are constants and they are 0.5 and 0.63, respectively. It can be seen that the frictional contribution for the frictional solid pressure and viscosity is dominant above α_{min}. The frictional model indicates that the normal stress includes solid frictional viscosity as a function of the internal angle of friction (set to $\varphi = 28.5°$).

The plastic potential theory can describe the phenomenon of particles dilatancy with the critical state approach. The dilatancy characteristics of particles are independence between the rate-of-strain tensor and the stress tensor. Assuming a slightly compressible and perfectly rigid-plastic system, the expression for the frictional bulk viscosity of particles may be written

$$\mu_{b,fr} = \frac{p_{fr}}{\sqrt{4\sin^2\varphi\Pi_{2D} + (\nabla\cdot\boldsymbol{u}_s)^2}} \qquad \alpha_s \ge \alpha_{min} \tag{1.119}$$

where Π_{2D} is the second invariant of the deviatoric stress tensor, and is expressed according to the rate of strain tensor D_{ij} of particles.

$$\begin{aligned}\Pi_{2D} &= D_{12}^2 + D_{23}^2 + D_{31}^2 \\ &\quad + \frac{1}{6}\left[(D_{11} - D_{22})^2 + (D_{22} - D_{33})^2 + (D_{33} - D_{11})^2\right]\end{aligned} \tag{1.120}$$

$$D_{ij} = \frac{1}{2}\left(\frac{\partial \boldsymbol{u}_{si}}{\partial \boldsymbol{x}_j} + \frac{\partial \boldsymbol{u}_{sj}}{\partial \boldsymbol{x}_i}\right) \tag{1.121}$$

As a consequence, the frictional shear viscosity and bulk viscosity are assumed to be proportional to the solid phase frictional pressure from [31, 51] correlations. For a detailed derivation of the solid phase friction models, see [31, 40, 46, 51].

1.5.2 Srivastava-Sundaresan Frictional Model

The individual particles interact with multiple neighbors through sustained contact at high solid volume fractions. Under such conditions, the normal forces and the associated tangential frictional forces at these sliding contacts are dominant. Based on the flow behavior of particles in the quasi-static regime, the frictional stresses manifest an order-zero dependence on the rate of strain. Srivastava and Sundaresan [49] suggested a value of sinφ for the dilation branch, to ensure that the granular assembly is not required to sustain tensile stress on the yield surface.

$$\mu_{fr} = \frac{p_{fr}\sqrt{2}\sin\varphi}{2\sqrt{\boldsymbol{S}:\boldsymbol{S}+\theta/d_p^2}}\left[n-(n-1)\left(\frac{p_{fr}}{p_{cr}}\right)^{1/(n-1)}\right] \tag{1.122}$$

$$\frac{p_{fr}}{p_{cr}} = \left(1-\frac{\nabla\cdot\boldsymbol{u}_s}{n\sqrt{2}\sin\varphi\sqrt{\boldsymbol{S}:\boldsymbol{S}+\theta/d_p^2}}\right)^{n-1} \tag{1.123}$$

$$n = \begin{cases} \frac{\sqrt{3}}{2\sin\varphi} & \nabla\cdot\boldsymbol{u}_s \ge 0 \\ 1.03 & \nabla\cdot\boldsymbol{u}_s < 0 \end{cases} \tag{1.124}$$

where p_{cr} is the critical state pressure. The expressions indicate that the frictional stresses of particles accounts for compressibility according to a coefficient n in dense granular flows. The coefficient n is an exponent that determines the shape of the yield distribution and is set to 1.03 at $\nabla\cdot\boldsymbol{u}_s < 0$. Apparently, these equations account for the dilatation ($\nabla\cdot\boldsymbol{u}_s > 0$ and $p_{fr} < p_{cr}$) and compaction ($\nabla\cdot\boldsymbol{u}_s < 0$ and $p_{fr} > p_{cr}$) of the particulate flow. Note that the condition $\nabla\cdot\boldsymbol{u}_s = 0$ is valid only for the critical state, in which there is a change of volume due to the solid phase deformation. In such a case the solid frictional pressure is equal to critical state pressure, i.e. $p_{fr} = p_{cr}$. Generally, the value of the coefficient n is close to unit, and the frictional viscosity can be expressed:

$$\mu_{fr} = \frac{p_{fr}\sqrt{2}\sin\varphi}{2\sqrt{\boldsymbol{S}:\boldsymbol{S}+\theta/d_p^2}} \tag{1.125}$$

The main difference between this frictional viscosity model and the one proposed by [46] is in the additional term ($\theta/{d_p}^2$) in the denominator. The role of this term is to take into account the strain rate fluctuations in the dense quasi-static flow. The advantage of such a formulation is that numerical singularity can be avoided where the term $\boldsymbol{S}$:$\boldsymbol{S}$ becomes zero in the particulate flows.

The critical state pressure p_{cr} in Eq. (1.122) represents the solid phase mean stress at the critical state. In general, p_{cr} increases monotonically with solid volume fractions, and is expected to become very large as the solid volume fraction approaches random close packing $\alpha_{s,max}$. The formulation of critical static solid pressure is proposed [43]:

$$p_{cr} = \frac{[(\alpha_{s,max} - \alpha_{s,min})\gamma_{sh} d_p]^2}{(\alpha_{s,max} - \alpha_s)^2} \tag{1.126}$$

where γ_{sh} is the shear rate. $\alpha_{s,min}$ is the lowest solids volume fraction at which the frictional contacts affect solid phase stress. The relationship between the frictional solid pressure p_{fr} and p_{cr} is

$$p_{fr} = (1 - \frac{\nabla \cdot \boldsymbol{u}_s}{n\sqrt{2}\sin\varphi\sqrt{\boldsymbol{S}:\boldsymbol{S} + \theta/d_p^2}})^{n-1} \frac{[(\alpha_{s,max} - \alpha_{s,min})\gamma_{sh} d_p]^2}{(\alpha_{s,max} - \alpha_s)^2} \tag{1.127}$$

Compared to Srivastava and Sundaresan's Eq. (1.121), this model only changes the formulation of p_{cr}, yet this formula affects both the solids pressure and solids viscosity.

A detailed explanation of the solid phase frictional model can be found in three papers by [40, 43, 49].

1.5.3 Inertial Number-Based Frictional Model

Granular flow in the low solid volume fractions of rapid flow regime is dominated by particle–particle collisions and the constitutive relations for the solid stress are obtained from the KTGF. In the high solid volume fractions of dense flow regime, on the other hand, particles interact via enduring contacts and the solid stress can be described with the so-called inertial rheology or $\mu(I_s)$ rheology [13, 17, 32, 41], where a phenomenological law relates the stress rate to the inertial number I_s.

$$I_s = \frac{2d_p|\text{dev}\boldsymbol{D}|}{\sqrt{p_{fr}/\rho_s}} \tag{1.128}$$

where d_p and ρ_s are the particle diameter and density, respectively. dev$\boldsymbol{D}$ denotes the deviatoric part of the solid phase rate of deformation tensor $\boldsymbol{D}$.

$$\mathrm{dev}\boldsymbol{D} = \boldsymbol{D} - \frac{1}{3}\mathrm{tr}(\boldsymbol{D})\boldsymbol{I} \tag{1.129}$$

The inertial number represents the ratio between the macroscopic deformation timescale and the microscopic inertial timescale, and it is a dimensionless number. A small inertial number implies that the macroscopic deformation is much smaller than the microscopic rearrangement, i.e., the quasi-static regime for the motion of a particle in physics. While a high value corresponds to the inertial state for binary collision of particles.

Considering granular materials as visco-plastic fluids, [32] assumed that the frictional shear stress is a function of the shear rate and the inertial number I_s. According to the rheological relationship between shear stress and solid pressure, the solid phase frictional viscosity is expressed as the function of inertial number.

$$\mu_{fr} = \left[\tan(20.9^\circ) + \frac{\tan(32.76^\circ) - \tan(20.9^\circ)}{0.279 + I_s} I_s\right] \frac{p_{fr}}{2|\mathrm{dev}\boldsymbol{D}|} \tag{1.130}$$

The frictional viscosity is an increasing function of the inertial number starting from the quasi-static flow at $I_s = 0$ up to the kinetic regime at large inertial numbers. I_s clearly accounts for the competing effects of particle inertia and confinement.

The inertial number-based rheology is assumed to be valid in the dense part of the granular flow, and extends the granular bed condition. An alternative expression of the granular frictional viscosity is expressed:

$$\mu_{fr} = \left[0.35 + \frac{0.97 - 0.35}{0.69 + I_s} I_s\right] \frac{p_{fr}}{2|\mathrm{dev}\boldsymbol{D}|} \tag{1.131}$$

From Eqs. (1.130) and 1.131), it can be seen that the frictional viscosity diverges as dev$\boldsymbol{D}$ tends to zero. It is also related to solid phase frictional pressure p_{fr}. According to the dilatancy law, the solid phase frictional pressure is

$$p_{fr} = 4\rho_s \left(\frac{0.2 d_p |\mathrm{dev}\boldsymbol{D}|}{\alpha_{s,max} - \alpha_s}\right)^2 \tag{1.132}$$

A detailed explanation of the inertial number-based frictional model can be found in four papers by [13, 17, 32, 41].

From the frictional rheology, it can be found that the frictional pressure and viscosity of particles relate to the inertial number I_s, volume fraction α_s, and particles diameter d_p and density ρ_s. The frictional stress model describes contact interactions in the dense granular flow regime, and is based on the dimensional analysis of the shear configuration. Note that both the frictional pressure and the frictional viscosity relate to particle diameter d_p. In contrast, the correlations proposed by [31, 49, 51] account for the volume fraction only. These expressions for frictional pressure and viscosity are monotonically increasing functions with increasing solid volume fractions.

1.5.4 Configurational Temperature-Based Frictional Model

In gas–solid fluidized beds, the flow of particles must go through an intermediate regime where both kinetic/collisional and frictional contributions play a role. With the increase of volume fraction of particles, the interparticle contact will be dominant. For the contact between two spheres the strain changes simultaneously as the particle is deformed, leading to the generation of contact force. The fluctuation velocity of particles $\boldsymbol{C}$ is decomposed into a translational velocity fluctuation $\boldsymbol{C}_t$ and a fluctuation velocity of deformation $\boldsymbol{C}_\delta$. The translational fluctuation kinetic energy is $m < \boldsymbol{C}_t{}^2 > /2$, and the deformation fluctuation kinetic energy is $m < \boldsymbol{C}_\delta{}^2 > /2$. The kinetic energy of the motion of particles is $m < \boldsymbol{C}^2 > /2$, and is the sum of the translational kinetic energy and the kinetic energy of deformation. Thus, the particle fluctuation kinetic energy is defined by

$$\frac{1}{2}m < \boldsymbol{C}^2 >= \frac{1}{2}m < \boldsymbol{C}_t^2 > + \frac{1}{2}m < \boldsymbol{C}_\delta^2 > \tag{1.133}$$

The first part on the right-hand side denotes the translational kinetic energy contribution by translational velocity fluctuations arising from the streaming of particles and direct collisions, and the second part represents the kinetic energy of deformation contribution by displacement fluctuation due to the contact interactions. From the definition of granular temperature, it is calculated from fluctuating translational velocity of particles as follows:

$$\theta_t = \theta = \frac{1}{3} < \boldsymbol{C}_t^2 > \tag{1.134}$$

From simulated instantaneous positions of particles using CFD-discrete element method (DEM), the instantaneous overlap δ_{ij} for a given collision pair is given by [34].

$$\delta_{ij}(t) = \begin{cases} \sqrt{\delta^2_{x,ij}(t) + \delta^2_{y,ij}(t) + \delta^2_{z,ij}(t)} & s_{ij} < (r_i + r_j) \\ 0 & s_{ij} \geq (r_i + r_j) \end{cases} \tag{1.135}$$

where r_i and r_j represent the radii of particle i and j, respectively. s_{ij} is the distance between the particle i and j, and is expressed according to the particle position x, y and z in three dimensions.

$$s_{ij} = \sqrt{(x_i - x_j)^2 + (y_i - y_j)^2 + (z_i - z_j)^2} \tag{1.136}$$

If the overlap δ is larger than zero (i.e., $\delta > 0$), there is a physical contact between particles. From instantaneous overlaps, the mean overlap of the cell is defined according to

$$\delta_m = \frac{1}{MN} \sum_{m=1}^{M} \sum_{i=1,j \neq i}^{N} \delta_{ij}(t) \tag{1.137}$$

where M is the total number of data over a given time period. N is the number of particles in the cell. The fluctuation velocity of deformation is expressed according to the time step Δt_p of particles.

$$C_{\delta,ij} = \frac{[\delta_{ij}(t) - \delta_m]}{\Delta t_p} \tag{1.138}$$

The variance of overlap is produced due to the deformation by the collision of inelastic particles. This is the principal characteristic of quasistatic flows of dense particles because of the deformation by enduring long sliding and rubbing contacts. Thus, the configurational temperature defines based on the fluctuation velocity of deformation of particles.

$$\theta_c = \frac{1}{3} < \boldsymbol{C}_\delta^2 > \tag{1.139}$$

In fluid–solid flow at low solid volume fractions, the interactions between particles occur via a series of successive collisions. The successive collisions are assumed to be instantaneous where the contact time is much smaller than the free motion time. Thus, the kinetic energy is relatively large and the elastic energy is close to zero. With increasing solid volume fractions, a large number of adjacent particles are engaged in force chains. These chains are dynamic structures, and the elastic energy varies greatly. However, the kinetic energy becomes low in comparison with the values at low solid volume fraction.

Similarly, we can define the generalized granular temperature of particles.

$$\theta_e = \frac{1}{3} < \boldsymbol{C}^2 > \tag{1.140}$$

From Eq. (1.134), the translational granular temperature applies to high solid volume fraction systems as long as no enduring contacts exist, but does not apply in systems dominated by force-chain networks where strong correlations exist among the particles. From Eq. (1.139), it describes fluctuations in elastic energy due to the contact interactions. Using these definitions, it can be seen that Eq. (1.133) combines the kinetic and configurational contributions, and can serve as a generalized granular temperature to be valid for all granular flows.

$$\theta_e = \theta_t + \theta_c \tag{1.141}$$

The definition of generalized granular temperature is more convenient because it is simply the sum of the variance of the translational fluctuation velocity and fluctuation velocity of deformation of particles. The distributions of generalized granular

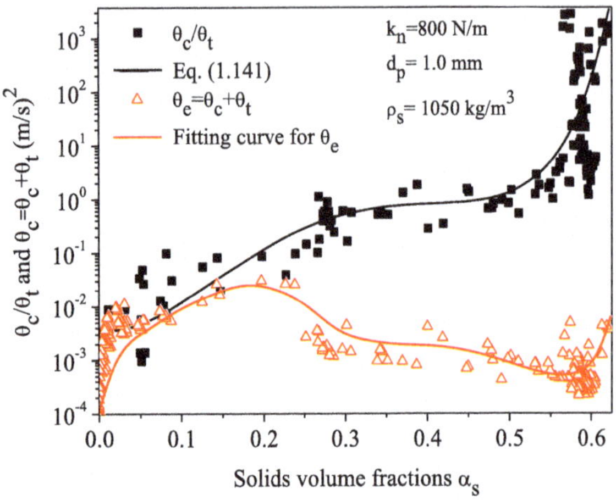

Fig. 1.3 Profile of ratio of configurational temperature to translational granular temperature

temperature of particles are given in Fig. 1.3 as a function of solids volume fractions in a gas–solid flat-base spout fluidized bed [34, 35]. The generalized granular temperature increases, reaches maximum, then decreases with increase of solids volume fraction, and finally increases at high solids volume fractions where the configurational temperature is larger than that of translational granular temperature. The collisional mechanism dominates, and the translational granular temperature is high and the configurational temperature is low at the low solids volume fractions. At the high solids volume fractions, the elastic deformation is caused by contact interactions between particles. The stronger interactive collisions have more active displacement field. This makes high fluctuation of overlaps, and leads to high configurational temperatures. This indicates the contact mechanism dominates at the high solids volume fractions. The generalized granular temperature approximates to translational granular temperature at low solids volume fraction, whereas it trends to configurational temperature at the high solids volume fractions.

From Eq. (1.141), it shows that the effect of kinetic energy of deformation appears in the form of an addition to the translational granular temperature. This addition is a measure of the local deviations from the continuum displacement field induced by the contact interaction of particles. From simulated configurational temperature and translational granular temperature, the ratio of the configurational temperature to the translational granular temperature, θ_c/θ_t, is calculated. The distribution of θ_c/θ_t increases with the increase of solids volume fractions. At low solids volume fractions, the configurational temperature is negligible. On the contrary, the configurational temperature is larger than translational granular temperature at high solids volume fractions. This indicates the contact interaction mechanism dominates at the high solids volume fractions because particles start to endure long sliding and rubbing contacts, which gives rise to a totally different form of energy dissipation and stress by friction. The following correlation is proposed for describing the relation between the configurational temperature and translational granular temperature as the function of solids volume fractions α_s, normal spring coefficient k_n and damping coefficient

η_n of particles [35].

$$\frac{\theta_c}{\theta_t} = \frac{[1.8657exp(-7.35 \times 10^{-4}k_n)][1.024 + 0.769\eta_n - 9.603\eta_n^2]}{1 - (\alpha_s/\alpha_{s,max})^{1/3}} \times 10^{(-5.2353+55.3782\alpha_s-357.9726\alpha_s^2+1176.8378\alpha_s^3-1854.8365\alpha_s^4+1126.0306\alpha_s^5)} \tag{1.142}$$

The correlation shows the ratio of the configurational temperature to translational granular temperature increases monotonously with the increase of solids volume fractions, and decreases with the increases of normal spring coefficient and damping coefficient of particles.

Since the configurational temperature is produced by contact interactions of particles, we anticipate a relationship between the particle pressure and the configurational temperature. Substituting generalized granular temperature for translational granular temperature in expression of particle pressure proposed by [20], the particle pressure is predicted by

$$p_s = \alpha_s \rho_s \theta_t + 2\alpha_s^2 \rho_s g_o (1 + e)\theta_e \tag{1.143}$$

The solids shear viscosity arises from particle momentum exchange due to translation and collision. Using Eq. (1.133) and replacing generalized granular temperature for translational granular temperature in Eq. (1.72), the solid phase shear viscosity is predicted by

$$\mu_s = \frac{10\alpha_s d_p \rho_s \sqrt{\pi \theta_t}}{96(1 + e)g_0}\left[1 + \frac{4}{5}(1 + e)\alpha_s g_0\right]^2 + \frac{4}{5}\alpha_s^2 \rho_s d_p g_0 (1 + e)\sqrt{\frac{\theta_e}{\pi}} \tag{1.144}$$

They are high at the large solids volume fraction due to high configurational temperature. It can be seen that the configurational temperature originates principally in the contact interactions of particles. The fluctuating overlaps due to the deformation of particles are large at high solids volume fractions. In view of the lack of a complete kinetic theory applicable to high volume fractions of particles, the generalized temperature which combines the configurational and translational granular temperatures allows to estimate solid pressure and viscosity in the complete range of solids volume fractions, and also provides a basis for potentially unifying granular flows over the entire range of regimes in fluidized beds and multiphase flows.

1.5.5 Stress-Blending Function

The solid phase stress includes contributions from the granular dilute regime and the dense regime. The stress-blending function blends the stress resulting from kinetic theory of granular flow with the frictional stress of particles when the solid volume fraction is close to the defined solid volume fraction of the packed bed. The goal of using the stress-blending function is to smooth the transfer between the dilute flow and the dense flow of particles. The solid phase stress is written in a continuous form as

$$\boldsymbol{\tau}_s = \phi_{str}(\boldsymbol{\tau}_k + \boldsymbol{\tau}_c) + (1 - \phi_{str})\boldsymbol{\tau}_{fr} \qquad \alpha_s \geq \alpha_{cr} \tag{1.145}$$

where the stress-blending function is expressed as (Guodong et al., 2010):

$$\phi_{str} = 0.5 - \frac{\arctan[25(\alpha_s - \alpha_{s,\min})(\alpha_{s,\max} - \alpha_{s,\min})^2}{\pi} \tag{1.146}$$

where $\alpha_{s,min}$ is the solid volume fraction at the critical state in which the frictional stress is contributed. $\alpha_{s,max}$ is the packed-bed solid volume fraction. It shows that the stress-blending function changes rapidly from 0.0 to 1.0 when the solid volume fraction changes from $\alpha_{s,min}$ to $\alpha_{s,max}$.

An alternative expression for the stress-blending function is represented using the hyperbolic tangent blending function, and it is defined as

$$\phi_{str} = \left[\tanh\left(\frac{2\pi(\alpha_f - \alpha_{f,min})}{\alpha_{2f} - \alpha_{1f}}\right) + 1\right]/2 \tag{1.147}$$

where $\alpha_{f,min}$ is the fluid volume fraction for the packed bed. $\alpha_{1f} = 0.99\alpha_{f,min}$, $\alpha_{2f} = 1.01\alpha_{f,min}$. The stress blending function varies from 0.0 to 1.0 with changing fluid volume fractions.

Using the inertial number, the regime transition from the granular dilute regime to the dense regime is realized by means of the stress-blending function [11]

$$\phi_{str} = \frac{1}{(0.2\alpha_s/I_s)^{1.5} + 1} \tag{1.148}$$

These indicate that the stress-blending function is a function of the inertial number I_s. The blending function is an increasing function of the inertial number. A high value of stress-blending function means that the momentum exchange among particles is dominated by kinetic motions and collisions and hence a higher contribution to solid stress. In contrast, a low value of stress-blending function means that the momentum exchange among particles is dominated by enduring frictional contacts, and hence a higher contribution to solid stress.

1.6 Boundary Conditions

Boundary conditions have to be specified at the wall and inlet. At the inlet and outlet, all properties should be defined based on the specific physics and assumption of the problem. No slip condition is used for the continuous fluid phase.

For the velocity field and the granular temperature, a zero momentum and turbulent kinetic energy flux at the wall are often formulated. The free slip condition of the solid phase volume fraction, velocity and granular temperature is customary to use.

$$\frac{\partial \alpha_s}{\partial \boldsymbol{n}} = \frac{\partial \boldsymbol{u}_s}{\partial \boldsymbol{n}} = \frac{\partial \theta}{\partial \boldsymbol{n}} = 0 \tag{1.149}$$

where $\boldsymbol{n}$ is the direction normal to the wall.

For the solid phase, the slip boundary condition is the recommended boundary conditions [31]:

$$\boldsymbol{\tau}_{sw} = -\frac{\pi}{6}\frac{\phi\alpha_s}{\alpha_{s,max}}\rho_s g_0\sqrt{3\theta}\boldsymbol{u}_{slip} \tag{1.150}$$

where u_{slip} is the particle slip velocity parallel to the wall. ϕ is the specularity coefficient between the particle and the wall, which is defined as the average fraction of relative tangential momentum transferred between the particle and the wall during a collision. The specularity coefficient varies from zero (smooth walls) to one (rough walls). A proper value based on the particles and wall properties should be assumed. For a specularity coefficient tending toward zero, a free slip boundary condition for the solids tangential velocity is imposed at a smooth wall boundary. [31] proposed the following wall boundary condition for the granular kinetic energy flux as

$$q_{sw} = \frac{\pi}{6}\sqrt{3}\phi\frac{\rho_s g_0\alpha_s}{\alpha_{s,max}}\sqrt{\theta}\boldsymbol{u}_{slip}\cdot\boldsymbol{u}_{slip} - \frac{\sqrt{3}\pi}{4}\frac{(1-e_w^2)\alpha_s}{\alpha_{s,max}}\rho_s g_0\theta^{3/2} \tag{1.151}$$

The dissipation of solids turbulent kinetic energy by collisions with the wall is specified by the particle–wall restitution coefficient e_w and the specularity coefficient ϕ between the particles and the wall. A high value of the specularity coefficient implies high production at the wall, and a value of e_w close to unity implies low dissipation of granular energy at the wall. It is expected that the specularity coefficient and the particle–wall restitution coefficient need to be calibrated for a given fluid/particle flow system because the specularity coefficient cannot be measured and the particle–wall restitution coefficient can be measured only with some difficulty.

From Eqs. (1.150) and (1.151), the boundary conditions for velocity and granular temperature could be written as [3, 26, 42, 49]

$$u_{sw} = -\frac{6\mu_{sw}\alpha_{s,max}}{\sqrt{3}\pi\phi\rho_s\alpha_s g_0\sqrt{\theta}}\frac{\partial u_{sw}}{\partial x} \tag{1.152}$$

$$\theta_w = -\frac{k_{sw}\theta}{\gamma_w}\frac{\partial\theta_w}{\partial x} + \frac{\sqrt{3}\pi\phi\rho_s\alpha_s u_{slip}^2 g_0\theta^{3/2}}{6\alpha_{s,max}\gamma_w} \tag{1.153}$$

where the expressions of solid viscosity μ_{sw} and solid thermal conductivity k_{sw} are deduced from Eqs. (1.72) and (1.93), and energy dissipation rate γ_w from Eq. (1.102) at the wall according to particle–wall collisional coefficient of restitution.

With increasing solid volume fraction at the wall, the frictional particle–wall contacts give rise to wall shear stresses in addition to particle–wall collisions. The frictional forces of wall-particles may be sufficiently high at the wall. The solid phase boundary condition is expressed according to slip velocity u_{slip} for particle–wall interactions.

$$\frac{\boldsymbol{u}_{slip}}{\left|u_{slip}\right|}\boldsymbol{\tau}_{sw}\cdot\boldsymbol{n} + \frac{\pi}{2\sqrt{3}}\frac{\phi\alpha_s\rho_s}{\alpha_{s,max}}g_0\sqrt{\theta}\left|\boldsymbol{u}_{slip}\right| + (\boldsymbol{\tau}_{fr}\cdot\boldsymbol{n})\tan\varphi_w = 0 \tag{1.154}$$

$$(k_{sw}\nabla\theta)\cdot\boldsymbol{n} - \frac{\pi\phi\alpha_s\rho_s g_0}{6\alpha_{s,max}}\sqrt{3\theta}\left|\boldsymbol{u}_{slip}\right|^2 + \frac{\pi\sqrt{3}\alpha_s}{4\alpha_{s,max}}(1-e_w^2)\rho_s g_0\theta\sqrt{\theta} = 0 \tag{1.155}$$

where $\boldsymbol{n}$ is the normal unit vector pointing into the wall. e_w and φ_w represent the coefficient of restitution and angle of internal friction for particle–wall collisions.

1.7 Summary

Using KTGF, the TFM approach has been given for the momentum and energy fluxes as functions of granular temperatures. These equations are almost free from empirical constants, only the restitution coefficient and specularity coefficient have to be given. The restitution coefficient depends, at least, on the material of the particles and the relative velocity between two colliding particles. In simulation and analysis, the restitution coefficient is constant, and its value varies for different cases. Low values of the restitution coefficient decrease the granular temperature, which in turn decreases solid viscosity and diffusivity. Furthermore, the limitations of the KTGF are due to the simple collision model with no particle rotation and no friction. Hence, more research is needed to derive a general model of the particle–particle interactions.

It should be emphasized that the model presented in TFM approach includes the influence of the fluid phase turbulence on the granular flow. For dilute flows, hydrodynamic interaction is important mechanism. For transport equations of fluid kinetic energy and energy dissipation rate, the values of the constants are not known for application to dense fluid-particles two-phase flow, and they may be different from the single-phase flow case, depending on solid volume fractions and turbulence properties.

In addition, one of the key issues of the Euler-Euler formulation is to derive an accurate model for the fluid-particle velocity covariance that is very important for momentum exchange. This term appears in the interfacial term of the transport equations for the fluctuating motion and can give production or destruction of the velocity fluctuations in both phases. Hence, more research is needed in this field for an accurate formulation of the transport equation of the fluid-particle velocity correlation tensor. As the suspension becomes denser, the effect of collisions should influence the fluid-particle velocity correlation tensor.

Nomenclature

$\boldsymbol{c}$	Instantaneous velocity, m/s
$\boldsymbol{C}$	Peculiar particle velocity, m/s
d_p	Diameter of particle, m
D_{iss}	Energy dissipation by means of friction, kJ/s
e	Restitution coefficient
E	Internal energy, J
f	Frequency distribution of velocities of particles
$\boldsymbol{F}$	Body force, N
g	Gravity acceleration, m/s^2
g_0	Radial distribution function at contact
h	Enthalpy per unit mass, J/kg
$\boldsymbol{I}$	Identity matrix
J_{pro}	Production of kinetic energy, kg/(m s^3)
J_{vis}	Dissipation of kinetic energy, kg/(m s^3)
k_f	Fluid turbulent kinetic energy of fluid, m^2/s^2
k_{fs}	Fluid-particle velocity covariance, m^2/s^2
k_s	Granular conductivity, kg/(m s)
k_n	Normal spring coefficient, N/m
K_i	Constants
m	Mass, kg
$m_i{}'$	Rate of production of phase i, kg/(s m^3)
n	Number of particles
p	Pressure, N/m^2
p_c	Collisional solid pressure, N/m^2
p_f	Frictional pressure, N/m^2
p_k	Kinetic pressure tensor, N/m^2
q	Rate of heat transfer, J
q_c	Collisional flux, J/(s m^2)
q_k	Kinetic flux, J/(s m^2)
$\boldsymbol{r}$	Position vector
$\boldsymbol{S}$	Deviatoric stress tensor
t	Time, s

$\boldsymbol{T}$	Stress tensor
T_0	Particle relaxation time, s
$\boldsymbol{u}$	Velocity vector, m/s
U	Internal energy per unit mass, J/kg
V	System volume, m^3

Greek Symbols

α_f	Fluid phase volume fraction
α_s	Solid volume fraction
θ	Granular temperature of solid phase, m^2/s^2
θ_c	Configurational temperature, m^2/s^2
θ_e	Generalized granular temperature, m^2/s^2
μ_b	Granular bulk viscosity, kg/(m s)
μ_{df}	Coefficient of dynamic friction
μ_f	Fluid phase dynamic viscosity, kg/(m s)
μ_{fr}	Frictional viscosity, kg/(m s)
μ_s	Solids viscosity, kg/(m s)
μ_t	Fluid turbulent viscosity, kg/(m s)
ρ_f	Density of the fluid phase, kg/m^3
ρ_s	Density of solids phase, kg/m^3
$\boldsymbol{\tau}_f$	Fluid phase stress tensor, Pa
$\boldsymbol{\tau}_{fr}$	Frictional stress, Pa
τ_{ts}	Dimensionless time scale
$\boldsymbol{\tau}_s$	Solids phase stress tensor, Pa
β	Momentum exchange coefficient, kg/(m^3 s)
γ	Dissipation rate due to particles collision, J/(m^3 s)
ξ	Collisional cooling rate, 1/s
ψ	Property of particle
ϕ	Specularity coefficient
Π	Second invariant of the deviatoric stress tensor
$\mathfrak{J}$	General property per unit volume
η_n	Normal damping coefficient, kg/s

Abbreviations

CFD	Computational fluid dynamics
DEM	Discrete element method
KT	Kinetic theory
KTGF	Kinetic theory of granular flow
TFM	Two-fluid model

References

1. Agrawal, K., Loezos, P. N., Syamlal, M., & Sundaresan, S. (2001). The role of meso-scale structures in rapid gas–solid flows. *Journal of Fluid Mechanics, 445*, 151–185.
2. Anderson, T. B., & Jackson, R. (1967). A fluid mechanical description of fluidized beds. *Industrial & Engineering Chemistry Fundamentals, 6*, 527–539.
3. Arastoopour, H., Gidaspow, D., & Abbasi, E. (2017). *Computational transport phenomena of fluid-particle systems*. Springer International Publishing.
4. Aris, R. (1989). *Vectors, tensors, and the basic equations of fluid mechanics*. Dover Publications Inc.
5. Benyahia, S., Syamlal, M., & O'Brien, T. J. (2012) *Summary of MFIX equations, Topical Report, National Technical Information Service*. Springfield.
6. Bird, R. B., Stewart, W. E., & Lightfoot, E. N. (2007) *Transport phenomena*. Wiley.
7. Brennen, C. E. (2005). *Fundamentals of multiphase flows*. University Press.
8. Campbell, C. S. (1990). Rapid granular flows. *Annual Review of Fluid Mechanics, 22*, 57–92.
9. Cao, J., & Ahmadi, G. (1995). Gas-particle two-phase turbulent flow in a vertical duct. *International Journal of Multiphase Flow, 21*, 1203–1228.
10. Chapman, S., & Cowling, T. G. (1970). *The mathematical theory of non-uniform gases*. Cambridge Mathematical Library.
11. Chialvo, S., & Sundaresan, S. (2013). A modified kinetic theory for frictional granular flows in dense and dilute regimes. *Physics of Fluids, 25*, 070603.
12. Crowe, C. T., Sommerfeld, M., & Tsuji, Y. (1998). *Multiphase flows with droplets and particles*. CRC Press.
13. Cruz, D. F., Emam, S., Prochnow, M., Roux, J. N., & Chevoir, F. (2005). Rheophysics of dense granular materials: Discrete simulation of plane shear flows. *Physical Review E, 72*, 021309.
14. Ding, J., & Gidaspow, D. (1990). A bubbling fluidization model using kinetic theory of granular flow. *AIChE Journal, 36*, 523–538.
15. Elghobashi, S. E. (1994). On predicting particle laden turbulent flows. *Applied Sciences Research, 52*, 309–329.
16. Ergun, S. (1952). Fluid flow through packed columns. *Chemical Engineering Progress, 48*, 89–94.
17. Forterre, Y., & Pouliquen, O. (2008). Flows of dense granular media. *Annual Review of Fluid Mechanics, 40*, 1–24.
18. Garzó, V. (2019). *Granular gaseous flows: A kinetic theory approach to granular gaseous flows*. Springer International Publishing.
19. Garzó, V., Tenneti, S., Subramaniam, S., & Hrenya, C. M. (2012). Enskog kinetic theory for monodisperse gas–solid flows. *Journal of Fluid Mechanics, 712*, 129–168.
20. Gidaspow, D. (1994). *Multiphase flow and fluidization, Continuum and kinetic theory description*. Academic Press Inc.
21. Gidaspow, D., & Jiradilok, V. (2009). *Computational techniques: The multiphase CFD approach to fluidization and green energy technologies*. Nova Science Publishers.
22. Guodong, L., Dan, S., Huilin, L., Bouillard, J., Yinhua, B., & Shuai, W. (2010). Computations of fluid dynamics of a 50 MWe circulating fluidized bed combustor. *Industrial & Engineering Chemistry Research, 49*, 5132–5140.
23. Hadinoto, K., & Curtis, J. (2004). Effect of interstitial fluid on particle−particle interactions in kinetic theory approach of dilute turbulent fluid-particle flow. *Industrial & Engineering Chemistry Research, 43*, 3604–3615.
24. Hrenya, C. M., & Sinclair, J. L. (1997). Effects of particle-phase turbulence in gas-solid flows. *AIChE Journal, 43*, 53–869.
25. Huilin, L., & Gidaspow, D. (2003). Hydrodynamics of binary fluidization in a riser: CFD simulation using two granular temperatures. *Chemical Engineering Science, 58*, 3777–3792.
26. Huilin, L., Gidaspow, D., & Shuyan, W. (2021). *Computational fluid dynamics theory of fluidization: Kinetic theory of granular flow and applications*. Springer International Publishing.

27. Huilin, L., Yurong, H., & Gidaspow, D. (2003). Hydrodynamic modelling of binary mixture in a gas bubbling fluidized bed using the kinetic theory of granular flow. *Chemical Engineering Science, 58*, 1197–1205.
28. Jackson, R. (1997). Locally averaged equations of motion for a mixture of identical spherical particles and a Newtonian fluid. *Chemical Engineering Science, 52*, 2457–2469.
29. Jackson, R. (2000). *The dynamics of fluidized particles*. University Press.
30. Jenkins, J., & Savage, S. B. (1983). Theory for the rapid flow of identical, smooth, nearly elastic, spherical particles. *Journal of Fluid Mechanics, 130*, 187–202.
31. Johnson, P. C., & Jackson, R. (1987). Frictional-collisional constitutive relations for granular materials, with application to plane shearing. *Journal of Fluid Mechanics, 176*, 67–93.
32. Jop, P., Forterre, Y., & Pouliquen, O. (2006). A constitutive law for dense granular flows. *Nature, 441*, 727–730.
33. Koch, D. L., & Sangani, A. S. (1999). Particle pressure and marginal stability limits for a homogeneous monodisperse gas fluidized bed: Kinetic theory and numerical simulations. *Journal of Fluid Mechanics, 400*, 229–263.
34. Liyan, S. (2016). *Research on kinetic theory of cohesive particle and numerical simulation of gas-solid flow*. Ph. D. Thesis. Harbin Institute of Technology.
35. Liyan, S., Weiguo, X., Huilin, L., Guodong, L., Qinghong, Z., Qing, T., & Tianyu, Z. (2015). Simulated configurational temperature of particles and a model of constitutive relations of rapid-intermediate-dense granular flow based on generalized granular temperature. *International Journal of Multiphase Flow, 77*, 1–18.
36. Louge, M. Y., Mastorakos, E., & Jenkins, J. T. (1991). The role of particle collisions in pneumatic transport. *Journal of Fluid Mechanics, 231*, 345–359.
37. Lun, C. K. K., Savage, S. B., Jeffrey, D. J., & Chepurniy, N. (1984). Kinetic theories for granular flow: Inelastic particles in Couette flow and slightly inelastic particles in a general flow field. *Journal of Fluid Mechanics, 140*, 223–256.
38. Lyczkowski, R. W., Gidaspow, D., & Solbrig, C. W. (1982) Multiphase flow models for nuclear, fossil and biomass energy production. *Advances in Transport Processes, II*, 198–351. Wiley Eastern Ltd.
39. Marchisio, D. L., & Fox, R. O. (2007). *Multiphase reacting flows: Modelling and simulation*. Springer.
40. MFIX. (2020). *MFiX-Exa documentation*. MFS Development Group, National Energy Technology Laboratory.
41. Midi, G. D. R. (2004). On dense granular flows. *The European Physical Journal E, 14*, 341–365.
42. Pannala, S., Syamlal, M., & O'Brien, T. J. (2015). *Computational gas-solids flows and reacting systems: Theory, methods and practice*. Engineering Science Reference.
43. Qiang, G., Yuxuan, Z., Azin, P., Kenan, X., Thomas, M. K., & Christopher, M. B. (2021). Dynamically structured bubbling in vibrated gas-fluidized granular materials. *Proceedings of the National Academy of Sciences, 118*, 1–8.
44. Rao, K. K., & Nott, P. R. (2009). *An introduction to granular flow*. Cambridge University Press.
45. Savage, S. B., & Jeffrey, D. J. (1981). The stress tensor in a granular flow at high shear rates. *Journal of Fluid Mechanics, 110*, 255–272.
46. Schaeffer, D. G. (1987). Instability in the evolution equations describing incompressible granular flow. *Journal of Difference Equations and Applications, 66*, 19–50.
47. Simonin, O. (1996) *Continuum modeling of dispersed two-phase flows in combustion and turbulence in two-phase flows*. Von Karman Institute of Fluid Dynamics Lecture Series.
48. Sinclair, J. L., & Jackson, R. (1989). Gas-particle flow in a vertical pipe with particle-particle interactions. *AIChE Journal, 35*, 1473–1486.
49. Srivastava, A., & Sundaresan, S. (2003). Analysis of a frictional-kinetic model for gas-particle flow. *Powder Technology, 129*, 72–85.
50. Syamlal, M., Musser, J., & Dietiker, J. F. (2017) In: Michaelides E. E., Crowe C. T., & Schwarzkopf J. D. (eds). *Two-fluid model in MFIX, multiphase flow handbook* (pp. 242–274). Boca Raton.

51. Syamlal, M., Rogers, W., & O'Brien, T. J. (1993). *MFIX documentation: Theory guide.* National Energy Technology Laboratory, Department of Energy, Technical Note DOE/METC-95/1013 and NTIS/DE95000031.
52. Wen, C. Y., & Yu, Y. H. (1966). Mechanics of fluidization. *American Institute of Chemical Engineers Symposium Series, 62*, 100–111.

Chapter 2
CFD-DEM Approach for Fluid-Particles Flow

Abstract This chapter provides an up-to-date description of the formulation and implementation of combined Computational Fluid Dynamics and Discrete Element Method (CFD-DEM) modeling. The different contact force models of particles are discussed along with their applicability and limitations. The couplings between solid and fluid momentum equations are described and the implementation of external forces is presented. The virtual calibration methods of the DEM input parameters are proposed to aid not only researchers but also engineers in CFD-DEM modelling industrial applications.

The Euler–Lagrange approach is a two-phase flow numerical modeling technique in which the Euler method is used for the fluid phase and the Lagrangian method for the discrete particles. Computational Fluid Dynamics-Discrete Element Method considers a combination of computational fluid dynamics (CFD) and the discrete element method (DEM), and is applied in simulations of fluid–solid processes. It is referred to as CFD-DEM approach. The DEM was first proposed by [7] to study the motion and interaction of blocky systems. In CFD-DEM approach, the fluid phase is assumed as a continuum and its meso-scale motion is described by the volume averaged Navier–Stokes equation, while the micro-scale discrete particle motion of the solid phase is described by Newton's second laws. Forces acting on a discrete particle includes contact interactions between colliding particles, fluid-particle interaction and interparticle forces which demonstrates the relation between micro- and meso-scales for modeling of a fluid-solid system. The two phases are coupled by a fluid-particle interaction force using a drag force correlation. In CFD-DEM approach, when equations of the fluid phase and discrete particles are coupled together, the DEM calculates the position and velocity data for each individual particle, and the CFD calculates the fluid profiles over fluid cells at each time step. These equations are presented for phenomena occurring with different length scales in the fluid-particles systems. CFD provides the simulated complex fluid flows with a quantitative and qualitative description of the temporal and spatial change of the flow field, while

H. Lu et al., *Computational Transport Phenomena of Multiphase Systems and Fluidization*, Fluid Mechanics and Its Applications 127,
https://doi.org/10.1007/978-981-96-0698-6_2

DEM provides distributions of interaction forces of particle–particle and particle–fluid in the system. Thus, CFD-DEM is a promising computational approach used in fluid–particle multiphase flow.

CFD-DEM approach groups into two different methods. One is the resolved CFD-DEM method in which the fluid cell size ΔL is much smaller than the particle size d_p (i.e., $\Delta L < d_p$). The other is known as the unresolved CFD-DEM method in which the fluid cell size is larger than the particle diameter (i.e., $\Delta L > d_p$). In the CFD, fluid properties are modelled in the fluid cell. Therefore, fluid properties are defined at a length scale as large as the size of fluid cell. In the resolved CFD-DEM method, fluid flow is fully resolved over the entire surface of particles since the size of fluid cell is much smaller than the particle diameter. Hence, this treatment needs a high computational cost of solving the fluid phase equations along particle surface, and is only feasible for multiphase flows containing no more than a few hundred particles. In the unresolved CFD-DEM method, the size of fluid cell is larger than the particle diameter. This makes possible to study multiphase flows containing thousands or even millions of particles. The unresolved CFD-DEM approach needs to consider the coupling between the discrete particles at particle scale and fluid phase at fluid cell scale.

2.1 Euler-Lagrangian Fluid–Solid Flow Model

2.1.1 Fluid Phase Conservation Equations

The fluid phase governing equations can be derived from the mass and momentum conservation laws. The isothermal fluid phase mass and momentum conservation equations are expressed on the basis of fluid velocity $\vec{u}$ and fluid pressure p as follows [6, 13, 17]:

$$\frac{\partial \alpha_f \rho_f}{\partial t} + \nabla \cdot (\alpha_f \rho_f \vec{u}_f) = 0 \tag{2.1}$$

$$\frac{\partial (\rho_f \alpha_f \vec{u}_f)}{\partial t} + \nabla \cdot (\rho_f \alpha_f \vec{u}_f \vec{u}_f) = -\alpha_f \nabla p + \nabla \cdot (\alpha_f \overline{\overline{\tau}}_f) - \vec{f}_{fp} + \alpha_f \vec{g} \rho_f \tag{2.2}$$

where α_f is the fluid volume fraction, and $\vec{f}_{fp}$ is the interaction term due to fluid drag force. For gas phase, the density ρ_f is calculated on the basis of idea gas law

$$\rho_f = \frac{p}{RT_f} \tag{2.3}$$

where R is the universal (or perfect) gas constant. T_f is the gas temperature. The fluid phase effective stress tensor is calculated as

$$\overline{\overline{\tau}}_f = (\mu_f + \mu_t)\{[(\nabla\vec{u}_f) + (\nabla\vec{u}_f)^{\mathrm{T}}] - \frac{2}{3}(\nabla \cdot \vec{u}_f)\} \tag{2.4}$$

where μ_f and μ_t are the fluid dynamic viscosity and turbulent viscosity, respectively. With increasing the fluid Reynolds number (i.e., increasing the inertia force compared to the viscous force), the flow regime changes from laminar, to transitional, and finally to turbulent flow. While the laminar flow is orderly, and the flow is significantly nonlinear and instable in the turbulent regime. Turbulence instability occurs when the time scale of the viscous damping of velocity fluctuations is much larger than the time scale of the convection. This happens at high Reynolds numbers. In other words, tendency of the instability to become damped by viscosity decreases with increasing Reynolds number. Despite initial impression about irregular and stochastic behavior of turbulent flow, turbulent fluctuations have coherent regular structures, named as eddies, with a wide range of length and time scales. In addition, eddies can stretch, rotate, and break up into smaller eddies or coalesce into larger ones. These eddies contain the turbulence energy and change over the time.

It is well known that the adoption of a suitable fluid turbulence model is the key to describe fluid hydrodynamics. Usually, the fluid turbulence models are grouped according to the required number of partial differential transport equations. These models include one-equation model (i.e., Spalart–Allmaras model) [46], two-equation model for the turbulent kinetic energy and the dissipation rate (i.e., k-ε and k-ω models) [30, 53], three-equation model for turbulent kinetic energy, laminar kinetic energy and dissipation rate (i.e., transition k-kl-ω model) [27, 50] and four equation model for transition shear stress transport (SST) model [33] as well as Reynolds-stress models [16, 19] to eddy resolved turbulent models, e.g., large eddy simulation (LES) [28, 45]. Technically, the turbulent viscosity is related to the fluid turbulence kinetic energy k and dissipation rate ε via the formulation given by

$$\mu_t = C_\mu \rho_f \frac{k^2}{\varepsilon} \tag{2.5}$$

where k and ε describe through the transport equations as follows

$$\frac{\partial(\rho_f \alpha_f k)}{\partial t} + \nabla \cdot (\rho_f \alpha_f \vec{u}_f k) = \nabla \cdot [\alpha_f (\mu_f + \frac{\mu_t}{\sigma_k})\nabla k] + \alpha_f G_k - \alpha_f \rho_f \varepsilon \tag{2.6}$$

$$\frac{\partial(\rho_f \alpha_f \varepsilon)}{\partial t} + \nabla \cdot (\rho_f \alpha_f \vec{u}_f \varepsilon) = \nabla \cdot [\alpha_f (\mu_f + \frac{\mu_t}{\sigma_\varepsilon})\nabla \varepsilon] + \alpha_f \frac{\varepsilon}{k}(C_1 G_k - C_2 \rho_f \varepsilon) \tag{2.7}$$

The constants C_μ, σ_k, σ_ε, C_1 and C_2 are 0.09, 1.0, 1.3, 1.44 and 1.92, respectively. It can be seen that the fluid phase volume fraction α_f is introduced into the transport equations, which is expected to obtain a more accurate prediction for fluid dynamics in fluid–solid systems. The term G_k represents the production of fluid turbulence kinetic energy, and evaluates in a manner consistent with the Boussinesq hypothesis.

$$G_k = \mu_t \frac{\partial \vec{u}_{fi}}{\partial x_j}\left(\frac{\partial \vec{u}_{fi}}{\partial x_j} + \frac{\partial \vec{u}_{fj}}{\partial x_i}\right) \tag{2.8}$$

2.1.2 Equations of Discrete Particles

In fluid–solid multiphase flows, the translational and rotational motions of discrete particles are highly dependent on type and strength of interactions of particles. The discrete particles interact with each other through their contact points/areas. Depending on the operating conditions of the fluid–solid system, interparticle interactions, such as van der Waals, liquid bridge and electrostatic forces, may exist between discrete particles even if they are not in contact. The fluid interacts with the discrete particles. The interaction between each particles and the fluid is mutual.

In the DEM approach, the discrete particles are considered either deformable or rigid. These two are mostly referred to as hard-sphere formulation and soft-sphere formulation. The hard–sphere model based on the event driven algorithm is first proposed by [2], and the soft–sphere model based on the spring–dashpot model is first proposed by [8, 54]. In the hard–sphere approach, collisions are binary and instantaneous, whereas the soft–sphere approach imposes no such restrictions, thus it is possible to have enduring, multi–particle contacts. In the hard–sphere (event driven) approach, the time step is determined by the minimum collision time between any one pair of particles, and is directly proportional to the mean free path or inversely proportional to the volume fraction of particles. In general, in a dense system, for which long contact times and multiple contacts exist between particles. Also, in dense regions, momentum transfer occurs more through enduring contacts (the so called quasi–static regime) than through binary collisions. In the soft-sphere (time-driven) approach, particles are allowed to overlap and their contact lasts for a certain period. This allows a particle to be in contact with more than one particle at a time. Therefore, the soft-sphere formulation covers more practical granular flow than the hard-sphere formulation, and is more appropriate in fluid-particles multiphase flow.

In the DEM method according to the soft-sphere formulation, the solid–phase is represented by N spherical particles with each particle having diameter d_p and density ρ_s. Accordingly, these particles are represented in a Lagrangian frame of reference at time t by $\{\vec{x}_i(t), \vec{v}_i(t), \vec{\omega}_i(t),\ i = 1,..., N\}$, where $\vec{x}_i(t)$ denotes the ith particle's position, $\vec{v}_i(t)$ and $\vec{\omega}_i(t)$ denote its linear and angular velocities. The position, linear and angular velocities of the ith particle evolve according to Newton's laws as [6, 21]:

$$\frac{d\vec{x}_i(t)}{dt} = \vec{v}_i(t) \tag{2.9}$$

$$m_i \frac{d\vec{v}_i(t)}{dt} = \sum\nolimits_{j=1, i\neq j}^{N} \vec{f}_{c,ij} + \vec{f}_{pf,i} + m_i \vec{g} \tag{2.10}$$

$$I_i \frac{d\vec{\omega}_i(t)}{dt} = \sum\nolimits_{j=1, i \neq j}^{N} \vec{M}_{ij} \tag{2.11}$$

where $\vec{f}_{pf}$ is the interaction force between fluid and particle in the cell. $\vec{f}_c$ is the contact force as a result of contact with other particles. $\vec{M}$ is the sum of all torques acting on the ith particle. $\vec{g}$ is the acceleration due to gravity. The mass m_i and moment of inertia I_i of the ith particle are according to particle diameter and density.

$$m_i = \rho_s \frac{\pi d_{pi}^3}{6} \text{ and } I_i = \frac{m_i d_{pi}^2}{10} \tag{2.12}$$

2.2 Contact Forces of Particles

In soft–sphere approach, the soft-sphere formulation is based on the following assumptions: (1) Particles are deformable, but the deformation is reversible and they retain their original shapes after the contact is released. (2) The interaction between contacting discrete particles is via the contact area. The overlap between particles is very small compared to the size of discrete particles. (3) The overlap progressively changes with time during a contact and its maximum is the function of physical properties and impact velocity of discrete particles. (4) The magnitude of contact force between colliding discrete particles is obtained from force-displacement laws. The force-displacement is a function of overlap, relative velocity, contact history, and shape and properties of discrete particles.

2.2.1 Force-Displacement Law

When there is a physical contact between any two particles, the force-displacement law is used to calculate collision properties as a function of normal and tangential overlaps, physical properties and collision history of the colliding particles. The force-displacement laws are applied to DEM. They can be classified into two main categories: continuous potential force-displacement model (which is mainly used in molecular dynamics) and non-continuous force-displacement model (which is widely used in DEM simulations of granular flow). The force model in the continuous potential force-displacement model is similar to the Lenard–Jones potential equation [31]. However, the continuous model has non-zero force at the beginning and the end of collision, which is physically unrealistic in granular flows. Non-continuous force-displacement model has zero repulsive force if particles are not in physical contact, which is more realistic in granular materials.

The overlap represents a physical contact between the two discrete particles with the radii R_i and R_j. The normal overlap of two particles is defined by

$$\delta_n = R_i + R_j - \left|\vec{x}_j - \vec{x}_i\right| \tag{2.13}$$

If the normal overlap is larger than zero, there is a physical contact between particle i and particle j. The collision force $\vec{f}_c$ between the two particles is comprised of normal collision force $\vec{f}_n$ and tangential collision force $\vec{f}_t$.

$$\vec{f}_{c,ij} = \vec{f}_{n,ij} + \vec{f}_{t,ij} \tag{2.14}$$

The normal vector is a vector points from the center of particle i to the center of particle j and is defined as:

$$\vec{n}_{ij} = \frac{\vec{x}_j - \vec{x}_i}{\left|\vec{x}_j - \vec{x}_i\right|} \tag{2.15}$$

The relative velocity between colliding discrete particles at the contact point and its normal and tangential components are defined as:

$$\vec{v}_{ij} = \vec{v}_i - \vec{v}_j + (R_i\vec{\omega}_i + R_j\vec{\omega}_j) \times \vec{n}_{ij} \tag{2.16}$$

$$\vec{v}_{ij,n} = (\vec{v}_{ij} \cdot \vec{n}_{ij})\vec{n}_{ij} \tag{2.17}$$

$$\vec{v}_{ij,t} = \vec{v}_{ij} - \vec{v}_{ij,n} \tag{2.18}$$

From the tangential velocity at contact point between two discrete particles, the tangential vector is calculated as:

$$\vec{t}_{ij} = \frac{\vec{v}_{ij,t}}{\left|\vec{v}_{ij,t}\right|} \tag{2.19}$$

The calculation of the tangential overlap is a bit different from the normal overlap. The tangential overlap is calculated from the tangential relative velocity of particles at the contact point and time step Δt of discrete particles.

$$\delta_t = \vec{v}_{ij,t} \min\left(\frac{|\delta_n|}{\left|\vec{v}_{ij} \cdot \vec{n}_{ij}\right|}, \Delta t\right) \tag{2.20}$$

Since each contact is processed for several time steps Δt, the tangential overlap of each collision should be saved at each time step.

2.2.2 Linear Viscoelastic Model

The linear spring-dashpot model introduced by [8] using the spring–dashpot contact law is widely used due to its applicability to the case of multiple collisions and non-spherical particles. A small value for the spring stiffness makes it possible to use larger time step for integration and accelerating calculations. However, it affects the maximum overlap and the magnitude of repulsive contact force between particles which may lead to unrealistic results.

The overlap between the two particles is represented as a system of springs and dashpots in both normal and tangential directions, as shown in Fig. 2.1. The spring causes the rebound off the colliding particles and the dashpot mimics the dissipation of kinetic energy due to inelastic collisions. The spring stiffness coefficients in the tangential and normal directions are k_t and k_n, respectively. Similarly, the dashpot damping coefficients in the tangential and normal directions are η_t and η_n, respectively. The collision force in the normal direction can be written as [8]

$$\vec{f}_{ij,n} = -(k_n \delta_n)\vec{n}_{ij} - (\eta_n \vec{v}_{ij} \cdot \vec{n}_{ij})\vec{n}_{ij} \tag{2.21}$$

The first term on the right-hand side represents an elastic force, and the second term denotes a viscous force. The elastic force is calculated by Hook's law, which relates the normal elastic force to the normal overlap by the proportionality factor of spring stiffness and conserves the kinetic energy of collision. In contrast, the viscous force is proportional to the relative velocity of particles and dissipates the kinetic energy of collision.

Analogous to the normal direction, the tangential collision force is defined by the following equation, assuming no micro-slip in the contact area of colliding particles with constant tangential spring stiffness:

$$\vec{f}_{ij,t} = -(k_t \delta_t)\vec{t}_{ij} - (\eta_t \vec{v}_{ij} \cdot \vec{n}_{ij})\vec{t}_{ij} \tag{2.22}$$

This equation is valid when the Coulomb's criterion is not violated. If the Coulomb's criterion is violated, gross sliding occurs and the tangential force is given by the model of [8]

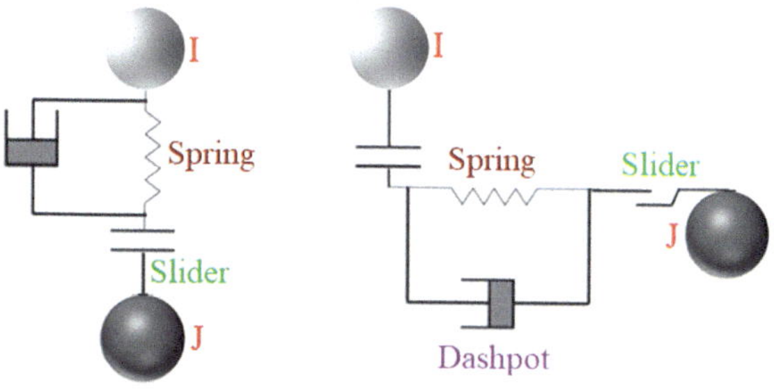

Fig. 2.1 Schematic of spring–dashpot system in soft–sphere approach

$$\vec{f}_{ij,t} = -\mu_{df}\vec{f}_{ij,n}\mathrm{sgn}(\delta_t)\vec{t}_{ij} \qquad \left|f_{ij,t}\right| \geq \mu_{df}\left|f_{ij,n}\right| \tag{2.23}$$

where μ_{df} stands for the coefficient of dynamic friction between particle i and j.

According to Hertzian contact theory [25], the normal and tangential spring stiffnesses can be calculated from the Young's modulus and Poisson ratio between contacting particle i and j.

$$k_n = \frac{4}{3}\frac{E_iE_j\sqrt{R_{eff}}}{E_i\left(1 - \nu_j^2\right) + E_j\left(1 - \nu_i^2\right)}\delta_n^{1/2} \tag{2.24}$$

$$k_t = \frac{16}{3}\frac{G_iG_j\sqrt{R_{eff}}}{G_i(2 - \nu_j^2) + G_j(2 - \nu_i^2)}\delta_n^{1/2} \tag{2.25}$$

where E_i and E_j denote the Young's moduli. and ν_i and ν_j represent the Poisson ratios for particle i and j, respectively. G_i is the shear moduli calculated as

$$G_i = \frac{E_i}{2(1 + \nu_i)} \tag{2.26}$$

The damping coefficients are related to the spring stiffness and restitution coefficient e, and are expressed as below.

$$\eta_n = \frac{2\sqrt{m_{eff}k_n}|\ln e|}{\sqrt{\pi^2 + \ln^2 e}} \tag{2.27}$$

and

$$\eta_t = \eta_n \tag{2.28}$$

where m_{eff} and R_{eff} are the effective mass and radius of particles.

$$m_{eff} = \left(\frac{1}{m_i} + \frac{1}{m_j}\right)^{-1} \tag{2.29}$$

and

$$R_{eff} = \left(\frac{1}{R_i} + \frac{1}{R_j}\right)^{-1} \tag{2.30}$$

The linear contact model is simple, making it very attractive in calculating collision parameters. It can be used for situations with multiple collisions. However, there are some unrealistic physical behavior associated with the linear viscoelastic model. One is that there is a non-zero force at the start and the end of the collision in viscoelastic models. Another is that the tangential collision is almost independent of the normal force (except the situation when the Coulomb's law is used). The tangential stiffness is constant and independent of normal overlap and collision history. Hence, the linear model should be used with caution, especially in cases in which particle–particle interactions are dominant in the flow of granular material.

For a more detailed discussion of the linear viscoelastic model, the reader is referred to [8, 10, 24].

2.2.3 Nonlinear Viscoelastic Model

The nonlinear viscoelastic contact force model proposed by [29] consists of two components, elastic force and viscous force. The theory of Hertz describes the elastic behavior of collision of spherical particles. The energy loss during a contact leads to different velocity of discrete particles before and after the collision. The energy dissipation is taken into account by introducing a damper. All nonlinear normal viscoelastic force-displacement models use the Hertz theory to describe the elastic behavior of collision. The normal component of the contact force is expressed as below [29].

$$\vec{f}_{ij,n} = \left(-\frac{4}{3}E_{eff}\sqrt{R_{eff}}\delta_n^{3/2}\right)\vec{n}_{ij} - (\eta_n\delta_n^{1/2}\vec{v}_{ij}\cdot\vec{n}_{ij})\vec{n}_{ij} \tag{2.31}$$

where E_{eff} is the effective Young's modulus and expresses as the function of Young's modulus E and Poisson's ratio v of particle i and j, respectively.

$$E_{eff} = \left(\frac{1-v_i^2}{E_i} + \frac{1-v_j^2}{E_j}\right)^{-1} \tag{2.32}$$

Tsuji et al. [49] proposed a nonlinear force model with a slightly different viscous term. The normal collision force is expressed as a function of effective modulus.

$$\vec{f}_{ij,n} = \left(-\frac{4}{3}E_{eff}\sqrt{R_{eff}}\delta_n^{3/2}\right)\vec{n}_{ij} - \left(\eta_n\delta_n^{1/4}\vec{v}_{ij}\cdot\vec{n}_{ij}\right)\vec{n}_{ij} \tag{2.33}$$

Both Eqs. (2.31) and (2.33) indicate that the normal collision forces consist of the elastic force of the first term and the viscous force of the second term of the right-hand side. It can be seen that the viscous forces are functions of normal deformation and normal relative velocity of particles. However, the viscous force is proportional to ${\delta_n}^{1/2}$ in Eq. (2.31), while it is proportional to ${\delta_n}^{1/4}$ in Eq. (2.33). This leads to different force-displacement curves and values for damping coefficients for these two equations.

The tangential collision force is expressed as a function of effective shear modulus according to the shear modulus G of particles.

$$\vec{f}_{ij,t} = -8G_{eff}\sqrt{R_{eff}}\delta_n^{1/2}\delta_t\vec{t}_{ij} \tag{2.34}$$

$$G_{eff} = \left(\frac{2-\nu_i}{G_i} + \frac{2-\nu_j}{G_j}\right)^{-1} \tag{2.35}$$

It can be seen that the nonlinear spring stiffness is continuously changing during a collision because the normal displacement is not constant. Since the real simulations involve micro-slip conditions and vary normal force at the contact surface, the use of such a model of the tangential force does not lead to accurate results.

For a more detailed application of the nonlinear viscoelastic models, the reader is referred to [10, 24, 29, 49].

2.2.4 Hertz-Mindlin Contact Model

The Hertz-Mindlin contact model is applied to model fluid-particles systems. In this model the normal force component is based on the Hertzian contact theory, and the tangential force model is based on Mindlin-Deresiewicz's work [37]. Both normal and tangential forces have damping components where the damping coefficient is related to the coefficient of restitution [49]. The tangential friction force follows the Coulomb law of friction model [8]. The rolling friction is implemented as the contact independent directional constant torque model.

Figure 2.2 shows the profile of normal force as a function of normal overlap. The normal contact force is expressed according to the effective Young's modulus E_{eff} and effective radius of particles R_{eff} [8, 37, 49].

$$\vec{f}_{ij,n} = (-\frac{4}{3}E_{eff}\sqrt{R_{eff}}\delta_n^{3/2})\vec{n}_{ij} - 2\sqrt{\frac{5}{6}}\beta\sqrt{m_{eff}S_n}(\vec{v}_{ij}\cdot\vec{n}_{ij})\vec{n}_{ij} \tag{2.36}$$

$$S_n = 2E_{eff}\sqrt{R_{eff}\delta_n} \tag{2.37}$$

The first term on the right-hand side represents the normal force component, and the second term is a damping force. The tangential force depends on the tangential overlap δ_t.

$$\vec{f}_{ij,t} = -8G_{eff}\sqrt{R_{eff}\delta_n}\delta_t\vec{t}_{ij} - 2\sqrt{\frac{5}{6}}\beta\sqrt{m_{eff}S_t}\vec{v}_{ij,t} \tag{2.38}$$

$$S_t = 8G_{eff}\sqrt{R_{eff}\delta_n} \tag{2.39}$$

The second term on the right-hand side represents a tangential damping component as the function of relative tangential velocity of particles. The tangential force is limited by Coulomb friction ($\mu_{fr}f_n$) where μ_{fr} is the coefficient of static friction.

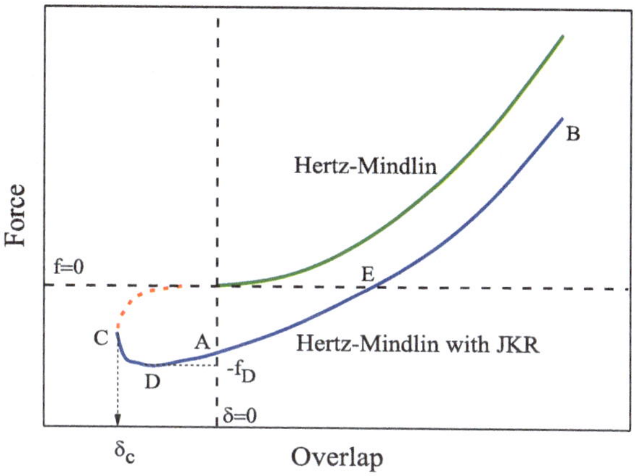

Fig. 2.2 Forces using Hertz-Mindlin model and Hertz-Mindlin with JKR model as a function of overlap

The parameter β denotes the normal damping ratio. This ratio takes a value $\beta = 0$ when there is no damping, and $\beta = 1$ when the collision of particles is damped.

$$\beta = \frac{\ln e}{\sqrt{\ln^2 e + \pi^2}} \tag{2.40}$$

Detailed derivation of the equations and applications can be found elsewhere [8, 10, 24, 37, 49].

2.2.5 *Hertz-Mindlin with JKR Model*

The cohesion or adhesion can be modeled by introducing an attractive force component to the contact model. The cohesion and adhesion terms are usually defined for attracting forces between similar and dissimilar materials respectively. The particle contact model of Hertz-Mindlin with JKR (Johnson-Kendall-Roberts) cohesion is a cohesion contact model that accounts for the influence of Van der Waals forces within the contact zone, and allows to model strongly adhesive systems, such as wet materials [26]. In this model, the implementation of normal elastic contact force is based on the Johnson-Kendall-Roberts theory. The model is also used to model materials where the adhesion is caused by capillary or liquid-bridge forces.

The profile of normal force of Hertz-Mindlin with JKR model is shown in Fig. 2.2 as a function of normal overlap. The forces using Hertz-Mindlin with JKR cohesion model are compared with Hertz-Mindlin (no slip) model results. Negative overlap is the gap between two separated particles. If the two particles are compressed until the contact force reaches a maximum (say point *B*) and then decompressed until all the work done during the loading stage has been recovered at point *A* during unloading. At this point, the contact area is not zero, the particles remain adhered together, and further work is required to break the contact. Then, the situation becomes unstable at the force $-f_D$ of the point D and the contact breaks. However, particles are always displacement driven in which case separation occurs at point C with the maximum overlap.

In Hertz-Mindlin with JKR model, the contact force is divided into a normal component and tangential component. The contact force in the normal direction consists of the damping force and elastic force. The normal force depends on overlap δ and surface energy γ in the following way:

$$\vec{f}_{ij,n} = \left[-4\sqrt{\pi \gamma E_{eff}} a^{3/2} + \frac{4E_{eff}}{3R_{eff}} a^3 \right] \vec{n}_{ij} - 2\sqrt{\frac{5}{6}} \beta \sqrt{2m_{eff} E_{eff} (R_{eff} \delta_n)^{0.5}} \vec{v}_{ij,n} \tag{2.41}$$

The attractive force can be introduced to the classic Hertz contact model, which acts over the circular contact area with the radius a.

$$\frac{a^2}{R_{eff}} = \delta + \sqrt{\frac{4\pi \gamma a}{E_{eff}}} \tag{2.42}$$

The second term on the right-hand side of Eq. (2.41) represents the damping force in the normal direction, and is related to the surface energy γ of particles and the normal relative velocity $\vec{v}_{ij,n}$ between the particle i and particle j during the contact.

The tangential force depends on the normal and tangential overlaps and the relative tangential velocity $\vec{v}_{ij,t}$ of particles.

$$\vec{f}_{ij,t} = \min \left[\mu_{df} \vec{f}_n, \left| -8E_{eff} \sqrt{R_{eff} \delta_n} \delta_t \vec{t}_{ij} - 2\sqrt{\frac{5}{6}} \beta \sqrt{8E_{eff} m_{eff} (R_{eff} \delta_n)^{0.5}} \vec{v}_{ij,t} \right| \right] \tag{2.43}$$

For two different objects, the free energy per unit area represents an adhesion free energy ΔG_{ij}, showing a negative adhesion work. It can be expressed as follows.

$$\Delta G_{ij} = \gamma_i + \gamma_j - \gamma_{ij} \tag{2.44}$$

where γ_i and γ_j denote the solid phase and liquid phase surface energies in contact with each other. γ_{ij} represents the JKR surface energy. On the other hand, the free energy can be expressed according to the general contact angle θ.

$$\Delta G_{sl} = \gamma_i (1 - \cos \theta) \tag{2.45}$$

where the subscript sl denotes solid–liquid phase. The contact angle can refer to the tangential line of the interface of two phases. Combining Eq. (2.44) with Eqs. (2.45) and (2.46) is obtained.

$$\gamma_{ij} = \gamma_j + \gamma_i \cos \theta \tag{2.46}$$

The equation indicates that the JKR surface energy relates with contact angle, similar to the repose angle of the accumulation of powder materials. These indicate that the JKR surface energy can be evaluated from contact angle between the liquid and the solid.

When the two surface materials are identical ($\gamma_i = \gamma_j = \gamma$), the interface energy is zero ($\gamma_{ij} = 0$), and the effective surface energy reduces to

$$\Delta G_{ij} = 2\gamma \tag{2.47}$$

The attractive force becomes:

$$F_{adh} = \pi a^2 \Delta G_{ij} = 2\pi \gamma a^2 \tag{2.48}$$

The JKR model provides attractive cohesion forces even if the particles are not in physical contact. The maximum contact patch radius a_c and maximum gap δ_c between particles with non-zero force are given by

$$a_c = \left[\frac{9\pi \gamma R_{eff}^2}{2E_{eff}} \left(\frac{3}{4} - \frac{1}{\sqrt{2}} \right) \right]^{1/3} \tag{2.49}$$

$$\delta_c = -\sqrt{\frac{4\pi \gamma a_c}{E_{eff}}} + \frac{a_c^2}{R_{eff}} \tag{2.50}$$

As the overlap δ is less than the maximum gap δ_c, the model returns zero force. The maximum value of the cohesion force occurs when particles are not in physical contact and the separation gap is less than δ_c. The maximum cohesion force, called pull-out force f_{po}, is given by

$$f_{po} = -\frac{3}{2}\pi \gamma R_{eff} \tag{2.51}$$

When two particles approach each other, the contact is formed. As two particles physically touch at $\delta_n = 0$ at the point *A*. At this point, the normal force suddenly jumps to

$$f_{n,A} = -\frac{8}{9}\pi \gamma R_{eff} \tag{2.52}$$

This attractive force will pull the discrete particles closer until the equilibrium point is reached (assuming no other forces are acting) where $f_n = 0$ at the point *E*. Further loading due to external forces will result in an increased positive overlap δ_n and an increase in the compressive forces.

Upon unloading, the force decreases down to a minimum value $f_{n,D}$ at the point *D*.

$$f_{n,D} = -3\pi\gamma R_{eff} \quad (2.53)$$

Although JKR model is designed for dry particles, it can be applied to model wet particles. The force needed to separate between the particle i and j depends on the liquid surface energy. For more detailed modeling and additional results, see [10, 24–26].

2.2.6 Hysteretic Spring Contact Model

The hysteretic spring contact model is based on the study of the [51] and accounts for the plastic deformation behaviour in the contact mechanics equations. The hysteretic spring contact model allows plastic deformation behaviors to be included in the contact mechanics equations, resulting in particles behaving in an elastic manner up to a predefined stress.

The hysteretic spring contact process includes a loading and unload/reload displacements, as shown in Fig. 2.3. In a loading, the force increases with the increase of displacement. The unloading force goes to zero before the displacement recovers to the initial contacting point. The quantity δ_0 represents the residual overlap that remains because of the plastic deformation (i.e., flattening) in the contact region. The exact location of each old contact spot is not remembered, so that particles become like new undeformed spheres once they separate. Subsequent contacts will load along a new loading path with slope K_1. Any reloading before separation follows the slope K_2 until the original loading is reached, where upon the lower slope is followed until unloading occurs. The normal contact force used in hysteretic spring model is expressed:

$$\vec{f}_{ij,n} = [-b_n\sqrt{\frac{4m_{eff}k}{1+\left(\frac{\pi}{\ln e}\right)^2}}\vec{v}_{ij,n}] - \vec{n}_{ij}\begin{cases} K_1\delta_n & K_1\delta_n < K_2(\delta_n - \delta_0) \\ K_2(\delta_n - \delta_0) & \delta_n > \delta_0 \\ 0 & \delta_n \le \delta_0 \end{cases} \quad (2.54)$$

where K_1 and K_2 are the loading and unloading stiffness, respectively. δ_n is the normal overlap and δ_0 is the residual overlap. b_n is a dimensionless parameter of damping factor. k is either K_1 or K_2. The loading stiffness K_1 relates to the yield strength Y_1 and Y_2 of each material participating in contact as follows

$$K_1 = 5R_{eff}\min(Y_1, Y_2) \quad (2.55)$$

where R_{eff} is the equivalent radius of the two particles participating in the contact. The following expression for K_2 is according to the coefficient of restitution of particles.

$$K_2 = e^{-2}K_1 \quad (2.56)$$

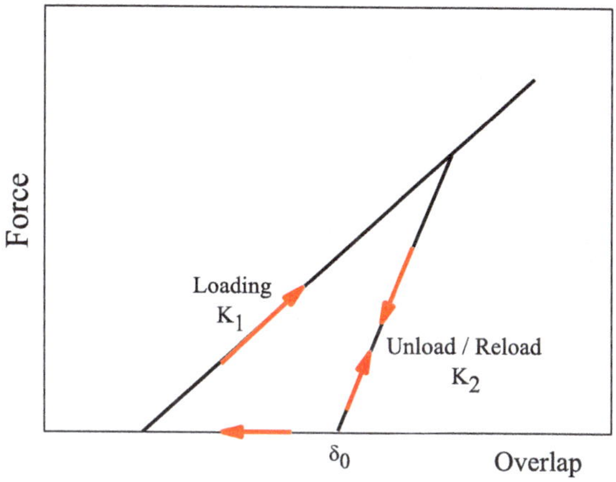

Fig. 2.3 Schematic of hysteretic spring contact force relationship

The first term on the right-hand side of Eq. (2.54) represents the damping force. The main energy dissipation mechanism is due to the difference in spring stiffness between loading and unloading phases. The need for an additional velocity dependent dissipation mechanism suppresses undamped low amplitude oscillations. This mechanism is implemented by a damping factor b_n. The dimensionless damping factor controls the amount of velocity dependent damping. Without this additional damping, small vibrations of particles can persist for a long time. Setting this parameter to zero suppresses only the velocity-dependent damping without affecting the energy loss mechanism. The recommended value for b_n is taken as 0.05 for small strain compressions.

The residual overlap δ_0 represents the residual displacement that remains because of the plastic deformation in the contact region. The residual overlap is updated every time step according to the following rule:

$$\delta_0 = \begin{cases} \delta_n(1 - \frac{K_1}{K_2}) & K_1\delta_n < K_2(\delta_n - \delta_0) \\ \delta_0 & \delta_n > \delta_0 \\ \delta_n & \delta_n \le \delta_0 \end{cases} \tag{2.57}$$

The yield strength is estimated from the Young's modulus E and the radius of the particle R. The possible value is calculated from:

$$Y = \frac{2}{525}\frac{E}{\sqrt{R}} \tag{2.58}$$

If this value is less than 100 Pa, it is replaced with a value of $0.003E$.

The tangential forces used in hysteretic spring model are expressed according to the dimensionless parameter of stiffness factor γ_t and the coefficient of dynamic friction μ_{df} as follows:

$$\vec{f}_{ij,t} = -\min[\mu_{df}\vec{f}_{ij,n}, \gamma_t K_1 \delta_t \vec{t}_{ij} + \vec{F}_{d,t}] \tag{2.59}$$

$$\vec{F}_{d,t} = -\sqrt{\frac{4m_{eff}\gamma_t k}{1+\left(\frac{\pi}{\ln e}\right)^2}}\vec{v}_{t,rel} \tag{2.60}$$

where $F_{d,t}$ represents the tangential damping force. The typical value of the stiffness factor parameter γ_t is in the range from 0.7 to 1.0.

For a detailed discussion and derivation of the contact forces of hysteretic spring contact model, see [10, 51].

2.2.7 Edinburgh Elasto-plastic Adhesion Contact Model

For cohesive particles, the flow characteristics show the manifestation of the unconfined yield strength that arises from the historical consolidation stress. The flow behavior of cohesive particles is strongly dependent on the consolidation stresses. Hence, the cohesive strength is stress-history dependent. Such stress-history dependent cohesive behavior must be captured if a numerical model is to successfully simulate the flow of cohesive particles.

The elasto-plastic adhesion model is a contact model to describe flow behavior of cohesive elasto-plastic particles [39, 48]. The adhesion contact model captures the history dependence and the key characteristic behavior of cohesive particles. The model comprises a nonlinear hysteretic spring model to account for the elastic–plastic contact deformation and an adhesive force component that is a function of the plastic contact deformation. When two particles are pressed together, they undergo elastic and plastic deformations. The pull-off strength (adhesion) increases with an increase of the plastic contact area.

Figure 2.4 shows the profile of elasto-plastic contact force as a function of displacement in the normal direction. The contact normal force f_n of cohesive particles is the sum of the normal hysteretic spring force and the normal damping force:

$$\vec{f}_{ij,n} = \vec{f}_{hys} + [-2\beta_n\sqrt{m_{eff}k_{tn}}\vec{v}_{ij,n}] \tag{2.61}$$

$$\vec{f}_{hys} = \begin{cases} f_0 + k_1\delta_n^m\vec{n}_{ij} & k_2(\delta_n^m - \delta_p^m) \geq k_1\delta_n^m \\ f_0 + k_2(\delta_n^m - \delta_p^m)\vec{n}_{ij} & k_1\delta_n^m > k_2(\delta_n^m - \delta_p^m) > -k_{adh}\delta_n^\chi \\ f_0 - k_{adh}\delta_n^\chi\vec{n}_{ij} & -k_{adh}\delta_n^\chi \geq k_2(\delta_n^m - \delta_p^m) \end{cases} \tag{2.62}$$

where m is the loading–unloading branch exponent, and assumes to be $m = 1.5$. χ is the adhesive branch exponent, and is assumed to have a typical value of $\chi = 1.5$. The first term on the right-hand side represents the normal hysteretic spring force, and the second term denotes the normal damping force. f_0 represents the pull-off force,

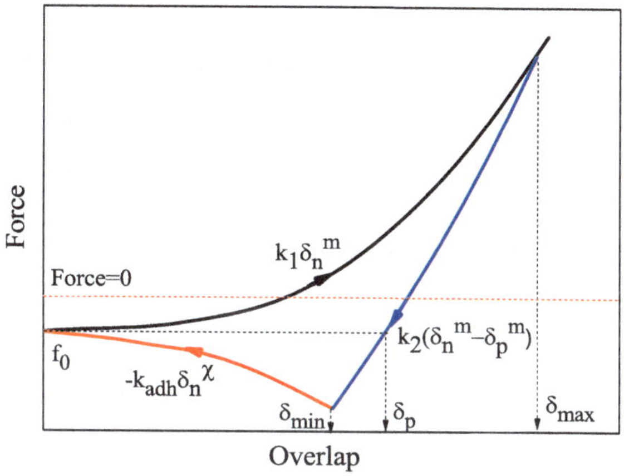

Fig. 2.4 Edinburgh elasto-plastic contact force in the normal direction

changing from negative infinity to zero. It shows the contact force at a normal overlap $\delta_n = 0$, which should take a negative value. The load–displacement function of cohesive particles consists of a loading, unloading, re-loading and adhesive branches, and are characterized by the constant adhesive strength f_0, adhesive stiffness parameter k_{adh}, loading stiffness parameter k_1, unloading and reloading stiffness parameter k_2, loading/unloading/reloading exponent m and the adhesive branch unloading exponent χ. During the initial loading, the contact model follows the loading path k_1, upon unloading of the contact. The maximum overlap δ_{max} is a history-dependent parameter and is updated and stored with the contact. After that, the contact of cohesive particles will switch from the loading path to the unloading/reloading path k_2. During the reloading stage, the contact force initially follows along the reloading k_2 path but switches to the loading k_1 path when the previous maximum loading force is reached. Unloading along the k_2 path below the plastic overlap δ_p results in the development of an adhesive force until the maximum adhesive force is reached. Further unloading past this point results in a reduction in both the normal overlap and the adhesive force until separation occurs ($\delta_n = 0$). If reloading of the contact occurs, the contact will reload along a k_2 path on the adhesion branch until the loading k_1 path is reached on further increase of load.

The normal critical damping ratio β_n is introduced. This ratio takes a value $\beta_n = 0$ when there is no damping, and $\beta_n = 1$ when the system is critically damped. The normal stiffness is given by:

$$k_{tn} = \begin{cases} mk_1\delta_n^{m-1} & k_1 \text{ path} \\ mk_2\delta_n^{m-1} & k_2 \text{ path} \\ \chi k_{adh}\delta_n^{\chi-1} & k_{adh} \text{ path} \end{cases} \tag{2.63}$$

The exponent m is used to switch between the linear and non-linear force-overlap relationships for the contact model. When $m = 1$, there is linear behavior, hence the model replicates a linear spring, whereas when $m = 1.5$ the model replicates the

Hertzian spring model. The adhesive branch exponent χ ranges from 1.0 to infinite. The stiffness k_1 is defined in terms of the effective Young's modulus in a format similar to that of the Hertz model.

$$k_1 = \frac{4}{3} E_{eff} \sqrt{R_{eff}} \tag{2.64}$$

The unloading and reloading stiffness parameter k_2 is defined in terms of the plasticity ratio λ_p of particles.

$$k_2 = \frac{k_1}{1 - \lambda_p} \tag{2.65}$$

When $\lambda_p = 0$, $k_2 = k_1$ and there is no plastic behavior. When λ_p trends to 1.0, k_2 goes to infinite, and the behavior is perfectly plastic. The adhesion stiffness is given by

$$k_{adh} = -\frac{3\pi a \gamma_{ij}}{2\delta_{min}^{\chi}} \tag{2.66}$$

where a is the contact radius. γ_{ij} is the contact effective surface adhesion energy based on the Johnson-Kendall-Roberts model for adhesive forces [26].

The pull-off force f_0 changes from a negative infinite to zero, and is available by default. The plastic overlap is defined as the overlap where the force on the k_2 branch is equal to the pull-off force f_0, and is given in terms of the maximum overlap and the plasticity ratio.

$$\delta_p = \left(1 - \frac{k_1}{k_2}\right)^{1/m} \delta_{max} \tag{2.67}$$

The contact tangential force of particles is the sum of tangential spring force $\vec{f}_{spr}$ and tangential viscous damping force $\vec{f}_{dam}$, as given by

$$\vec{f}_{ij,t} = \vec{f}_{spr} + \vec{f}_{dam} \tag{2.68}$$

The tangential spring force is expressed in incremental terms:

$$\vec{f}_{spr} = f_{spr,0} + 8k_{sf} G_{eff} \sqrt{R_{eff}\delta_n} \Delta\delta_t \vec{t}_{ij} \tag{2.69}$$

where $f_{spr,0}$ is the tangential spring force at the previous time step. G_{eff} is the effective shear modulus, and k_{sf} is a scaling factor. The shear stiffness scaling factor changes from zero to infinite, and can be allowed to be nearly one (i.e., $k_{sf} = 1.0$). $\Delta\delta_t$ is the shear displacement increment.

The viscous damping in the shear direction is given by the vector, and can be obtained from:

$$\vec{f}_{dam} = -2\beta_s\sqrt{m_{eff}k_{tn}}\vec{v}_{ij,t} \tag{2.70}$$

where β_s is the shear critical damping ratio of particles, and changes from 0.0 to 1.0. $v_{ij,t}$ is the relative shear translational velocity.

The limiting tangential friction force $\vec{f}_{ct}$ is calculated using the Coulombic friction criterion with the normal force modified by the adhesives forces given by:

$$\vec{f}_{ct} \leq \mu_{fc}(\left|\vec{f}_{hys} + k_{adh}\delta^m - f_0\right|) \tag{2.71}$$

where μ_{fc} is the friction coefficient of particles.

For a more detailed explanation for the normal and tangential forces of elasto-plastic adhesion contact model, see [10, 24, 39, 48].

2.2.8 Hertz-Mindlin with Bonding Model

The Hertz Mindlin with bonding contact model fully considers the bonding characteristics between particles, and the bond is established between particles withstanding the effects of tension, compression and shear. In Hertz-Mindlin with bonding model, the normal force component is based on Hertzian contact theory [36]. The Mindlin-Deresiewicz model forms the foundation for tangential forces. The damping coefficient of both normal and tangential forces is connected to the coefficient of restitution [49]. The tangential friction force follows the Coulomb law of friction model [8]. From Fig. 2.5, a linear softening path is described by the force–displacement law in the normal bond component. The compression and stretch of the bond are taken as positive. The force reaches maximum f_{max} at δ_1. The normal bond breaks at δ_2. Once a failure condition is reached, the forces in the bond are immediately reduced to zero.

The particle normal force $\vec{f}_n$ relates to the normal elastic force, normal damping force and normal adhesion force of the bonding at the time t_n [43].

$$\vec{f}_{ij,n} = \begin{cases} \frac{3}{4}E_{eff}\sqrt{R_{eff}}\delta_n^{3/2}\vec{n}_{ij} - 2\sqrt{\frac{5}{6}}\beta_n\sqrt{m_{eff}S_n}\vec{v}_{ij,n} & t \leq t_{bond} \\ f_{n0} - \pi R_B^2 S_n \vec{v}_{ij,n}\delta t & t > t_{bond} \end{cases} \tag{2.72}$$

where t_{bond} represents the bonding time. The time step is $\delta t = t_n\text{-}t_{n\text{-}1}$. R_{B} denotes the bonded disk radius. f_{n0} represents the normal force at the time $t_{n\text{-}1}$. The bonding time should exceed the stabilization time of the filling spheres to ensure the relative stability of the formed parallel bonds. The normal forces show that it comprises the normal elastic force of the first term and the normal damping force of the last term before bonding. Parallel bonds are formed when simulation time reaches the preset bonding time t_{bond}. After particles are bonded, the normal elastic force and damping force are to be zero [43]. The normal contact force of bounding particles represents

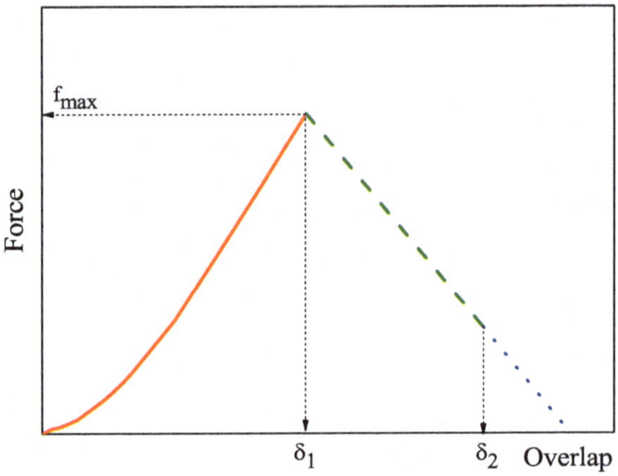

Fig. 2.5 Force–displacement contact law for the normal bond

the normal adhesion force that is progressively changed every time step. The particle normal stiffness S_n are expressed as the function of effective Young's modulus E_{eff}.

$$S_n = 2E_{eff}\sqrt{R_{eff}\delta_n} \tag{2.73}$$

Similarly, when the particles are not actually in contact, the tangential force $\vec{f_t}$ encompasses the tangential elastic force and the tangential damping force, respectively. Furthermore, the tangential force is constrained by Coulomb friction. When simulation time reaches the preset bonding time t_{bond}, the bonds of particles are formed. The tangential force is represented by the tangential adhesion forces of the bonding after particles are bonded at the time t_n.

$$\vec{f}_{ij,t} = \begin{cases} -8G_{eff}\sqrt{R_{eff}}\delta_t^{3/2}\vec{t}_{ij} - 2\sqrt{\frac{5}{6}}\beta_t\sqrt{k_t m_{eff}}\vec{v}_{ij,t} & t \le t_{bond} \\ f_{t0} - \pi R_B^2 S_t \vec{v}_{ij,t}\delta t & t > t_{bond} \end{cases} \tag{2.74}$$

where f_{t0} represents the tangential force at the time $t_{n\text{-}1}$. The particle tangential stiffness S_t is expressed as the function of effective shear modulus G_{eff}.

$$S_t = 8G_{eff}\sqrt{R_{eff}\delta_n} \tag{2.75}$$

From Eqs. (2.72) and (2.74), both β_t and β_n are the damping ratios, and they vary from 0.0 to 1.0. When the bonding model is used to calculate contact force, the force will be decomposed into the elastic force in the expected direction and the shear damping in the tangential direction. The bonding force between the time step δt is a fixed value formed during the initial loading process. The interactions between the time step δt before the bonding follow the Hertz–Mindlin (no-slip) contact model. The normal direction will exhibit a tensile force that prevents from moving away from each other. When the tensile force or the shear force in any direction exceeds the strength limit, the tangential tensile force and the tangential shear force generated by the bonding model will immediately disappear, and subsequent properties will be consistent with the linear model.

The bonding contact model indicates that the normal and tangential adhesion forces are added into the contact model of particles. Due to the fact that the bonds in this model can act even when the particles are no longer physically connected, the contact radius should be greater than the radius of the spheres. The bond fractures when the normal and tangential shear stresses exceed critical values.

$$\sigma_{max} < \frac{-f_n}{\pi R_B^2} + \frac{8M_t}{\pi R_B^3} \tag{2.76}$$

$$\tau_{max} < \frac{-f_t}{\pi R_B^2} + \frac{8M_n}{\pi R_B^3} \tag{2.77}$$

where M_n and M_t represent the normal and tangential torques, respectively. Typically, the contact radius R_B should be about (1.2–1.3)R that is larger than the physical radius of discrete particles. These indicate that the Hertz–Mindlin contact model with bonding accounts for bonds formed among particles in fluid-particle multiphase flow.

For a more detailed explanation of bonding contact interactions for the Hertz-Mindlin with bonding model, see [8, 10, 43, 49].

2.3 Momentum Coupling and Decomposition

In the numerical simulation of CFD–DEM, high grid resolution is required because of the need to obtain accurate flow field data. The solid phase information is available at the particle scale while the fluid phase information is available at the fluid cell scale. Therefore, a special attention is required when particle scale information is used to calculate a fluid cell property (e.g., volume fraction) and the fluid cell information is used to calculate the particle scale property (e.g., Reynolds number of particle). The momentum coupling between phases within a control volume exists between the fluid phase and discrete particles by comparing the total drag force of discrete particles with the momentum exchange of fluid phase.

Two sets of formulations, referred to as model A and model B, have been proposed for describing the governing equations of fluid and solid phases [17]. If the pressure is shared between fluid and solid phases, it is referred to as model A, and if the pressure is considered for the fluid phase alone, it is referred to as model B. The fluid phase momentum conservation equation of model A is expressed as below [17].

$$\frac{\partial(\rho_f \alpha_f \vec{u}_f)}{\partial t} + \nabla \cdot (\rho_f \alpha_f \vec{u}_f \vec{u}_f) = -\alpha_f \nabla p + \alpha_f \nabla \cdot (\overline{\overline{\tau}}_f) - \vec{f}_{fp}^A + \alpha_f \rho_f \vec{g} \tag{2.78}$$

The fluid phase momentum conservation equation of model B is expressed as below.

$$\frac{\partial(\rho_f \alpha_f \vec{u}_f)}{\partial t} + \nabla \cdot (\rho_f \alpha_f \vec{u}_f \vec{u}_f) = -\nabla p + \nabla \cdot (\overline{\overline{\tau}}_f) - \vec{f}_{fp}^{B} + \alpha_f \rho_f \vec{g} \tag{2.79}$$

where $\vec{f}_{fp}$ represents the volumetric fluid–particle interaction in model A and B, respectively. This term represents the volumetric mean value of forces acting on particles by surrounding fluid. These equations show that the true coupling can be achieved when the same source terms are used for fluid phase and discrete particles. Although these two models are named differently, they are physically the same.

2.3.1 Fluid–Particle Interaction Force

The fluid–particle interaction force in the unresolved surface method can be obtained according to the resolved surface method [3, 6, 13]:

$$\vec{f}_{fp,i} = \vec{f}_{d,i} + \vec{f}_{u,i} + \vec{f}_{p,i} + \vec{f}_{\tau,i} + \vec{f}_{lf,i} \tag{2.80}$$

where $\vec{f}_d$ is the steady drag force. $\vec{f}_u$ is the unsteady force, $\vec{f}_p$ is the pressure gradient force. $\vec{f}_\tau$ is the viscous force. $\vec{f}_{lf}$ is the lift force. The unsteady force includes unsteady history Basset force $\vec{f}_{his}$ and added mass force $\vec{f}_{ad}$:

$$\vec{f}_{u,i} = \vec{f}_{his,i} + \vec{f}_{ad,i} \tag{2.81}$$

The pressure gradient force and viscous force are obtained from the fluid phase momentum conservation equation.

$$\vec{f}_{p,i} = -V_i \nabla p \tag{2.82}$$

$$\vec{f}_{\tau,i} = -V_i (\nabla \cdot \overline{\overline{\tau}}_f) \tag{2.83}$$

where V_i represents the volume of a particle in the cell.

The lift force includes Saffman and Magnus forces:

$$\vec{f}_{lf,i} = \vec{f}_{sa,i} + \vec{f}_{ma,i} \tag{2.84}$$

The fluid–particle interaction force acting on the particle is simulated at the particle scale while the analogous term in the fluid phase momentum equation, which is the volumetric fluid–particle interaction force acting on the fluid, is simulated in each fluid cell. This difference in the length scale of fluid interaction forces results in development of different coupling methods [14].

Coupling method I: The average force acting on the fluid phase due to the motion of particles is simulated in each fluid cell by the average gas and particle velocities and average particle diameter, while the force acing on a particle is determined based on the velocity of each particle from the fluid phase.

Coupling method II: The force acting on the fluid phase due to the motion of particles is first simulated as in Scheme 1, then it is distributed among individual particles based on an averaging method of surface area or volume of particles.

Coupling method III: The fluid–particle interaction force acting on individual particles is first simulated in a fluid cell, then this force is summed on all particles to estimate the fluid–particle interaction force in the fluid cell.

In view of the conservation laws of fluid, the force of the fluid phase acting on the solid phase should be equal to the force of the solid phase acting on the fluid phase in the opposite direction. Therefore, the coupling methof III is recommnded to calculate the volumetric fluid interaction force term.

For model A, the volumetric fluid–particle interaction is expressed

$$\vec{f}^{A}_{fp,i} = \frac{1}{V_{cell}} \sum_{i=1}^{m} (\vec{f}_{d,i} + \vec{f}_{u,i} + \vec{f}_{lf,i}) \tag{2.85}$$

where m is the number of particles in each fluid cell of volume V_{cell}, and the summation is performed on all particles in the fluid cell. For model B, the volumetric fluid–particle interaction is given by

$$\vec{f}^{B}_{fp,i} = \frac{1}{\alpha_f V_{cell}} \sum_{i=1}^{m} (\vec{f}_{d,i} + \vec{f}_{u,i} + \vec{f}_{lf,i}) - \frac{1}{V_{cell}} \sum_{i=1}^{m} (\rho_f V_i \vec{g}). \tag{2.86}$$

2.3.2 Drag Force

In all fluid-particle flows, the aerodynamic drag plays an important role on the motion of particles. Thus, an accurate calculation of drag force is important for successful simulation of fluid-particle flows. From the theoretical point of view, the exact relation of the drag force acting on a spherical particle is the Stokes–Einstein relationship [11], which is valid for an isolated sphere in the creeping flow regime where Reynolds number is less than 0.1. In many practical applications, it is convenient to express the drag force based on a proportionality dimensionless constant that is called the drag coefficient. The drag coefficient can be obtained as a function of Reynolds number of particle.

For flows of fluid-particle mixture, the drag force exerted on a particle must be determined in the presence of other particles. Correct theoretical formula for estimating drag forces in such systems is not available yet. Thus, empirical correlations have been developed for this purpose. Care must be taken when using the empirical correlations because sometimes the total force exerted on a particle is also referred to as the drag force. The drag force is represented by the product of the fluid-particle momentum transfer coefficient and the slip velocity, describing the drag force is affected by fluid acting on particles.

$$\vec{f_d} = \beta_{fs}(\vec{u}_f - \vec{v}) \tag{2.87}$$

where β_{fs} is the fluid-particle inter-coefficient. The different expressions for the drag function for particulate flows are based on the relative interstitial velocity, including [17, 47] models. For solid concentrations greater than 0.2, the Ergun correlation of fluid-particles momentum transfer coefficient β_E [12] is used, whereas for values less than or equal to 0.2, the [52] expression of momentum transfer coefficient β_{WY} is employed. To avoid discontinuity of the two correlations, a switch function φ is introduced to give a smooth from the dilute regime to the dense regime [23]. The fluid-particle inter-coefficient is given by

$$\beta_{fs} = (1 - \varphi)\beta_E + \varphi\beta_{WY} \tag{2.88}$$

$$\beta_E = 150\frac{(1-\alpha_f)^2\mu_f}{(\alpha_f d_p)^2} + 1.75\frac{\rho_f(1-\alpha_f)\left|\vec{u}_f - \vec{v}\right|}{\alpha_f d_p} \quad \alpha_f \le 0.8 \tag{2.89}$$

$$\beta_{WY} = \frac{3}{4}C_D\frac{\rho_f(1-\alpha_f)\left|\vec{u}_f - \vec{v}\right|}{d_p}\alpha_f^{-2.65} \quad \alpha_f > 0.8 \tag{2.90}$$

$$\varphi = \frac{\arctan[150 \times 1.75(\alpha_f - 0.8)]}{\pi} + 0.5 \tag{2.91}$$

where the drag coefficient C_D for an isolated particle depends on the particle Reynolds number as

$$C_D = \begin{cases} \frac{24}{Re}(1 + 0.15Re^{0.687}) & Re < 1000 \\ 0.44 & Re \ge 1000 \end{cases} \tag{2.92}$$

$$Re = \frac{\rho_f \alpha_f d_p(\vec{u}_f - \vec{v})}{\mu_f}. \tag{2.93}$$

2.3.3 *Unsteady Force and Lift Force*

The unsteady forces are generated by the non-stationary motion of particle and fluid. It is caused by the lagging time response of the boundary layer (the time required for penetration of the generated momentum into the body of fluid) to accelerate particles. In other words, the unsteady force is generated by temporal development of the viscous region near particles and provides a prompt flow resistance. Particles, therefore, experience an unsteady force due to the instability of the boundary layer. The unsteady drag force is also referred to as Basset force or Boussinesq–Basset force [6]. Since this force depends on the penetration history [3], it is called history force and is defined based on the integral of all past particle accelerations. The Basset force is negligible in gases and at low particle accelerations. The unsteady history Basset force has the following form for an isolated spherical particle at low Reynolds numbers in a fluid at rest [6]:

$$\vec{f}_{his,i} = \frac{3}{2} d_{pi}^2 \sqrt{\pi \rho_f \mu_f} \int_0^{\tau_p} \frac{1}{\sqrt{\tau_p - t}} \left(\frac{d\vec{v}_i}{dt} \right) dt \quad (2.94)$$

where ρ_f and μ_f are the fluid density and viscosity. τ_p is the particle momentum response time. The derivative velocity represents the appropriate frame-indifferent acceleration of particles.

When a particle is accelerated in a fluid, it can also accelerate a part of its surrounding fluid. The corresponding force of displaced fluid is called added or virtual mass force. This is due to the fact that it is equivalent to adding a virtual mass to the particle. The added mass is insignificant at small density ratios in flow of gas-particles mixtures. For a particle moving in an inviscid fluid at rest, the added mass force can be analytically derived as the function of fluid and particle velocities [41].

$$\vec{f}_{ad,i} = \frac{1}{2} C_{ad} \rho_f V_i \left[\left(\frac{\partial \vec{u}_f}{\partial t} \right) + \vec{u}_f (\nabla \cdot \vec{u}_f) - \frac{d\vec{v}_i}{dt} \right] \quad (2.95)$$

where V_i is the volume of particle i. C_{ad} is the added-mass coefficient, and the value may be equal to unity.

The lift force is generally normal to the relative motion of fluid and particle and is due to particle rotation and fluid shear stress. This force involves the shear-dependent Saffman lift force, caused by the fluid velocity gradient, and the rotational-dependent Magnus lift force, imposed by particle contact and rebound from a surface.

The Saffman force is caused by the pressure difference on a non-rotating particle under a non-uniform shear velocity field. The Saffman force also tends to lift up particles from the wall region. The correlation for evaluating the Saffman lift force is expressed according to particle rotation velocity $\vec{\omega}_i$ [9, 32].

$$\vec{f}_{sa,i} = 1.615 C_{sa} \vec{\omega}_i d_p^2 \sqrt{\rho_f \mu_f} \left| \vec{w}_f \right|^{-0.5} (\vec{\omega}_i \times \vec{w}_f) \quad (2.96)$$

where $\vec{w}_f$ is the curl of velocity vector of fluid. C_{sa} is the Saffman lift coefficient.

$$C_{sa} = \begin{cases} e^{-0.1Re_{sa}} + 0.3314\sqrt{0.5|\vec{w}_f|d_p|\vec{\omega}_i|^{-1}}(1 - e^{-0.1Re_{sa}}) & Re_{sa} \le 40 \\ 0.0524\sqrt{0.5|\vec{w}_f|d_p Re_{sa}|\vec{\omega}_i|^{-1}} & Re_{sa} > 40 \end{cases} \tag{2.97}$$

$$\vec{w}_f = \nabla \times \vec{u}_f \tag{2.98}$$

where Re_{sa} is the shear Reynolds number based on the fluid velocity difference between the bottom and the top of the particle:

$$Re_{sa} = \frac{\rho_f d_p^2 |\nabla \vec{u}_f|}{\mu_f} \tag{2.99}$$

The Magnus lift force is caused by asymmetric pressure distribution around the particle due to velocity difference between the bottom and the top of the particle. Particle rotation is caused by sources other than the velocity gradient. The Magnus lift force is related to the particle rotation in a uniform flow [44].

$$\vec{f}_{ma,i} = \frac{\pi}{8} C_{ma} \vec{\omega}_i^2 d_p^2 \rho_f \sqrt{\rho_f \mu_f} \frac{(\vec{w}_r \times \vec{\omega}_i)}{|\vec{w}_r||\vec{\omega}_i|} \tag{2.100}$$

where C_{ma} is the Magnus lift coefficient

$$C_{ma} = d_p \frac{|\vec{w}_r|}{|\vec{\omega}_i|} \begin{cases} 1 & Re_{sa} \le 1 \\ (0.178 + 0.822 Re_{sa}) & Re_{sa} > 1 \end{cases} \tag{2.101}$$

$$\vec{w}_r = \frac{1}{2}\vec{w}_f - \vec{\omega}_i \tag{2.102}$$

This indicates that the Magnus lift force affects the motion of discrete particles in both radial and axial directions.

2.3.4 Bridge Force and van der Waals Force

Adhesion is the tendency of different particles to adhere to each other. When the particle size is very small, particle motion will be influenced by the adhesion force. Such adhesion forces are part of the short-range surface contact forces. Both van der Waals forces and bridge forces are manifestations of adhesion forces.

The van der Waals forces, also known as intermolecular forces, are divided into dispersion force, orientation force and induction force. The calculation of the van der Waals force on the particle i with the position $\vec{r}_i$ is the summation of van der Waals forces exerted on this particle [15, 20].

$$F_{vdw,ij} = \frac{A_{ha} d_{pi}^2}{24} \sum_j \frac{\vec{r}_i - \vec{r}_j}{\left|\vec{r}_i - \vec{r}_j\right|^3} \tag{2.103}$$

where A_{ha} is the Hamaker constant which depends on the material properties.

The van der Waals forces make the fine particle flow significantly different from the conventional coarse particle flow in terms of contact mechanics, collision mechanics, kinetic behavior, and macroscopic laws and phenomena of particle motion. The van der Waals force is the main factor affecting the adhesion strength of fine particles agglomerating with other particles, or forming physical contact after deposition on the wall in multiphase flow and gas-fine particles fluidized beds. The magnitude of van der Waals force F_{vdw} between particles i an j is modeled by the following

$$F_{vdw}(s) = -\frac{A_{ha}}{3} \frac{2R_i R_j (R_i + R_j + s)}{s^2 (2R_i + 2R_j + s)^2} \left[\frac{s(2R_i + 2R_j + s)}{(R_i + R_j + s)^2 - (R_i - R_j)^2} - 1 \right]^2 \tag{2.104}$$

where s is the distance between the particle surfaces. R_i and R_j are the radius of particle i and j. It is assumed that the force saturates at a minimum separation distance s_{min} (corresponding to the inter-molecular spacing). Since the magnitude of the van der Waals force quickly decreases as the distance between the surfaces increases, a maximum cutoff distance $s_{max} = (R_i + R_j)/4$ is used to accelerate the simulation. For $s > s_{max}$, the van der Waals force is not accounted for simulations of fluid-particles flows [15, 38].

$$\vec{f}_{vdw,ij} = \begin{cases} -F_{vdw}(s)\vec{n}_{ij} & s_{min} < s < s_{max} \\ -F_{vdw}(s_{min})\vec{n}_{ij} & s < s_{min} \end{cases} \tag{2.105}$$

where n_{ij} represents the unit normal vector pointing from particle i to particle j. The van der Waals force between the particle i and the wall can be modeled as,

$$\vec{f}_{vdw,iw} = \begin{cases} -\frac{A_{ha} R_i}{6s^2}\vec{n}_{iw} & s_{min} < s < s_{max} \\ -\frac{A_{ha} R_i}{6s_{min}^2}\vec{n}_{iw} & s < s_{min} \end{cases} \tag{2.106}$$

The Hamaker constant A depends on the particles material and surface condition. This constant varies from 1.0×10^{18} to 1.0×10^{21} J and lies between 1.0×10^{19} and 1.0×10^{20} J. It can be also estimated from the equation of the van der Waals force between two particles being equal to the weight of one particle.

The liquid bridge force is generally considered to be the external manifestation of surface tension or capillary forces in liquids due to the presence of humidity in the gas, and it causes the formation of a free liquid film layer on the particles in fluidized bed dryers. During the drying process of particles, the moisture content decreases, the free water on the surface and the bound water inside the particle progressively migrates to the surface. the particle surface is covered with free moisture. The drying rate relates to water diffusion. The free water forms a wetted film on the particle surface, generating a liquid bridge when the particles come into contact within fluidized bed dryers and polyolefin fluidized bed reactors.

The liquid bridge force is composed of the capillary force $\vec{f}_{caf}$ and the viscous force $\vec{f}_{vis}$. There is only the capillary force in static conditions, but the viscous force acts in the bridge elongation. Depending on the speed of the two particles at the time of the collision, two situations can be expected after the collision. If the velocity of particles at the collision is low, the predominant force is attractive, while at high speed of particles, the predominant force is repulsive. In the latter case, the liquid bridge is stretched until the critical rupture distance is reached and the particles are separated. Generally, the dimensionless capillary number is used to examine the contribution of both viscosity and surface tension to the liquid bridge force [35].

Assuming a smooth surface of particles and a uniform film on them, all the liquid on the particles contributes to the liquid bridge force. The capillary force is calculated as follows [35]

$$\vec{f}_{caf} = -\pi \gamma_w \sqrt{(R_i R_j)} \left[\exp\left(-\frac{1.1s}{V_w^{0.53} R_m} + A \right) + B \right] \vec{n}_{ij} \tag{2.107}$$

$$A = (-0.34 \ln V_w - 0.96)\vartheta^2 - 0.019 \ln V_w + 0.48 \tag{2.108}$$

$$B = 0.0042 \ln V_w + 0.078 \tag{2.109}$$

where $V_w = V/R^3$. V represents the volume of the liquid on the particle surface. γ_w is the surface tension of the liquid. s is the distance from the particle surface. R_m is the largest radius of particle i and particle j. ϑ denotes the contact angle of the particle with the liquid. For particle–wall collisions, the particle–wall capillary force is:

$$\vec{f}_{caf} = -\pi \gamma_w R_i \left[\exp\left(-\frac{1.9s}{V_w^{0.51} R_i} + A \right) + B \right] \vec{n}_{iw} \tag{2.110}$$

$$A = (-0.016 \ln V_w - 0.76)\vartheta^2 - 0.12 \ln V_w + 1.2 \tag{2.111}$$

$$B = 0.013 \ln V_w + 0.18 \tag{2.112}$$

The normal component $f_{vis,n}$ and tangential component $f_{vis,t}$ of the viscous force are calculated as follows

$$\vec{f}_{vis,n,ij} = 6\pi \mu_w s^{-1} R_{eff}^2 \vec{u}_{n,ij} \tag{2.113}$$

$$\vec{f}_{vis,t,ij} = 6\pi \mu_w R_{eff} \vec{u}_{t,ij} \left[\frac{8}{15} \log_{10}\left(\frac{R_{eff}}{s} \right) + 0.9588 \right] \tag{2.114}$$

where $\vec{u}_{n,ij}$ and $\vec{u}_{t,ij}$ are the normal and tangential relative velocities of particles at the contact point, and μ_w is the viscosity of the liquid. Strictly speaking, the viscous forces could become arbitrarily large for infinitely small separation distance s. In practice, the distance limits the minimum separation distance to some constant value s_{min}. Generally, a touching particles are assumed to be no closer than $s_{min} = 10^{-6}$ m in simulations of fluid-particles flows. For a more detailed derivation and application on liquid bridge force expressions, see [35].

2.3.5 Magnetic and Electrostatic Forces

The magnetized fluidized bed or magnetofluidized bed is a corollary with the development of fluid-particles fluidized bed and electromagnetism technologies that particles are magnetizable or magnetically susceptible. In a magnetized fluidized bed, the magnetizable particles are fluidized under the action of an external magnetic field that controls interparticle forces. A magnetized fluidized bed shares a number of advantages. The magnetic forces suppress the formation of bubbles beyond minimum fluidization velocity and its contacting efficiency is improved. Therefore, the bed behavior depends on magnetic field intensity. The magnetic field can work from the outside of the fluidized bed, and it is not affected by the chemical corrosion condition inside the bed, keeping a much longer service life.

Depending upon the position of particles and relating to their dipole moment orientation, the interparticle magnetic force can be attractive or repulsive. If the particles approach each other along the line of the dipole moments, the force is attractive. If particles approach perpendicular to the dipole moments, the force is repulsive. For simplification, it is assumed that two identical magnetically susceptible particles are positioned side-by-side in a uniform magnetic field, and separated by a distance r_{ij}, as shown in Fig. 2.6. Here, γ represents the dipole angular orientation, and θ denotes the polar angle. α is the angle difference (θ–γ). The forces acting on the particles do not depend on the azimuthal angle. Therefore, it is adequate to locate particle on the x–z plane. Furthermore, it is assumed that these two particles have the same diameter, and can be considered ideal dipoles under the influence of a uniform external magnetic field.

The radial and angular magnetic forces between the particle i and the particle j can be expressed according to the dipole moment m and the magnetic susceptibility μ_0 of permeability of free space [42, 55].

$$\vec{f}_{mag,r,ij} = -\frac{\mu_0}{4\pi}\left\{\frac{m^2}{r_{ij}^3}[-6\cos\alpha\sin\alpha]\gamma_r + [1 - 3\cos^2\alpha]\left(-\frac{3m^2}{r_{ij}^4} + \frac{2m}{r_{ij}^3}m_r\right)\right\}(\vec{n}_r)_{ij} \tag{2.115}$$

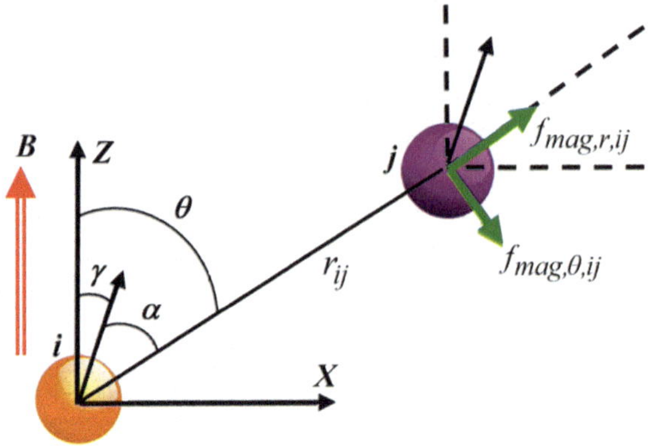

Fig. 2.6 Repulsive and attractive magnetic forces of two diploes

$$\vec{f}_{mag,\theta,ij} = -\frac{\mu_0}{4\pi r_{ij}^4} \left\{6m^2 \cos\alpha \sin\alpha(1-\gamma_\theta) + 2m\left[1-3\cos^2\alpha\right]m_\theta\right\}(\vec{n}_\theta)_{ij} \tag{2.116}$$

where r_{ij} is the distance between the particle i and j. $\vec{n}_r$ and $\vec{n}_\theta$ are unit vectors in radial and angular directions, which vary with respect to relative position of particles i and j.

For a sphere, the magnetic dipole moment m is given according to the particle magnetic permeability μ_p and magnetic field strength B distribution.

$$m = \frac{V_p}{\mu_0}\left[\frac{3(\mu_p - \mu_0)}{\mu_p + 2\mu_0}\right]B \tag{2.117}$$

The external magnetic field B is the sum of the applied external magnetic field and the induced field due to the magnetization of the surrounding particles. The gradient of the diploe angular orientation along radial direction γ_r and angular direction γ_θ expresses as the function of effective magnetic susceptibility χ and the volume of particle V_p.

$$\gamma_r = -\frac{72\pi r_{ij}^2 \chi V_p \sin 2\theta}{\left\{\left[8\pi r_{ij}^3 + \chi V_p(3\cos 2\theta - 1)\right]^2 + (3\chi V_p \sin 2\theta)^2\right\}} \tag{2.118}$$

$$\gamma_\theta = \frac{2(3\chi V_p \sin 2\theta)^2 + 6\chi V_p \cos 2\theta\left[8\pi r_{ij}^3 + \chi V_p(3\cos 2\theta - 1)\right]}{\left[8\pi r_{ij}^3 + \chi V_p(3\cos 2\theta - 1)\right]^2 + (3\chi V_p \sin 2\theta)^2} \tag{2.119}$$

The effective magnetic susceptibility is different depending on the magnetic susceptibility of ferrite particles. From the Clausius–Mossotti relationship which links the mixture susceptibility with the particle magnetic susceptibility χ_p.

$$\chi = \frac{\chi_p}{1 + \chi_p/3} \tag{2.120}$$

Similarly, the gradient of the diploe moment of each particle along radial direction m_r and angular direction m_θ are

$$m_r = \frac{(8\pi r_{ij}^3)^3 \chi V_p B}{\mu_0 \left\{ 8\pi r_{ij}^3 \cos\gamma - \chi V_p \left[\cos\gamma + 3\cos(2\theta - \gamma) \right] \right\}^2} \left\{ \left[\left(8\pi r_{ij}^3 - \chi V_p \right) \sin\gamma + 3\chi V_p \sin(2\theta - \gamma) \right] \gamma_r - 3 r_{ij}^{-3} \chi V_p \left[\cos\gamma + 3\cos(2\theta - \gamma) \right] \right\} \tag{2.121}$$

$$m_\theta = \frac{(8\pi r_{ij}^3)^3 \chi V_p B}{\mu_0 \left[\left(8\pi r_{ij}^3 - \chi V_p \right) \cos\gamma - 3\chi V_p \cos(2\theta - \gamma) \right]^2} \left\{ \left[\left(8\pi r_{ij}^3 - \chi V_p \right) \sin\gamma + 3\chi V_p \sin(2\theta - \gamma) \right] \gamma_\theta - 6\chi V_p \sin(2\theta - \gamma) \right\} \tag{2.122}$$

The radial magnetic force and the angular magnetic force relate with the distance and the angular position between particles. The angle between magnetic field and center to center line of particles has $\theta = \alpha + \gamma$. From Eq. (2.115) the radial magnetic force is reduced when the distance between the particles increases. The radial force is more significant when the particles are located at $\theta = 0.0$ than they are located at $\theta = \pi/2$. From Eq. (2.116), we see that the angular magnetic force has a maximum at $\theta = \pi/4$, and becomes zero at $\theta = 0$ and $\theta = \pi/2$. The angular magnetic force will vanish as the particles move away each other.

The total magnetic force is evaluated by adding the radial vector force and the angular vector force between the particles. Both magnetic forces are added into the equations of motion of discrete particles. Therefore, we have to know the particle position of r and θ, particle susceptibility and external magnetic field B for the numerical evaluation of the interparticle magnetic forces.

For a detailed explanation and derivation of the magnetic force model, see [42].

As particles are fluidized in the fluidized bed reactor, it is associated with particles rub against each other and against the wall. These circumstances lead to electrostatics through triboelectric charging or tribo-electrification, which is the process of charge transfer between two materials. When charged surfaces are close to one other, a charged state is described by the Columbic attraction between the oppositely charged surfaces stabilizes the charged system. Accumulation of electrostatic charges can be attributed to entrainment of charged fine particles, leaving net positive or negative charges behind within the fluidized bed. Electrostatic charges can change the forces on particles, thereby inducing particle–wall adhesion, inter-particle cohesion and agglomeration of fine particles. Hence, the electrostatic forces generated by charged particles can influence the hydrodynamics of fluidized beds and other granular systems.

The electrostatic interaction between point charges is governed by the Coulomb's law. As a first approximation, the charge is assumed to distribute uniformly on the surface of a spherical particle and cannot be polarized. Then, the electrostatic force between charged particles is governed by the Coulomb's law as [22]:

$$\vec{f}_{ef,ij} = \frac{q_i}{4\pi\varepsilon_0}\sum_j \frac{q_j}{r_{ij}^2}\vec{n}_{ij} \tag{2.123}$$

where q_i and q_j are the charge of particles i and j, ε_0 is the permittivity of a vacuum, r_{ij} is the distance between the centers of the particle i and particle j. When a charged particle approaches a conducting surface, a re-distribution of the charge on the surface is induced. The charge on the particle and the induced charge on the surface cause the so-called image (charge) force. The electrostatic force between the particle i and the induced charge wall is determined as follows

$$\vec{f}_{ef,iw} = \frac{q_i q_w}{4\pi\varepsilon_0(2r_{iw})^2}\vec{n}_{iw} \tag{2.124}$$

where q_w is the charge on the wall, which is the product of the surface area and the surface charge density of the wall.

In order to consider the influence of other charged particles near the target particle, an alternative expression of the electrostatic force is expressed as [1]

$$\vec{f}_{ef,ij} = \frac{q_i q_j}{4\pi\varepsilon_0}\left(\frac{k}{r_{ij}} + \frac{1}{r_{ij}^2}\right)e^{-kr_{ij}} \tag{2.125}$$

where, k is the inverse of the Debye length, and is based on the local charge concentration (where n is the number of particles of charge z) around a particle.

$$k = q_e\sqrt{\left(\frac{1}{\varepsilon_r\varepsilon_0 K_B T}\sum_i n_i z_i^2\right)} \tag{2.126}$$

where q_e represents the charge of the electron (1.602×10^{-19} C), ε_r denotes the relative permittivity, and T and K_B are the gas temperature in Kelvin and Boltzmann constant (1.38×10^{-23} J/K).

In order to obtain the electrostatic forces, an effective radius is set to determine whether two particles interact with each other. When the distance between two particles is larger than the effective radius, the electrostatic force is negligible. When the distance between two particles is shorter than or equal to the effective distance, the electrostatic interactions between the two particles occurs.

2.3.6 Vibration and Acoustic Fields

Vibration is usually done by vertically vibrating the vessel walls or air distributor of fluidized beds. This technique is also used in hoppers and packing to improve mobility of particles. Vibration affects kinetic energy of particles.

In order to incorporate the vibration induced forces into a DEM model, we assume a sinusoidal vibration. The displacement and velocity of the walls can be expressed by the following equations:

$$\vec{x}_w = \vec{x}_{w0} + A\sin(2\pi ft) \tag{2.127}$$

$$\vec{v}_w = 2\pi fA\cos(2\pi ft) \tag{2.128}$$

where A and f are the amplitude and frequency of vibration, and $\vec{x}_{w0}$ is the position of the wall at rest.

Acoustic fields are used in many fluid-particles multiphase processes to affect the flowability of particles. An acoustic radiation force occurs when momentum is transferred from a sound wave to the surrounding medium. The sound wave can be expressed by a periodic function. The sound velocity of an acoustic field can be given by:

$$\vec{U}_{sound} = \vec{U}\sin(2\pi ft) \tag{2.129}$$

where f is the sound frequency. $\vec{U}$ is the absolute magnitude of sound velocity that can be calculated from the sound pressure level [40].

$$SPL = 20\log(\frac{\left|\vec{U}\right|}{\sqrt{2}U_{ref}}) \tag{2.130}$$

where U_{ref} is the reference velocity at reference pressure. For air, the reference pressure is chosen to be the value at the threshold of human hearing, 2×10^{-5} Pa, and the value of U_{ref} is 4.83×10^{-8} m/s.

As the sound wave attenuates through a gas–solid suspension, the sound pressure level follows exponential attenuation:

$$SPL = L^{-2}\exp(-2a_c L)SPL_0 \tag{2.131}$$

where SPL_0 is the inlet sound wave pressure (in decibel) and SPL is the sound wave pressure at the distance L within the gas–solid mixture having an attenuation coefficient a_c.

$$a_c = \frac{2b(\pi f)^2}{(\alpha_f\rho_f + \alpha_s\rho_s)c_0^3} \tag{2.132}$$

where c_0 is the sound speed of the acoustic medium (air), and b is a constant. It suggests that they are 340 m/s and 10^{-5} kg/(m s) in gas–solid fluidized beds. Taking the attenuation into account, the sound velocity at the distance L in flows of gas-solid mixture is:

$$\vec{U}_{sound}(L, t) = \vec{U} \exp(-a_c L) \sin\left(2\pi ft - \frac{2\pi f}{c_0}L\right) \tag{2.133}$$

The acoustic force acting on a particle arises from the relative velocity between the velocity of particle and the oscillatory component of gas (sound velocity). Therefore, the acoustic force exerted on particle i can be evaluated from:

$$\vec{f}_{ac,i} = \frac{\beta_{fs} V_i}{1 - \alpha_f}\left[\vec{U}_{sound}(L, t) - \vec{v}_i\right] \tag{2.134}$$

where β_{fs} is the interphase momentum transfer coefficient, and V_i is the particle volume. It shows that the slip velocity is the velocity difference between the sound velocity and particle velocity due to the acoustic field.

2.3.7 Torque Expressions

The torque acting on contacting particles is comprised of two contributions: rotational torque caused by interparticle contacts in the tangential direction and rolling resistant torque that arises from the unevenness of the contact pressure distribution in the contact area. The former causes particles to rotate while the latter resists against the rotations of particles. The rotational torque of particles is defined as:

$$\vec{M}_{ij} = R_i \vec{n}_{ij} \times \vec{f}_{ij,n} \tag{2.135}$$

Since the contact force is transformed from the contact point to the mass center of the particle, the outcome is the rotational torque on the particle.

The rolling resistance torque is defined as [56]:

$$\vec{M}_{ij} = -\mu_r R_{eff} \left|\vec{f}_{ij,n}\right| \frac{\vec{\omega}_i - \vec{\omega}_j}{\left|\vec{\omega}_i - \vec{\omega}_j\right|} \tag{2.136}$$

where μ_r is the dimensionless coefficient of rolling resistance. In this model, a constant resistant torque is applied on colliding particles. The minus sign is used to emphasize that this torque opposes the relative rotation of the particle.

2.4 Initial and Boundary Conditions

The initial and boundary conditions play a very important role in the simulations of dynamic behavior and status of fluid-discrete particles multiphase flows. To start the simulation of fluid phase and integration of equations of motion of discrete particles, it is necessary to define initial values and boundary conditions for all variables involved in the mass and momentum conservation equations.

For fluid phase, the initial fluid velocity is set at zero throughout the entire system. A constant fluid pressure is also specified in the entire system.

For discrete particles, the initial conditions define values for position x_i, velocity v_i, rotational velocity ω_i for all particles. The initial values of particles are assigned to each variable differ from case to case. However, the dynamic solution usually starts from a condition in which all particles are at rest and packed. There are various methods to obtain a packed assembly of particles. A very important question that we must answer is whether or not the initial condition affects the simulation results. In general, when we are going to study the long-term dynamic characteristics of a multiphase flow, the initial condition has minimal effects. However, the initial condition does affect the results when we are going to study the transient behavior of the system (from stationary/initial to long-term dynamic behavior) or when the simulation is static or quasi-static.

One option for initial positioning of particles in the system is to pack particles randomly within the system boundaries. The first method of random packing is to position all particles randomly in the simulation domain and allowing them to settle down by gravity. Coordinates of each new particle can be generated by a random number generator engine. For initial positioning of particles, we must ensure that any new inserted particle does not overlap with other particles.

For fluid phase inlet condition, a uniform velocity profile is applied. A value for the fluid pressure is specified at the outlet. For the fluid phase, no-slip and non-penetrating wall conditions are considered.

For discrete particles, they are confined in the system boundaries via the existence of walls. Walls are stationary in most multiphase flows and the interaction of walls with discrete particles affects simulations of discrete particles. Two different methods are used to define the wall boundary condition. One assumes that the wall is built up from particles called wall particles. Another is the real wall definition.

In the wall particles definition, each wall particle has its own size and mechanical properties. The interaction forces between a wall particle and a discrete particle are calculated based on the force-displacement models. The wall particles are locked according to the wall configuration. Thus, the size of wall particles and their mechanical properties influence the dynamic interaction between the wall particle and the discrete particles. The advantage is that the implementation of particulate wall into discrete element simulation is relatively straightforward. In addition, there is no limitation on the wall shape of the arc, circular and elliptic that can be defined in the simulation.

In the real wall definition, the real shape or geometry of the wall is given by an analytical equation. In this way, the wall is considered particle j with an infinite radius. Physical properties of the real wall are replaced by physical properties of particle j. The coordinate of the real wall is first transformed into the local wall coordinates and the distance is calculated. Then, the contact point is determined using the force-displacement models.

In both the wall particles definition and the real wall definition, the normal and tangential spring stiffnesses between the particle i and the wall is calculated according to the wall Young's modulus E_w, Poisson's ratio ν_w and wall shear modulus G_w as follows [6, 21]

$$k_n = \frac{4}{3}\frac{E_i E_w \sqrt{R_i}}{E_i\left(1-\nu_w^2\right)+E_w\left(1-\nu_i^2\right)}\delta_n^{1/2} \tag{2.137}$$

$$k_t = \frac{16}{3}\frac{G_i G_w \sqrt{R_i}}{G_i\left(2-\nu_w^2\right)+G_w\left(2-\nu_i^2\right)}\delta_n^{1/2} \tag{2.138}$$

The effective mass m_{eff} and the effective radius R_{eff} of the walls are equal to the mass m_i and the radius R_i of the particle i. The effective Young's modulus and the effective shear modulus of the walls are expressed as the function of the wall elastic modulus E_w and wall shear modulus G_w.

$$E_{eff} = \left(\frac{1-\nu_i^2}{E_i}+\frac{1-\nu_w^2}{E_w}\right)^{-1} \tag{2.139}$$

$$G_{eff} = \left(\frac{2-\nu_i}{G_i}+\frac{2-\nu_w}{G_w}\right)^{-1} \tag{2.140}$$

In some cases, the periodic boundary condition is applied instead of using a real wall. Using the periodic boundary condition, we can mimic the dynamic behavior of the whole system with a smaller simulation domain and less number of particles. Applying the periodic boundary condition, each particle which leaves the system from one side is reinserted from the other side of the boundary. Consequently, interactions between particles at the opposite side are considered in the simulations.

2.5 Calibration of DEM Parameters

For simulations of flow of discrete particles using CFD-DEM approach, it involves discrete element parameters used in simulations. The determination of the DEM model parameters affects to produce satisfactory quantitative predictions. Hence, the choice of discrete element parameters is the first step in the numerical simulations using CFD-DEM approach [4, 5].

Two types of discrete element parameters are involved in CFD-DEM approach. One is the contact parameters, including modulus and Poisson ratio. These parameters relate to the properties of the contact materials, and can be measured through experimental measurements. The other is the model parameters, including restitution coefficient, friction coefficient and surface energy. These contact parameters can be calibrated using virtual simulations in the absence of experimental measurements.

The virtual simulations of contact parameters used in DEM approach are categorized into several methodologies according to the use of a specific simulation procedures for calibrations. The virtual simulations are predominately used as a way to correlate the DEM input variables. In addition, virtual simulation can also be used for optimal selection of input parameters in simulations.

2.5.1 Dropping Test of Virtual Angle of Repose

The angle of repose (AoR) is an important physical property used for characterization of the bulk material. The angle of repose is used as an indicator of the flowability of bulk solids. It is a quick method that allows to get an idea of the flowability of particles. The angle of repose is described as the highest angle that is created between the horizontal and conical surfaces when the particles are allowed to flow freely on the flat platform. The angle of repose is the slope angle of a loose particles at rest. The lower the angle of repose, the more flowable are the discrete particles.

Using DEM method, the virtual AoR test is performed by means of a dropping test in a funnel device, as shown in Fig. 2.7 with the particle diameter and density of 0.4 mm and 1500 kg/m^3. The Young's modulus and Poisson's ratio are 1.7×10^7 Pa and 0.23. The restitution coefficient and friction coefficient are 0.9 and 0.5, respectively. The funnel device consists of two parts. One is a horizontal flat plate located at the bottom. The other is a storage bin located at the top.

The virtual AoR test of particles includes three phases using DEM method. The first phase is the creation of particles packing during the generating process. Particles are generated randomly in the storage bin. The second phase is the dropping process of particles. The outlet of the storage bin is in the open state. Particles fall toward the horizontal flat plate under the action of gravity. The third phase is the accumulation

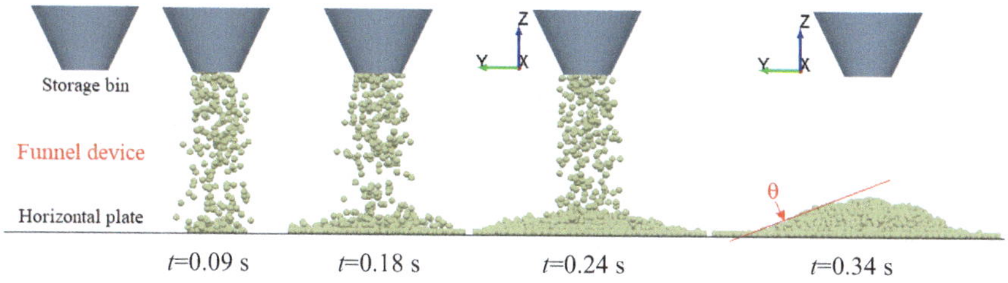

Fig. 2.7 Creation and repose of particles accumulation in the funnel device

of particles during the collection process. Particles accumulate gradually on the horizontal flat plate, and a pile of particles is formed. The accumulation angles θ are determined from the particles pile.

A series of virtual AoR test are conducted with the change of contact parameters of restitution coefficients e and friction coefficients μ_{df} of non-cohesive particles using DEM method at the specified modulus and Poisson's ratio. For cohesive particles, the surface energy γ is also an input parameter. Based on simulated virtual AoR, the regression model describing the relationship between the AoR and contact parameters can be described as

$$\theta = a_0 + a_1 e + a_2 \mu_{df} + a_3 \gamma + b_1 e \mu_{df} + b_2 e \gamma + b_3 \gamma \mu_{df} \tag{2.141}$$

where a_i is the regression coefficients of the contact parameters. b_i is the regression coefficients of the intersection terms. For non-cohesive particles, the surface energy is zero. From the measured AoR of particles, the contact parameters of restitution coefficient, friction coefficient and surface energy are determined using the method of the multi-objective optimization of parameters.

2.5.1.1 Lifting Cylinder Test for Virtual Repose Angle

The lifting cylinder test is applied to measure the AoR of non-cohesive and cohesive particles. The lifting cylinder test can represent a quite slow bulk material flow. The apparatus of the lifting cylinder test consists of an open-ended cylinder and a horizontal plate. The open-ended cylinder is placed on the plate, as shown in Fig. 2.8 using DEM. The particle diameter and density are 0.2 mm and 1200 kg/m^3. The Young's modulus and Poisson's ratio are 1.8×10^7 Pa and 0.2. The restitution coefficient and friction coefficient are 0.95 and 0.2, respectively. The virtual AoR is obtained from the final particles pile.

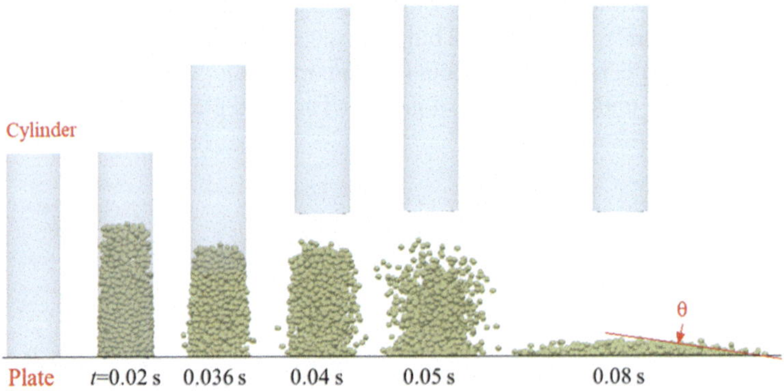

Fig. 2.8 Repose angle of particles accumulation in the lifting cylinder test

In the lifting cylinder test using DEM, the simulation of the AoR of particles consists of the generation of particles, the settling stage, the resting stage, the cylinder rising stage, the particles settling stage under gravity, and resting pile. In the stage of generation of an assembly of particles, two methods of particle generation are used. One is that particles are positioned in layers starting from the plate at 1.01d center to center in the cylinder, where d is the particle diameter. They are all assigned an initial velocity of zero. The other is particles randomly within the cylinder. In the settling and resting stages, the settlement of the particles is performed under gravity until a minimum particle velocity is reached. Then, the cylinder is raised upwards with a specified lifting velocity u_r in the rising stage.

$$u_r = \begin{cases} \text{constant} & t \le t_s \\ 0 & t > t_s \end{cases} \tag{2.142}$$

where t_s is a fixed limit time. When the cylinder is lifted, particles begin to flow out of the created gap in the settling stage. The column of particles finally collapsed under its own weight. The upward movement of the cylinder will terminate as the time exceeds the limit time. A pile of particles is formed at the resting stage. The virtual AoR of particles is determined after the particles are settled on the plate.

2.5.2 Ring Shear Tests for Virtual Shear Stress

The direct shear tests, both translational and rotational (ring), have been used to calibrate contact parameters of non-cohesive materials and slightly cohesive materials. The ring shear tests using DEM are applied to study the mechanical behavior of particle assembly in ring shear process. The ring shear model using DEM is composed of a cylinder annulus with an inner radius R_{in}, outer radius R_{ou} and a shear cell height h. The ring shear model is divided into upper and lower platens. The upper and lower platens are surrounded by walls, which are smooth and frictionless. The wall of the bottom plate rotates in a counterclockwise direction around the central axis of the cylinder at a specified angular velocity ω_{wb} in the ring shear test

$$\omega_{wb} = \text{constant} \tag{2.143}$$

Thus, the particles wrapped by the upper and lower platens are sheared circumferentially. Finally, the virtual shear stresses τ_{ss} of particles are determined from DEM simulations.

The virtual particle–particle stresses using the ring shear tester is depicted in Fig. 2.9. The force chain patterns describe stress distribution due to the contact interactions through the rotation of the bottom plate. The inner height of the shear cell is 10 mm, and its outer and inner radius are 5 and 10 mm, respectively. The particle diameter and density are 0.2 mm and 1200 kg/m^3. The ring shear test consists of an

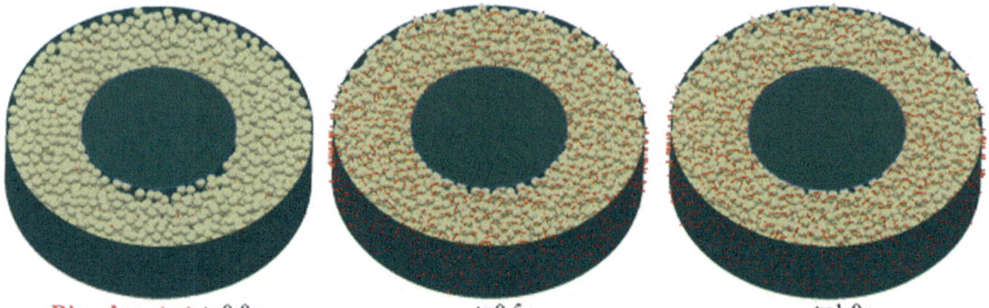

Fig. 2.9 Simulation setup and contact force chain of particles in the virtual ring shear test

annular chamber which contains the sample. The bottom lid rotates with a constant velocity of 100 rad/s to create a steady-state shear flow by moving the particles. The top lid moves downward to release the applied pressure. After reaching the steady-state shear stress, the top lid is kept fixed in all degrees of freedom except vertical direction in the virtual ring shear tester.

In the ring shear test using DEM, the simulation of the virtual tangential shear stress of particles includes the particle generation stage, the consolidation stage and the shearing stage. In the first stage of the generation of particles, particles are released randomly within the shear cell under gravity. Particles filled in the cell are settled at the specified time step. The top lid moves downward to create the required vertical pressure in the pre-shear stage between the particles and the lids. It expects to replicate the lid motion in the simulation adequately. A static condition is reached. In the consolidation stage, the top platen moves downward to create the required vertical pressure in the pre-shear stage between particles and the platen. Particles are consolidated, and the specified downward axial pressure p_{ap} is applied to the top wall of the upper platen.

$$p_{ap} = \text{constant} \tag{2.144}$$

In the shearing stage, the bottom cell rotates with a constant velocity to create a steady-state shear flow by moving the particles. The force chain is formed from the contact interactions of particles. The characteristics of force chain is the key to the granular dynamics. After reaching the steady-state shear stress, the virtual shear stress τ_{ss} is calculated from the axial component of the torque exerted over the bottom surface.

From DEM simulations, the virtual tangential shear stress τ_{ss} at steady state of flow is determined with the change of restitution coefficient e and friction coefficient μ_{df}. The second-order regression equation of the virtual shear stress and contact parameters is correlated at the specified modulus and Poisson's ratio.

$$\tau_{ss} = a_0 + a_1 e + a_2 \mu_{df} + a_3 e \mu_{df} \tag{2.145}$$

where a_i represents the corresponding regression coefficients. The optimal solution is determined according to the optimized second-order regression equation in comparison with experiments. The contact parameters can be used to simulate and predict the fluid-particles flow processes.

2.5.3 Compression Test for Virtual Modulus

The contact forces between particles are the function of Young's modulus and Poisson's ratio. Thus, defining the accurate values of the contact parameters is vital to simulate the behavior of discrete particles. The uniaxial compression test and the confined compression test using DEM method are applied to determine compressibility of particles, and study the mechanical behavior of a granular assembly subjected to vertical loading. Hence, the contact forces relate to the stress–strain behavior of granular materials, and the bulk material strain is associated with the stiffness parameter in DEM simulations.

Using DEM method, the uniaxial compression and confined compression tests are conducted to determine modulus. The virtual compression tester consists of the particles container, a top cover and bottom cover. The container has a box or cylinder shape, as shown in Fig. 2.10. The bottom cover and the container walls are stationary. The top cover moves downwards at a specified loading rate during the simulations. A faster loading rate enables shorter total simulation times. The box size in the virtual uniaxial compressive test is 1.0 mm × 1.0 mm × 2.0 mm to simulate the compaction process. The particle diameter and density are 200 μm and 1200 kg/m^3. The discrete particle restitution coefficient e and friction coefficient μ are 0.85 and 0.05. The top cover plate moves at a fixed speed of 0.002 mm/s to simulate constant loading. It is particularly noted that the discrete particles are first generated to ensure sufficient space for subsequent particles.

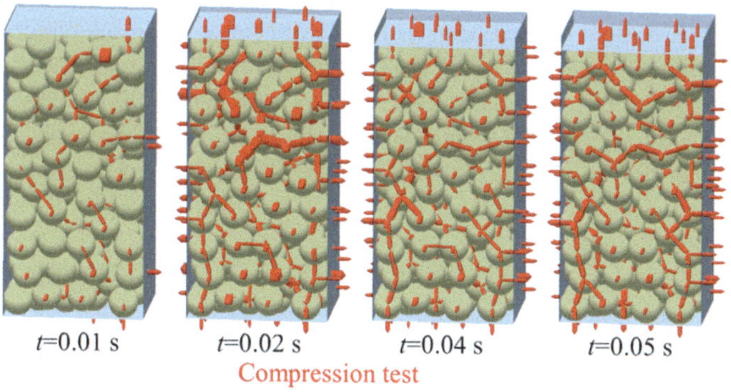

Fig. 2.10 Profile of force chains at four times in virtual compression test

The simulation of virtual compression test includes the particle packing stage and the loading stage. In the packing stage, particles are randomly generated, and packed in the container. Particles are allowed to fall under gravity to achieve the initial filled state. Then, a vertical load is applied to the top cover, and forced on bulk materials confined in the container. During the loading stage, the bulk material is compressed along the vertical direction. The movements of particles are occurred in this stage. Particles are subjected to compression until a pseudo-steady state is achieved. From the distribution of force chain within discrete particles, the contact forces become large as the top cover plate moves downward along vertical direction. It can be seen that some of the contact points are located on the thick force chain and some are located on the thin force chain. And more do not constitute a contact force chain. The greater the contact force, the thicker the force chain within the particles.

A series of simulations of the virtual modulus using DEM approach is determined at the different restitution coefficient and friction coefficient, and the virtual compressive stress σ_s is correlated at the specified modulus and Poisson's ratio.

$$\sigma_s = a_0 + a_1 e + a_2 \mu_{df} + a_3 e \mu_{df} \tag{2.146}$$

where a_i is the corresponding regression coefficients. Comparing to the experimental results, the contact parameters calibrated by the uniaxial confined compression test and the confined compression test can be used in DEM simulations of fluid-particles multiphase flow.

2.6 Summary

The coupling of computational fluid dynamics and discrete element method (CFD-DEM) has become an important method for studying fluid-particles multiphase flows. Since CFD can qualitatively and quantitatively describe the fluid evolution process and DEM can track the motion behavior of each particle individually, hydrodynamics using CFD–DEM method have been realized from small scale to large engineering scale. The detailed discussion of CFD-DEM approach is summarized mainly from the CFD modeling theory (fluid turbulence model), DEM modeling theory (particle–particle interaction models and interparticle interaction models, such as van der Waals, liquid bridge and electrostatic forces even if they are not in contact), CFD–DEM coupled methods (particle–fluid interaction model) to calibration of DEM parameters.

The CFD-DEM approach has shortcomings in terms of fluid-particle velocity correlation tensor that is the partitioning of the particle velocity fluctuations to the particle turbulent kinetic energy and granular temperature. Great care must be used when the fluid flow is turbulent because the Reynolds averaged Navier–Stokes (RANS) turbulence models cannot be fully used for CFD-DEM approach. This is because DEM simulation requires the instantaneous fluid velocity at each particle location at each time step, and this information is not given by a RANS CFD simulation. Furthermore, the fluid-particle velocity correlation appears as unclosed term in the fluid kinetic energy transport equation, representing the fluctuating velocity of the fluid phase seen by the particle phase. The fluid-particle velocity correlations are determined from a transport equation for the fluid-particle velocity correlation tensor. Assuming that particles are spherical and follow Stokes' law of resistance in the homogeneous turbulence of fluid, an algebraic expression for the fluid-particle velocity correlation is expressed as the function of particle relaxation time and the time of interaction between particle motion and continuous phase fluctuations. An algebraic expression for the fluid-particle velocity correlation tensor is proposed to study the dispersion of discrete particles in a turbulent flow field. Hence, CFD-DEM models need to be improved to capture the essential physical mechanisms of turbulent fluid-particle flow in multiphase flows.

Nomenclature

a_i	Coefficients
A	Coefficients
b_i	Coefficients
B	Magnetic field, T
C_d	Drag coefficient
d_i	Diameter of particle *i*, m
E_{eff}	Effective Young's modulus of particles, Pa
E_i	Young's modulus of particle *i*, Pa
$\vec{f}_{ad}$	Added mass force on particle *i*, N
$\vec{f}_d$	Drag force on particle *i*, N
$\vec{f}_i$	Force acting on particle *i*, N
$\vec{f}_{fp}$	Interaction force between fluid and particle *i*, N
$\vec{f}_{his}$	History Basset force on particle *i*, N
$\vec{f}_{lf}$	Lift force on particle *i*, N
$\vec{f}_{ma}$	Magnus force on particle *i*, N
$\vec{f}_n$	Normal force of particle, N
$\vec{f}_p$	Pressure gradient force on particle *i*, N
$\vec{f}_{sa}$	Saffman force on particle *i*, N
$\vec{f}_t$	Tangential force of particle, N
$\vec{f}_\tau$	Viscous force on particle *i*, N
G_{eff}	Effective shear modulus, Pa
G_i	Shear modulus of particle *i*, Pa
k	Spring stiffness coefficient, and turbulent kinetic energy, m^2/s^2
m	Dipole moment magnitude, $A \cdot m^2$
$\vec{M}_{ij}$	Torque acting between particles *i* and *j*, N m
n	Normal unit vector

N	Number of particles
p	Pressure, Pa
R_i	Radius of particle i, m
R_{eff}	Reduced radius of particles, m
Re	Reynolds number
T_f	Temperature of fluid, K
$\vec{u}$	Fluid velocity, m/s
$\vec{v}_i$	Velocity of particle i, m/s
$\vec{v}_{n,ij}$	Normal relative velocity between particles i and j, m/s
$\vec{v}_{t,ij}$	Tangential relative velocity between particles i and j, m/s
V_{cell}	Volume of fluid computational cell, m^3
V_i	Volume of particle i, m^3
V_p	Particle volume, m^3
x_{ij}	Center to center distance of particles, m

Greek Symbols

α	Angle difference, rad
α_f	Fluid volume fraction
β	Interphase momentum transfer coefficient
χ	Empirical coefficient
ε	Dissipation rate of turbulent kinetic energy per unit mass, m^2/s^3
μ_f	Fluid dynamic viscosity, Pa s
μ_p	Particle magnetic permeability, T m/A
μ_t	Fluid turbulent viscosity, Pa s
μ_0	Magnetic permeability of free space, T m/A
v	Poisson ratio
τ_f	Fluid stress tensor, N/m^2
ρ_f	Density of fluid, kg/m^3
ρ_s	Density of particles, kg/m^3
$\vec{\omega}_i$	Angular velocity of particle i, rad/s
γ	Dipole angular orientation, rad
θ	Polar angle, rad

References

1. Adamson, A. W., & Gast, A. P. (1976). *Physical chemistry of surfaces*. Wiley-Interscience.
2. Allen, M. P., & Tildesley, D. J. (1989). *Computer simulation of liquids*. Oxford University Press.
3. Bird, R. B., Stewart, W. E., & Lightfoot, E. N. (2002). *Transport phenomena*. Wiley.

4. Carr, M. J., Katterfeld, A., Roessler, T., & Wheeler, C. A. (2023). Calibration procedure of DEM method parameters for wet and sticky bulk materials. *Powder Technology, 429*, 118919.
5. Coetzee, C. J. (2017). Review: Calibration of the discrete element method. *Powder Technology, 310*, 104–142.
6. Crowe, C. T., Schwarzkopf, J. D., Sommerfeld, M., & Tsuji, Y. (2011). *Multiphase flows with droplets and particles*. Taylor & Francis.
7. Cundall, P. A. (1971). A computer model for simulating progressive large-scale movements in blocky system. In *Proceedings of the International Symposium on Rock Mechanics* (pp. 8–12).
8. Cundall, P. A., & Strack, O. D. L. (1979). A discrete numerical model for granular assemblies. *Geotechnique, 29*, 47–65.
9. Drew, D. (1976). Two-phase flows: Constitutive equations for lift and Brownian motion and some basic flows. *Archive for Rational Mechanics and Analysis, 62*, 149–164.
10. EDEM. (2020). *EDEM theory guide, DEM Solutions*. Altair Engineering Inc.
11. Einstein, A. (1956). *Investigation on the theory of Brownian movement. Edited by R.* Dover Publications.
12. Ergun, S. (1952). Fluid flow through packed columns. *Chemical Engineering Progress, 48*, 89–94.
13. Fan, L. S., & Zhu, C. (1998). *Principles of gas-solid flows*. Cambridge University Press.
14. Feng, Y. Q., & Yu, A. B. (2004). Assessment of model formulations in the discrete particle simulation of gas-solid flow. *Industrial & Engineering Chemistry Research, 43*, 8378–8390.
15. Galvin, J. E., & Benyahia, S. (2014). The effect of cohesive forces on the fluidization of aeratable powders. *AIChE Journal, 60*, 473–484.
16. Gibson, M. M., & Launder, B. E. (1978). Ground effect on pressure fluctuations in the atmospheric boundary layer. *Journal of Fluid Mechanics, 86*, 491–511.
17. Gidaspow, D. (1994). *Multiphase flow and fluidization*. Academic Press.
18. Gidaspow, D., & Jiradilok, V. (2009). *Computational techniques*. Nova Science.
19. Hagerty, W. W., & Shea, J. F. (1955). A study of the stability of plane fluid sheets. *Journal of Applied Mechanics, 22*, 509–514.
20. Hamaker, H. C. (1937). The London–van der Waals attraction between spherical particles. *Physica, 4*, 1058–1072.
21. Hamid, R. N., Reza, Z., Rahmat, S. G., & Navid, M. (2016). *Coupled CFD-DEM modeling: Formulation, implementation and application to multiphase flows*. Wiley.
22. Hogue, M. D., Calle, C. I., Weitzman, P. S., & Curry, D. R. (2008). Calculating the trajectories of triboelectrically charged particles using Discrete Element Modeling (DEM). *Journal of Electrostatics, 66*, 32–38.
23. Huilin, L., & Gidaspow, D. (2003). Hydrodynamics of binary fluidization in a riser: CFD simulation using two granular temperatures. *Chemical Engineering Science, 58*, 3777–3792.
24. ITASCA. (2019). *PFC: Contacts and contact models*. ITASCA Consulting Group Inc.
25. Johnson, K. L. (1985). *Contact mechanics*. Cambridge University Press.
26. Johnson, K. L., Kendall, K., & Roberts A. D. (1971) Surface energy and the contact of elastic solids. *Proceedings of the Royal Society of London. Series A, Mathematical and Physical Sciences, 324*, 301–313.
27. Keith, D. W., & Davor, C. (2008). A three-equation eddy-viscosity model for Reynolds-averaged Navier-stokes simulations of transitional flows. *Journal of Fluids Engineering, 130*, 121401.
28. Kim, W. W., & Menon, S. (1997). *Application of the localized dynamic subgrid-scale model to turbulent wall-bounded flows*. Technical Report AIAA-97–0210, American Institute of Aeronautics and Astronautics, 35th Aerospace Sciences Meeting, Reno, NV.
29. Kuwabara, G., & Kono, K. (1987). Restitution coefficient in collision between two spheres. *Japanese Journal of Applied Physics, 26*, 1230–1233.
30. Launder, B. E., & Spalding, D. B. (1972). *Lectures in mathematical models of turbulence*. Academic Press.
31. Lennard-Jones, J. E. (1931). Cohesion. *Proceedings of the Physical Society, 43*, 461–482.

32. Mei, R. (1992). An approximate expression for the shear lift force on a spherical particle at finite Reynolds number. *International Journal of Multiphase Flow, 18*, 145–147.
33. Menter, F. R., Langtry, R. B., Likki, S. R., Suzen, Y. B., Huang, P. G., & Volker, S. (2004). A correlation based transition model using local variables Part 1-Model formulation. ASME-GT2004–53452.
34. MFIX. (2020). *MFiX-Exa documentation*. MFS Development Group, National Energy Technology Laboratory.
35. Mikami, T., Kamiya, H., & Horio, M. (1998). Numerical simulation of cohesive powder behavior in a fluidized bed. *Chemical Engineering Science, 53*, 1927–1940.
36. Mindlin, R. D. (1953). Elastic spheres in contact under varying oblique forces. *Journal of Applied Mechanics, 20*, 327–344.
37. Mindlin, R. D., & Deresiewicz, H. (1953). Elastic spheres in contact under varying oblique force. *ASME Journal of Applied Mechanics, 20*, 327–344.
38. Molerus, O. (1982). Interpretation of Geldart's type A, B, C and D powders by taking into account interparticle cohesion forces. *Powder Technology, 33*, 81–87.
39. Morrisey, J. P., Thakur, S. C., & Ooi, J. Y. (2014). *EDEM contact model: Adhesive elasto-plastic model*. Technical report, School of Engineering, University of Edinburgh.
40. Morse, P. M., & Ingard, K. U. (1968). *Theoretical acoustics*. McGraw-Hill.
41. Odar, F., & Hamilton, W. S. (1964). Forces on a sphere accelerating in a viscous fluid. *Journal of Fluid Mechanics, 18*, 302–315.
42. Pinto-Espinoza, J. (2002). *Dynamic behavior of ferromagnetic particles in a liquid-solid magnetically assisted fluidized bed (MAFB): Theory, experiment, and CFD-DPM simulation*. PhD thesis, Oregon State University.
43. Potyondy, D. O., & Cundall, P. A. (2004). A bonded-particle model for rock. *International Journal of Rock Mechanics and Mining Sciences, 41*, 1329–1364.
44. Rubinow, S. I., & Keller, J. B. (1961). The transverse force on a spinning sphere moving in a viscous fluid. *Journal of Fluid Mechanics, 11*, 447–460.
45. Smagorinsky, J. (1963). General circulation experiments with the primitive equations. I. The basic experiment. *Monthly Weather Review, 91*, 99–164.
46. Spalart, P., & Allmaras, S. (1992). *A one-equation turbulence model for aerodynamic flows*. Technical Report AIAA-92–0439, American Institute of Aeronautics and Astronautics.
47. Syamlal, M., & O'Brien, T. J. (1989). Computer simulation of bubbles in a fluidized bed. *AIChE Symposium Series, 85*, 22–31.
48. Thakur, S. C., Morrissey, J. P., Sun, J., Chen, J. F., & Ooi, J. Y. (2014). Micromechanical analysis of cohesive granular materials using discrete element method with an adhesive elasto-plastic contact model. *Granular Matter, 16*, 383–400.
49. Tsuji, Y., Tanaka, T., & Ishida, T. (1992). Lagrangian numerical simulation of plug flow of cohesionless particles in a horizontal pipe. *Powder Technology, 71*, 239–250.
50. Walters, D. K., & Cokljat, D. (2008). A three-equation eddy-viscosity model for Reynolds-averaged Navier-stokes simulations of transitional flows. *Journal of Fluids Engineering, 130*, 121401.
51. Walton, O. R., & Braun, R. L. (1986). Inelastic frictional disks viscosity, granular-temperature and stress calculations for shearing assemblies of inelastic, frictional disks. *Journal of Rheology, 30*, 949–980.
52. Wen, C. Y., & Yu, Y. H. (1966). Mechanics of fluidization. *Chemical Engineering Progress Symposium Series, 62*, 100–111.
53. Wilcox, D. C. (1998). *Turbulence modeling for CFD*. DCW Industries Inc.
54. Williams, J. R., & Mustoe, G. G. (1987). Modal methods for the analysis of discrete systems. *Computers and Geotechnics, 4*, 1–19.
55. Zhenghua, H., Xiang, L., Huilin, L., Guodong, L., Yurong, H., Shuai, W., & Pengfei, X. (2010). Numerical simulation of particle motion in a gradient magnetically assisted fluidized bed. *Powder Technology, 203*, 555–564.
56. Zhou, Y. C., Wright, B. D., Yang, R. Y., Xu, B. H., & Yu, A. B. (1999). Rolling friction in the dynamic simulation of sand pile formation. *Physica A: Statistical Mechanics and its Applications, 269*, 536–553.

Chapter 3
Fluid–Solid Two-Phase CFD-KTGF-DEM Approach

Abstract This chapter presents an innovative CFD-KTGF-DEM method in multiphase flows and fluidization. It is common knowledge that the quantitative difference exists between CFD-DEM and TFM-KTGF simulations due to the uncoordinated input parameters of collision and turbulence models. In CFD-KTGF-DEM method, the collisional parameters of the KTGF are simultaneously acquired from DEM simulations, and the turbulence kinetic energies of the DEM are simultaneously acquired from KTGF simulations using the momentum and energy coupling strategies. The CFD-KTGF-DEM method is free from input empirical constants, and produces a multiscale continuum-discrete modeling linked the particles scale and the grid scale in multiphase flows.

In numerical simulations of fluid-particles processes, two formulations are possible: Eulerian formulation, known as Euler-Euler TFM-KTGF approach, and Lagrangian formulation, known as Euler-Lagrangian CFD-DEM approach. Table 3.1 lists the main governing equations used in Euler-Euler TFM-KTGF approach and Euler-Lagrangian CFD-DEM approach in multiphase flows. Both formulations consist of two different phases. One represents the primary phase, and the other is known as the secondary phase.

In the Eulerian formulation (see Chap. 1 for more details), the statistical treatment of the primary phase and secondary phase is referred to as Euler-Euler two-fluid models (TFM). Both primary phase and secondary phase are modeled by applying laws of momentum and mass conservations in each fluid cell, leading to averaged Navier–Stokes equations with the turbulence model of fluid turbulent kinetic energy k_f, granular phase turbulent kinetic energy k_s and as well as the fluid-particle velocity covariance k_{fs}. Capability of the TFM in capturing the secondary phase hydrodynamics greatly depends on the closure laws that must be modeled using kinetic theory of granular flow (KTGF). This treatment of the secondary phase uses classical results from the kinetic theory of dense gases [4] in order to predict the transport equations of momentum and kinetic energy as the function of granular temperature. The solid pressure and viscosity correlate to granular temperature which is predicted

H. Lu et al., *Computational Transport Phenomena of Multiphase Systems and Fluidization*, Fluid Mechanics and Its Applications 127,
https://doi.org/10.1007/978-981-96-0698-6_3

Table 3.1 Main conservation equations in Euler-Euler and Euler-Lagrangian approaches

Phases	TFM-KTGF approach	CFD-DEM approach
Primary Phase	Momentum equation	Momentum equation
	Transport equations k_f with k_{fs}	With or without transport equations k_f, not k_{fs}
Secondary Phase	Momentum equation	Motion equation
	Granular temperature equation with empirical restitution coefficient e	Contact model with physical properties of Young's modulus E and Poisson's ratio ν

from granular fluctuating energy conservation equation [8, 14, 18, 20, 26]. The granular constitutive correlations of KTGF are almost free from empirical constants: only the restitution coefficient e and specularity coefficient ϕ have to be given. The specularity coefficient ranges from 0.0 for the perfect specular collisions to 1.0 for the perfectly diffuse collisions. The restitution coefficient varies from 0.0 to 1.0. Low values of the restitution coefficient decrease the granular temperature, which in turn decreases viscosity and diffusivity, affecting on numerical simulations of fluid-particles multiphase flows.

In Euler-Lagrangian approach (see Chap. 2 for more details), the computational fluid dynamic (CFD) coupled with discrete element method (DEM) is a computational approach used to model fluid–particle systems and fluidized beds [5, 6, 45]. In the CFD-DEM approach, the primary phase is assumed as a continuum, and fluid hydrodynamics are described by the volume averaged Navier–Stokes equation. Either constant fluid dynamic viscosity method or turbulent viscosity method using turbulent kinetic energy k_f and turbulent dissipation rate ε_f are adopted. For the secondary phase, the motion of discrete particles is described by Newton's second law. Forces acting on a discrete particle are gravity, contact between colliding particles and interparticle forces. In the CFD-DEM, each discrete particle is tracked, and gives an accurate description of the motion of a single discrete particle (rotation and collision with another particle) as the function of Young's modulus E and Poisson's ratio ν of discrete particles [7, 44, 49].

From Table 3.1, we retain the momentum equations of primary and secondary phases in Euler-Euler approach and the motion equation of the secondary phase in Euler-Lagrangian approach. We abolish the momentum equation of the primary phase in Euler-Lagrangian approach. Thus, the hydrodynamics of primary phase and secondary phase are described by the momentum equations of Euler fluid and Euler granular phases and the Lagrangian discrete particles motion equation. Here, it is known as the CFD-KTGF-DEM approach in fluid-particles multiphase flow. Figure 3.1 shows a classification of TFM-KTGF, CFD-DEM and CFD-KTGF-DEM methods. In the CFD-KTGF-DEM approach, the CFD provides the fluid phase flow profiles, and the KTGF provides granular flow properties and the DEM provides the motions of discrete particles. The restitution coefficient and specularity coefficient used in KTGF are modelled from DEM. The turbulence used in DEM is modelled

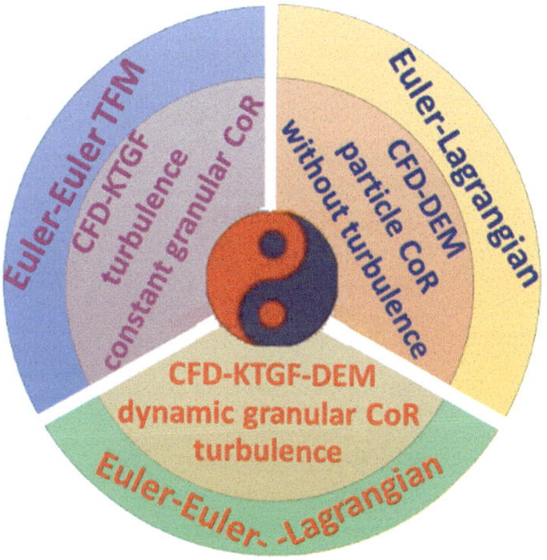

Fig. 3.1 TFM-KTGF, CFD-DEM and CFD-KTGF-DEM approaches

from CFD and KTGF. Thus, the CFD-KTGF-DEM approach is almost free from model empirical constants combining with the fluid phase-granular phase turbulence and collisions of discrete particles from the microscale to mesoscale in fluid-particles multiphase flows.

From Fig. 3.1, we see that the TFM-KTGF methods are the most commonly used to solve problems at the macroscale, at which the hydrodynamic behavior of fluid and solid phases can be described by continuum mechanics. In contrast, the CFD-DEM methods describe dynamics of discrete particles at the microscale. Multiphase flow and fluidization are multiscale in nature. Therefore, the CFD-KTGF-DEM approach describes collisions of discrete particles at the microscale and hydrodynamics of fluid and granular phases at the macroscale. The microscale parameters communicate directly and instantly with the macroscale parameters through coupling procedure. Thus, CFD-KTGF-DEM approach allows to study multi-scale simulations in multiphase flow and fluidized beds.

3.1 CFD-KTGF-DEM Approach

In fluid-particles multiphase flows, it assumes that the control volume consists of Euler fluid phase as the primary phase, Euler granular phase and Lagrangian discrete particles as the secondary phase, and is shown in Fig. 3.2 in the control volume dV (i.e., $dx \times dy \times dz$) of CFD-KTGF-DEM approach. The Euler fluid phase flow field is described by volume fraction α_f, velocity $\vec{u}_f$ and fluid pressure p. The Euler granular phase flow field is described by volume fraction α_s, velocity $\vec{u}_s$ and granular temperature θ. The Lagrangian discrete particle motion is represented by velocity $\vec{v}_p$. Both the granular phase and the discrete particles have the same particle diameter d_p and density ρ_s. These indicate that the solid materials are represented by both Euler

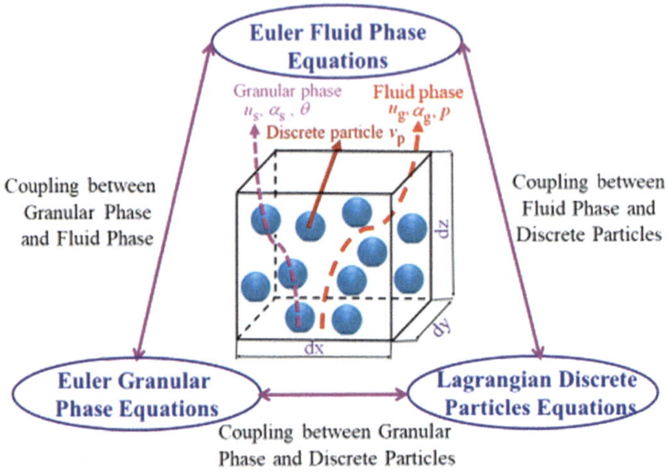

Fig. 3.2 Fluid phase, granular phase and discrete particles in CFD-KTGF-DEM approach

granular phase and Lagrangian discrete particles in CFD-KTGF-DEM approach. The Euler fluid phase and granular phase are modelled by Eulerian method which is a way of looking at the motion of fluid and particles from a continuum point of view. The discrete particles are modelled by Lagrangian approach in which the motion of individual particles is tracked in space and time. Both granular phase and discrete particles are coupled to satisfy the balance of momentum and forces, and link to fluid phase to satisfy the balance of momentum transfer. CFD-KTGF-DEM approach is a coupling approach between continuum and discrete models, and it is also a multiscale approach for fluid-particles multiphase flows and fluidized bed reactors.

3.1.1 Euler Fluid Phase

For flow of Euler fluid phase, fluid properties are defined in the fluid cells, which cover the whole simulation domain. Therefore, fluid properties are defined at a length scale as large as the size of fluid cells. The size of fluid cell is larger than the size of particle. The coupling between the granular phase at particle scale and the fluid phase at fluid cell scale is done via drag force with empirical or theoretical correlations developed for the flow across a single particle or an assembly of particles.

From the continuum viewpoint, the Euler fluid phase mass and momentum conservation equations are expressed according to fluid volume fraction, velocity and viscous stress tensor [14, 18, 24].

$$\frac{\partial}{\partial t}(\alpha_f \rho_f) + \frac{\partial}{\partial x_i}(\alpha_f \rho_f \vec{u}_{fi}) = 0 \tag{3.1}$$

$$\frac{\partial}{\partial t}(\alpha_f \rho_f \vec{u}_{fi}) + \frac{\partial}{\partial x_j}(\alpha_f \rho_f \vec{u}_{fi}\vec{u}_{fj}) = -\alpha_f \frac{\partial p}{\partial x_i} + \frac{\partial \overline{\tau}_{fij}}{\partial x_j} + \vec{I}_{sf} + \alpha_f \rho_f \vec{g}_i \tag{3.2}$$

$$\overline{\tau}_{fij} = \alpha_f(\mu_f + \mu_t)\left(\frac{\partial \vec{u}_{fi}}{\partial x_j} + \frac{\partial \vec{u}_{fj}}{\partial x_i}\right) - \frac{2}{3}\alpha_f \delta_{ij}\left[\rho_f k_f + (\mu_f + \mu_t)\frac{\partial \vec{u}_{fk}}{\partial x_k}\right] \tag{3.3}$$

where $\vec{I}_{sf}$ represents the aerodynamic drag forces between the Euler fluid phase and Euler granular phase.

The fluid phase effective viscosity consists of dynamic viscosity and turbulent viscosity. The former is obtained according to fluid temperature and pressure. In order to predict fluid turbulent viscosity, the turbulent correlations are modeled. Although the Euler fluid phase Reynolds stress does appear formally in the momentum balance, its description is not crucial since the pressure drop and drag terms dominate in fluid-particles flows. Thus, this correlation is described using a two-equation turbulence model to represent the effect of granular phase Reynolds stress. It is acknowledged that single phase fluid turbulence models contain a certain amount of empiricism, and this empiricism will not be the same for fluid turbulence with granular phase turbulence. However, the physical mechanisms governing the turbulence should be the same. The fluid phase eddy viscosity μ_t is defined as

$$\mu_t = \rho_f C_\mu \frac{k_f^2}{\varepsilon_f} \tag{3.4}$$

where $C\mu$, is an empirical model constant, and its value is 0.09. The transport equations of fluid turbulent kinetic energy k_f and its dissipation rate ε_f using two-equation k_f-ε_f turbulence model [3, 36].

$$\alpha_f \rho_f\left(\frac{\partial k_f}{\partial t} + \vec{u}_{fj}\frac{\partial k_f}{\partial x_j}\right) = \frac{\partial}{\partial x_i}\left(\alpha_f \frac{\mu_t}{\sigma_k}\frac{\partial k_f}{\partial x_j}\right) + \alpha_f \overline{\tau}_{fij}\frac{\partial \vec{u}_{fi}}{\partial x_j} + \vec{\Pi}_{kf} - \alpha_f \rho_f \varepsilon_f \tag{3.5}$$

$$\alpha_f \rho_f\left(\frac{\partial \varepsilon_f}{\partial t} + \vec{u}_{fj}\frac{\partial \varepsilon_f}{\partial x_j}\right) = \frac{\partial}{\partial x_i}\left(\alpha_f \frac{\mu_t}{\sigma_\varepsilon}\frac{\partial \varepsilon_f}{\partial x_j}\right) + \alpha_f \frac{\varepsilon_f}{k_f}\left(C_{1\varepsilon}\overline{\tau}_{fij}\frac{\partial \vec{u}_{fi}}{\partial x_j} - C_{2\varepsilon}\rho_f \varepsilon_f\right) + \vec{\Pi}_{\varepsilon f} \tag{3.6}$$

The terms $\vec{\Pi}_{kf}$ and $\vec{\Pi}_{\varepsilon f}$ represent the influence of the granular phase on the fluid turbulence and are defined by the following equations according to the momentum exchange coefficient β between the fluid phase and granular phase:

$$\vec{\Pi}_{kf} = \beta[k_{fs} - 2k_f + (\vec{u}_{si} - \vec{u}_{fi})\vec{u}_{di}] \tag{3.7}$$

$$\vec{\Pi}_{\varepsilon f} = C_3 \frac{\varepsilon_f}{k_f} \vec{\Pi}_{kf} \tag{3.8}$$

where $C_{1\varepsilon}$, $C_{2\varepsilon}$, $C\mu$, σ_k and σ_ε are the empirical model constants, and they are taken as 1.44, 1.92, 1.0 and 1.3, respectively. C_3 is 1.2. u_d represents the fluid phase drift velocity. k_{fs} is the cross-correlation of fluid phase and granular phase fluctuating velocities. The interaction term Π_{kf} represents the energy necessary to accelerate the particles, or the energy transferred from the particles to the fluid phase if the particles have high fluctuation velocities in a region where the fluid phase turbulent kinetic energy is low.

The drift velocity represents the correlation between the fluid phase fluctuating velocity and the granular phase spatial distribution of particles. According to [36], this velocity represents the dispersion of the discrete particles by the large turbulent scales of the fluid phase. The drift velocity u_d is expressed by $u_d = u_f - u_m$, here u_f is the fluid phase velocity, and u_m is the volume-weighted velocity of fluid-granular mixture. It can be computed according to fluid phase–granular phase velocity covariance k_{fs}, as

$$\vec{u}_{di} = \frac{1}{3} k_{fs} \tau_{12t} \left(\frac{1}{\alpha_f} \frac{\partial \alpha_f}{\partial x_i} - \frac{1}{\alpha_s} \frac{\partial \alpha_s}{\partial x_i} \right) \tag{3.9}$$

where the crossing trajectory effect is accounted for by τ_{12t}. The fluid phase Lagrangian integral time-scale, computed along the discrete particle trajectory (the eddy-particle interaction time), is related to the fluid phase turbulent time-scale, representing the crossing trajectory effect.

$$\tau_{12t} = \frac{1.5 C_\mu k_f}{\varepsilon_f \sqrt{1 + 1.5 C_\beta \left| \vec{U}_r \right|^2 / k_f}} \tag{3.10}$$

$$\vec{U}_r = \vec{u}_{si} - \vec{u}_{fi} - \vec{u}_{di} \tag{3.11}$$

where k_f is the gas phase turbulent kinetic energy. The constant $C\mu$ is given by $C\mu = 0.09$. The definition of the relative velocity magnitude proposed by [36] is used in the formulation of the drag term. C_β is a constant, which depends on the type of flow. The constant is recommended as 0.45 for dispersion parallel to the stream and 1.80 for dispersion perpendicular to the flow direction in fluid–solid multiphase flows.

The Euler fluid phase turbulent transport has a relatively great effect on the mean flow field in dilute fluid-particle flows, including in the freeboard and gas injectors in fluidized bed reactors. Gas feed is injected using nozzles to promote lateral mixing of gas and particles. Typical examples are secondary air injection in circulating fluidized bed combustors, attrition nozzles in the fluid coking reactor and stripping with gas jets in thermal cracking processes, gas perforated plate and bell-type air nozzle distributors, and cyclone separators in fluidized bed reactors. These imply that the simulations may require a model simultaneously taking into account fluid turbulent transport.

3.1.2 Euler Granular Phase

The granular flow model is formulated by applying the classical Eulerian continuum concept for the continuous phase, while the granular constitutive correlations are modelled in accordance with the principles of KTGF. It is postulated that the granular phase can be represented considering a collection of identical, smooth, rigid spheres with particle diameter d_p. The balance describes the rate of change of the granular phase distributions with respect to position and time. [14] presents the details of the derivation of the granular phase conservation equations for momentum and granular energy.

Kinetic theory is used for deriving expressions for momentum transfer that occurs at a length scale of particle size because of particle collisions. The stresses resulting from kinetic theory can be viewed as laminar stresses and analogous to single phase flow, and the Reynolds stresses in the granular fluid may be viewed as resulting from turbulence. The extension of the turbulence model to dense flows includes modifications to the collisional part of the granular stresses [36]. The main feature of this approach is the formulation of the dissipative time scale τ_{12t} and the particle collisional time τ_{2c}. Both collisions and fluid-particle interactions contribute to granular stresses due to drag.

The Euler granular phase mass and momentum conservation equations are expressed according to Euler granular phase stress tensor $\vec{\tau}_s$ [14].

$$\frac{\partial}{\partial t}(\alpha_s \rho_s) + \frac{\partial}{\partial x_i}(\alpha_s \rho_s \vec{u}_{si}) = 0 \tag{3.12}$$

$$\frac{\partial(\alpha_s \rho_s \vec{u}_{si})}{\partial t} + \frac{\partial(\alpha_s \rho_s \vec{u}_{si} \vec{u}_{sj})}{\partial x_j} = -\alpha_s \frac{\partial p}{\partial x_i} - \frac{\partial p_s}{\partial x_i} + \frac{\partial \vec{\tau}_{sij}}{\partial x_j} + \vec{I}_{fs} + \alpha_s \rho_s \vec{g}_i \tag{3.13}$$

where $\vec{I}_{fs}$ represents the fluid-granular interphase momentum exchange. The granular phase stresses are defined as the function of granular shear viscosity μ_s and granular bulk viscosity μ_b.

$$\vec{\tau}_{sij} = \alpha_s \mu_s \left[\frac{1}{2} \left(\frac{\partial \vec{u}_{si}}{\partial x_j} + \frac{\partial \vec{u}_{sj}}{\partial x_i} \right) - \frac{2}{3} \delta_{ij} \frac{\partial \vec{u}_{sk}}{\partial x_k} \right] + \alpha_s \mu_b \frac{\partial \vec{u}_{sk}}{\partial x_k} \tag{3.14}$$

The granular pressure p_s is expressed according to granular temperature θ from the KTGF and the hybrid theories of granular flow [1].

$$p_s = \begin{cases} p_{s,kc} & \alpha_s < \alpha_{min} \\ p_{s,kc} + p_{s,fr} & \alpha_s \geq \alpha_{min} \end{cases} \tag{3.15}$$

$$p_{s,kc} = \alpha_s \rho_s \theta + 2\alpha_s^2 \rho_s g_0 (1 + e_s)\theta \tag{3.16}$$

$$p_{s,fr} = 0.1\alpha_s \frac{(\alpha_s - \alpha_{min})^q}{(\alpha_{max} - \alpha_s)^p} \tag{3.17}$$

where the coefficients q and p are to be 2.0 and 3.0 [22, 40]. The values of α_{min} and α_{max} are assumed to be 0.5 and 0.63, respectively. The turbulent kinetic energy k_s of granular phase can be written with granular temperature θ.

$$\theta = \frac{2}{3} k_s \tag{3.18}$$

The granular viscosity consists of the kinetic part derived from dilute kinetic theory of gases and the collisional part due to the inelastic collisions as well as the frictional part due to the sustained frictional contacts [39, 40].

$$\mu_s = \begin{cases} (\mu_{s,ki} + \mu_{s,co}) & \alpha_s < \alpha_{\min} \\ (\mu_{s,ki} + \mu_{s,co}) + \mu_{s,fr} & \alpha_s \geq \alpha_{\min} \end{cases} \tag{3.19}$$

$$\mu_{s,fr} = \frac{p_{fr} \sin \varphi}{\sqrt{4\Pi_{2D} + (\nabla \cdot \vec{u}_s)^2}} \tag{3.20}$$

where φ is the angle of internal friction. Π_{2D} is the second invariant of the deviatoric stress tensor. The granular phase shear viscosity is expressed as the function of granular kinetic energy k_s and fluid-granular covariance k_{fs} [36].

$$\mu_{s,ki} = \frac{2\rho_s}{3} \frac{[k_{fs}(\tau_{12t}/\tau_{12x}) + k_s(1 + 0.4(1 + e_s)(3e_s - 1)\alpha_s g_0)]}{[(2/\tau_{12x}) + 0.2(1 + e_s)(3 - e_s)/\tau_{2c}]} \tag{3.21}$$

$$\mu_{s,co} = \frac{4}{5}\alpha_s \rho_s g_0 (1 + e_s)\left(\mu_{s.ki} + d_p \sqrt{\frac{2k_s}{3\pi}}\right) \tag{3.22}$$

The discrete particle relaxation time represents the characteristic time scale of interactions between the fluid phase and discrete particles. The discrete particle collision time represents the time between two binary collisions for a given particle. The discrete particle relaxation time and the discrete particle–particle collision time are defined by:

$$\tau_{12x} = \frac{4 d_p \rho_s}{3\rho_f C_D \left|\vec{U}_r\right|} \tag{3.23}$$

$$\tau_{2c} = \frac{d_p}{24\alpha_s g_0}\sqrt{\frac{\pi}{\theta}} \tag{3.24}$$

where C_D is the averaged drag coefficient for a single discrete particle in a suspension. It is calculated according to the mean Reynolds number Re which is defined by

$$Re = \frac{d_p \rho_f \left|\vec{U}_r\right|}{\mu_f} \tag{3.25}$$

It can be seen that the granular velocity fluctuations combine the fluid turbulent eddy viscosity with the kinetic theory for granular material. In addition, the granular bulk viscosity μ_b is expressed as follows.

$$\mu_b = \begin{cases} \mu_{b,kc} & \alpha_s < \alpha_{\min} \\ \mu_{b,kc} + \frac{p_s}{\sqrt{4\sin^2\varphi \Pi_{2D} + (\nabla\cdot\vec{u}_s)^2}} & \alpha_s \geq \alpha_{\min} \end{cases} \tag{3.26}$$

$$\mu_{b,kc} = \frac{4}{3}\alpha_s g_0 \rho_s (1+e_s)\left(\mu_{s,ki} + d_p\sqrt{\frac{2k_s}{3\pi}}\right) \tag{3.27}$$

The granular phase turbulence includes the effects of turbulence in the fluid phase and collisions of granular phase [10, 11, 30, 34, 37, 38]. The transport equations for the granular turbulent kinetic energy k_s and the cross-correlation of fluid phase and granular phase fluctuating velocities k_{fs} consists of k_s-k_{fs} two-equation model with additional terms taking into account the interfacial turbulent momentum transfer [2, 36], referred to as the SIMONIN model in [28].

$$\begin{aligned} \alpha_s\rho_s\left[\frac{\partial k_s}{\partial t} + \vec{u}_{sj}\frac{\partial k_s}{\partial x_j}\right] &= \frac{\partial}{\partial x_i}\left(\alpha_s\rho_s K_s \frac{\partial k_s}{\partial x_j}\right) \\ &+ \alpha_s\rho_s\overline{\tau}_{sij}\frac{\partial \vec{u}_{si}}{\partial x_j} + \Pi_{ks} - \frac{8\rho_s\alpha_s^2 g_0(1-e_s^2)}{d_p}\sqrt{\frac{2}{3}\frac{k_s}{\pi}}k_s \end{aligned} \tag{3.28}$$

$$\begin{aligned} \alpha_s\rho_s\left[\frac{\partial k_{fs}}{\partial t} + \vec{u}_{sj}\frac{\partial k_{fs}}{\partial x_j}\right] &= \frac{\partial}{\partial x_i}\left(\alpha_s\rho_s \frac{\nu_{fs}}{\sigma_k}\frac{\partial k_{fs}}{\partial x_j}\right) \\ &+ \alpha_s\rho_s\overline{\tau}_{fsij}\left(\frac{\partial \vec{u}_{si}}{\partial x_j} + \frac{\partial \vec{u}_{sj}}{\partial x_i}\right) + \Pi_{kfs} - \rho_s\alpha_s\frac{k_{fs}}{\tau_{12t}} \end{aligned} \tag{3.29}$$

where e_s is the granular phase coefficient of restitution. The constant used in the k_s-k_{fs} model is chosen as $\sigma_k = 1.0$. The last term on the right-hand side of Eq. (3.28) is the dissipation rate of kinetic energy due to collisions. The turbulence exchange terms Π_{ks} and Π_{kfs} are expressed as the function of granular phase turbulent kinetic energy and fluid phase-granular phase velocity covariance.

$$\Pi_{ks} = -\frac{\alpha_s \rho_s}{\tau_{12x}}(2k_s - k_{fs}) \tag{3.30}$$

$$\Pi_{kfs} = -\frac{\alpha_s \rho_s}{\tau_{12x}}\left[\left(1 + \frac{\alpha_s \rho_s}{\alpha_f \rho_f}\right)k_{fs} - 2k_f - 2\frac{\alpha_s \rho_s}{\alpha_f \rho_f}k_s\right] \tag{3.31}$$

The turbulent fluid phase-granular phase viscosity is given according to the interaction time τ_{12t} between the granular phase and fluid phase fluctuations. This time scale depends mainly on the loss of correlation due to the crossing trajectories effect.

$$\nu_{fs} = \frac{k_{fs}\tau_{12t}}{3} \tag{3.32}$$

The granular phase conductivity K_s is expressed in terms of its kinetic and collisional contributions [2, 36]:

$$K_s = K_{s,kin} + K_{s,col} \tag{3.33}$$

$$K_{s,kin} = \frac{0.6\tau_{12t}\tau_{2c}k_{fs} + (2/3)\tau_{2c}\tau_{12x}k_s[1 + 0.01(1 + e_s)^2(2e_s - 1)\alpha_s g_0]}{[1.6\tau_{2c} + 0.4(1 + e_s)(3e_s - 1)\tau_{12x}]} \tag{3.34}$$

$$K_{s,col} = \frac{6}{5}\alpha_s g_0(1 + e_s)\left(K_{s,kin} + \frac{5}{9}d_p\sqrt{\frac{2k_s}{3\pi}}\right) \tag{3.35}$$

The granular phase conductivity is consistent with the two limits: for flows of low granular volume fractions, these parameters are controlled by fluid phase turbulence, and in the high granular volume fraction cases, they reduce to the kinetic theory expressions for granular flow [14, 26]. In the intermediate regimes, these two mechanisms contribute according to the characteristic dissipation time scales.

For a detailed explanation and derivation of the fluid phase-granular phase k_s-k_{fs} two-equation model, see [2, 29, 36] and [28].

For the granular phase, the slip boundary condition is applied. [19] derived a theory to calculate the rate at which momentum and energy are supplied to the flow per unit area of the wall. The shear stress and the kinetic energy flux at the wall can be related to the wall-discrete particle collisional properties of normal restitution coefficient e_w and tangential restitution coefficient β_w. However, in the model of Johnson and Jackson [22], colliding discrete particles (distinction is made between colliding and sliding discrete particles) are characterized by a specularity coefficient ϕ. The recommended granular boundary condition is given by [22, 43]:

$$\vec{u}_{s,w} = -\frac{6\alpha_{s,max}[\mu_{s,fr}\tan\varphi_w + \alpha_s \rho_s(\mu_{s,kin} + \mu_{s,col})]}{\sqrt{3}\pi\phi\alpha_s\rho_s g_0\sqrt{\theta}}\frac{\partial \vec{u}_{s,w}}{\partial x} \tag{3.36}$$

$$\theta_w = \left[-K_s\theta\frac{\partial \theta_w}{\partial x} + \frac{\sqrt{3}\pi\phi\rho_s\alpha_s g_0\theta^{3/2}u_{s,slip}^2}{6\alpha_{s,\max}}\right]\frac{4\alpha_{s,\max}}{\sqrt{3}\pi(1 - e_w^2)\alpha_s\rho_s g_0\theta^{3/2}} \tag{3.37}$$

where φ_w represents the angle of internal friction. e_w is the normal restitution coefficient between the wall and granular phase. ϕ denotes the specularity coefficient between the granular phase and the wall.

3.1.3 Lagrangian Discrete Particles

The motion of individual discrete particles is tracked in space and time using the Lagrangian approach. The discrete element method (DEM) refers to a type of modeling approach in which the dynamic/time transient behavior of a system comprising a large number of discrete particles is studied. The translational and rotational motions of each discrete particle are formulated in the model and are highly dependent on the type and strength of interactions of the discrete particles with the surrounding medium through contact and interparticle effects. Particles interact with each other through their contact points/areas. Depending on the operating conditions of the system, interparticle interactions may exist between discrete particles even if they are not in contact.

The motion equation of a discrete particle i is expressed according to linear velocities $\vec{v}_p$ and spin velocities $\vec{\omega}_p$ with particle diameter d_p [5, 6, 13].

$$m_i \frac{d\vec{v}_{pi}}{dt} = \sum_{j=1, i \neq j}^{N} \vec{f}_{cij} + \vec{f}_{di} - V_{pi} \nabla p + V_{pi} \nabla \cdot \overline{\boldsymbol{\tau}}_f + m_i \vec{g} \tag{3.38}$$

$$I_{pi} \frac{d\vec{\omega}_{pi}}{dt} = \sum_{j=1, i \neq j}^{N} \vec{T}_{pij} \tag{3.39}$$

where V_{pi} is the discrete particle volume located in the cell. The contact force $\vec{f}_c$ is dissolved into the normal force component $\vec{f}_{nij}$ and tangential force component $\vec{f}_{tij}$, and they are modeled according to the overlap between colliding discrete particles. $\vec{T}_{pij}$ is the sum of all torques acting on the ith particle. I_p is the moment of inertia of particle.

The contact forces of discrete particles are modelled using the force–displacement laws. The nonlinear viscoelastic contact force model consists of two components, elastic and viscous forces. The theory of Hertz accurately describes the elastic behavior of collision of spherical particles. The energy loss during a contact leads to different velocity of discrete particles before and after the collision. By introducing a damper, the energy dissipation is introduced. Accordingly, the collision force in normal direction is expressed using the normal spring stiffness of linear spring k_n and normal damping coefficient η_n. Analogous to the normal direction, the tangential collision force is expressed using the tangential spring constant k_t and damping coefficient η_t. The contact force is expressed as follows [42, 48].

$$\vec{f}_{nij} = (-k_n \delta_{nij}^{3/2} - \eta_n \vec{v}_{rij} \cdot \vec{n}_{ij})\vec{n}_{ij} \tag{3.40}$$

$$\vec{f}_{tij} = \begin{cases} -k_t \delta_{tij} - \eta_t \vec{v}_{tij} & \left|\vec{f}_{tij}\right| < \mu_{df} \left|\vec{f}_{nij}\right| \\ -\mu_{df} \left|\vec{f}_{nij}\right| \vec{v}_{tij} / \left|\vec{v}_{tij}\right| & \left|\vec{f}_{tij}\right| \geq \mu_{df} \left|\vec{f}_{nij}\right| \end{cases} \tag{3.41}$$

where μ_{df} stands for the coefficient of dynamic friction between the discrete particle i and j. $\vec{n}$ represents the unit vector, and calculates according to the position vectors $\vec{x}_i$ and $\vec{x}_j$ of two discrete particles.

$$\vec{n}_{ij} = \frac{\vec{x}_j - \vec{x}_i}{\left|\vec{x}_j - \vec{x}_i\right|} \tag{3.42}$$

The normal overlap of two discrete particles is defined in terms of the particle radii R_i and R_j.

$$\delta_n = R_i + R_j - \left|\vec{x}_j - \vec{x}_i\right| \tag{3.43}$$

The relative velocity between colliding discrete particles at the contact point and its normal and tangential components are defined as:

$$\vec{v}_{ij} = \vec{v}_i - \vec{v}_j + (R_i \vec{\omega}_i + R_j \vec{\omega}_j) \times \vec{n}_{ij} \tag{3.44}$$

$$\vec{v}_{n,ij} = (\vec{v}_{ij} \cdot \vec{n}_{ij})\vec{n}_{ij} \tag{3.45}$$

$$\vec{v}_{t,ij} = \vec{v}_{ij} - \vec{v}_{n,ij} \tag{3.46}$$

The normal and tangential spring coefficients k_n and k_t are calculated according to the Poisson's ratio v and elasticity moduli E of discrete particles [5, 45]:

$$k_n = \frac{4}{3}\sqrt{\frac{d_{pi} d_{pj}}{2(d_{pi} + d_{pj})}} \left[\frac{E_j E_i}{E_i(1 - \nu_j^2) + E_j(1 - \nu_i^2)}\right] \delta_{n,ij}^{1/2} \tag{3.47}$$

$$k_t = \frac{16}{3}\sqrt{\frac{d_{pi} d_{pj}}{2(d_{pi} + d_{pj})}} \left[\frac{E_i E_j}{E_j(2 - \nu_i)(1 + \nu_i) + E_i(2 - \nu_j)(1 + \nu_j)}\right] \delta_{n,ij}^{1/2} \tag{3.48}$$

The normal dashpot damping coefficient η_n is expressed according to the mass of discrete particles, spring coefficient and discrete particle restitution coefficient e_{ij}.

$$\eta_n = \frac{2\sqrt{m_i m_j k_n} \left|\ln e_{ij}\right|}{\sqrt{(m_i + m_j)(\pi^2 + \ln^2 e_{ij})}} \tag{3.49}$$

where m is the mass of a discrete particle. A similar expression can be written for the tangential dashpot damping coefficient η_t according to the tangential spring coefficient. These indicate that the collisional properties of discrete particles relate to Poisson's ratio and elasticity moduli of discrete particles.

3.2 CFD-KTGF-DEM Coupling

For Lagrangian discrete particles, the motion of discrete particles is based on discrete mechanics, in which the interaction between discrete particles is described by classical Newtonian mechanics. The variables associated with this class of methods are only defined in the discrete particle positions. In contrast, the motion of granular phase is based on continuum mechanics, in which the studied domain is assumed to be continuous and completely filling the space. The mechanical behavior of this domain is described by granular constitutive laws involving continuous field variables.

To model fluid–particle flows, CFD-KTGF-DEM approach provides detail hydrodynamics of Euler fluid and Euler granular phases and Lagrangian discrete particles in multiphase flow systems and fluidized bed reactors. In CFD-KTGF-DEM approach, both fluid and granular phases are considered as continuum phases, and discrete particles are represented by the Lagrangian approach. Inclusion of constitutive equations relating to fluid phase and granular phase as well as discrete particles using the proper coupling strategy in CFD-KTGF-DEM allow to study momentum and mass transports in almost any detail on the particle scale. Hence, CFD-KTGF-DEM approach involves the interphase coupling of mass, momentum and energy from the Euler to Lagrangian domain, including the momentum exchanges of the Euler fluid phase-Euler granular phase and the Euler fluid phase-Lagrangian discrete particles. These are caused by the collisional interactions between the Euler granular phase and Lagrangian discrete particles and the turbulent kinetic energy transfer among the Euler fluid and Euler granular phases and Lagrangian discrete particles.

3.2.1 *Coupling of Interfacial Momentum*

Momentum coupling between phases within a control volume can be addressed by comparing the total drag force due to granular phase with the convective momentum flux of fluid phase as well as drag force acting on discrete particles. The fluid-particle interaction force acting on the particle is estimated at the particle scale while the analogous term in the fluid phase momentum equation, which is the volumetric fluid–particle interaction force acting on the fluid, should be calculated in each fluid cell [12]. The exchange of momentum between the Euler granular phase and the Euler fluid phase is calculated by means of the coefficient of interphase momentum exchange β. The momentum exchange is computed using the relative velocity between the fluid phase and the granular phase and the drag coefficient which is calculated according to Huilin-Gidaspow drag model [13, 17].

$$\vec{I}_{fs} = \beta(\vec{u}_f - \vec{u}_s) \tag{3.50}$$

$$\beta = (1-\varphi)\left[150\frac{\alpha_s(1-\alpha_f)\mu_f}{\alpha_f d_p^2} + 1.75\frac{\alpha_s\rho_f\left|\vec{u}_f - \vec{u}_s\right|}{d_p}\right] + \varphi\frac{3}{4}C_d\frac{\alpha_s\alpha_f\rho_f}{d_p}\left|\vec{u}_f - \vec{u}_s\right|\alpha_f^{-2.65} \tag{3.51}$$

$$\varphi = \frac{\arctan[150 \times 1.75(\alpha_f - 0.8)]}{\pi} + 0.5 \tag{3.52}$$

For a discrete particle residing in ith cell, a cell-centered value of u_{fi} is used. Similarly, a local cell averaged velocity u_{si} is used. With this simplification, the drag force f_{di} in Eq. (3.38) is computed in terms of the drag coefficient β from Eq. (3.51) according to the interaction force decomposition scheme.

$$\vec{f}_{di} = \frac{\beta V_{pi}}{\alpha_{si}}(\vec{u}_{fi} - \vec{u}_{si}) \tag{3.53}$$

The momentum interactions are modelled according to the Newton's third law of motion for the Euler fluid phase and Euler granular phase.

$$\vec{I}_{sf} = -\vec{I}_{fs} \tag{3.54}$$

Thus, all Eqs. (3.50), (3.53) and (3.54) give a coupling relationship of the drag force $\vec{f}_d$ used in the determination of velocities of Lagrangian discrete particles and the inter-phase momentum transfer $\vec{I}_{fs}$ used in the momentum conservation Eqs. (3.2) and (3.13) of Euler fluid phase, Euler granular phase and Lagrangian discrete particles [33].

3.2.2 Coupling of Pressure and Shear Stress

The pressure gradient and viscous forces are applied to each particle on an individual basis, but that they manifest themselves directly in the Euler fluid equations. The viscous (shear) forces are needed because of turbulence effect of Euler fluid phase and Euler granular phase. The interaction forces between the fluid phase and discrete particles include fluid pressure gradient force and viscous force due to the fluid stress tensor.

$$\vec{f}_{fd} = -V_{pi}\nabla p + V_{pi}\nabla \cdot \overline{\tau}_f \tag{3.55}$$

The volumetric mean force is obtained from:

$$\vec{F}_{fd} = \frac{1}{V_{cel}} \sum_{i=1}^{N} (-V_{pi} \nabla p + V_{pi} \nabla \cdot \overline{\tau}_f) \tag{3.56}$$

where N is the number of particles in the cell volume V_{cel}.

According to the definition of volume fraction of discrete particles, the mean fluid volumetric forces are

$$\vec{F}_{fd} = -\alpha_p \nabla p + \alpha_p \nabla \cdot \overline{\tau}_f \tag{3.57}$$

where α_p is the volume fraction of discrete particles in the cell. Replacing the volume fraction of discrete particles with the granular volume fraction, it is also expressed as follows:

$$\vec{F}_{fd} = -\alpha_s \nabla p + \alpha_s \nabla \cdot \overline{\tau}_f \tag{3.58}$$

All Eqs. (3.55), (3.57) and (3.58) give a coupled relationship of forces among the Euler fluid phase and Euler granular phases as well Lagrangian discrete particles in the cells.

3.2.3 Coupling of Interactions of Collisions

The kinetic theory of granular flow is fairly free from empirical constants. The only constant that appears directly in the classical KTGF is the restitution coefficient e and specularity coefficient ϕ. The coefficient of restitution captures the elasticity of normal particle impact. As far as the restitution coefficient is concerned, its value affects the results significantly in numerical simulations of fluid-particles multiphase flows. For dense suspensions, a decrease of the restitution coefficient e causes the granular temperature to decrease. This in turn causes a decrease of the viscosity and diffusion, resulting in larger gradients in the calculated results.

From Eq. (3.44) for the relative velocity v_{ij} between the discrete particle i and j, the definition of discrete particles restitution coefficient is expressed according to the relative velocity before impact v and the rebound velocity v' during the binary collision. A collision of discrete particles may be described by normal restitution coefficient e_n and tangential restitution coefficient e_t. The restitution coefficients relate the post- and pre-collisional values of the normal relative velocities and tangential relative velocities during the contact between the particle i and particle j.

$$e_{n,ij} = \left| \frac{v'_{n,i} - v'_{n,j}}{v_{n,i} - v_{n,j}} \right| \tag{3.59}$$

$$e_{t,ij} = -\frac{(v'_{t,i} - v'_{t,j})}{(v_{t,i} - v_{t,j})} \tag{3.60}$$

The normal restitution coefficient of discrete particles is in the range $0 \le e_n \le 1.0$, where $e_n = 0$ denotes a fully plastic collision and $e_n = 1.0$ a fully elastic collision wherein the normal velocity is conserved (with reversed sign). The tangential restitution coefficient e_t is in general defined in the range $-1.0 \le e_t \le 1.0$, whereby the case $e_t = 0$ denotes the total loss of tangential relative velocity after the impact, while $e_t \in [+1.0, -1.0]$ describes two elastic limits, i.e. smooth and rough particles, respectively. Smooth particles are described by $e_t = 1.0$ as the tangential relative velocity and thus the angular relative velocity of the interacting particles is not changed during the collision. Rough particles are represented by $e_t = -1.0$, which leads to a reversal of the tangential relative velocities at impact. These indicate that normal and tangential restitution coefficients depend on the discrete particle material parameters, the mass, the particle size as well as the relative impact velocity.

The colliding discrete particle number is N in the grid, and the mean value can be calculated according to discrete particle coefficient of restitution. The granular phase restitution coefficient e_s and tangential restitution coefficient β_s are defined as

$$e_s = \frac{1}{N} \sum_{j=1, i \neq j}^{N} e_{n,ij} \tag{3.61}$$

$$\beta_s = \frac{1}{N} \sum_{j=1, i \neq j}^{N} e_{t,ij} \tag{3.62}$$

We see that Eqs. (3.61) and (3.62) give a relationship between the granular phase restitution coefficient e_s and β_s and the discrete particles restitution coefficient $e_{n,ij}$ and $e_{t,ij}$, indicating that the granular phase restitution coefficient e_s and β_s of KTGF relates to granular volume fractions from the number of discrete particles N in the grid. Both of them give a coupled relation of interactions of collisions between the Euler granular phase and the Lagrangian discrete particles in the grids.

The granular phase wall boundary condition proposed by [22] involves two important parameters, the granular phase–wall restitution coefficient e_w and the specularity coefficient ϕ. The former describes the dissipation of the granular phase kinetic energy with the wall by collisional interactions. The latter specifies the shear condition at the walls. A value of e_w close to unity implies very low dissipation of granular kinetic energy at the wall. Similarly, we can obtain the equations for the discrete particle–wall restitution coefficient according to the normal and tangential relative velocities between the discrete particle i and the wall during the contact.

$$e_{n,iw} = \left| \frac{v'_{n,iw}}{v_{n,iw}} \right| \tag{3.63}$$

$$e_{t,iw} = -\frac{v'_{t,iw}}{v_{t,iw}} \tag{3.64}$$

The granular phase-wall restitution coefficient and tangential restitution coefficient are determined according to the discrete particle number N in the grid.

$$e_w = \frac{1}{N}\sum_{i=1}^{N} e_{n,iw} \tag{3.65}$$

$$\beta_w = \frac{1}{N}\sum_{i=1}^{N} e_{t,iw} \tag{3.66}$$

These equations link the granular phase-wall energy loss used in KTGF and the kinetic energy loss per unit area due to the collisions of discrete particles from DEM.

The specularity coefficient ϕ is an important parameter used to quantify the nature of particle–wall collisions. In other words, it characterizes the average fraction of relative tangential momentum flux transfer during a particle–wall collision [23]. The specularity coefficient is an indicator of the smoothness of the wall, and it ranges from zero to unity. Generally speaking, the shear stress is also known as the flux of momentum transfer. The relative tangential momentum flux of each particle is simply written as (mu_s), where m and u_s are the particle mass and mean slip velocity between the particles and the wall, respectively. A specularity coefficient is defined as [16]

$$\phi = \frac{\sigma_w}{mu_s} \tag{3.67}$$

where σ_w is the tangential momentum flux transferred to a unit area of the frictional flat surface. It is clear that the specularity coefficient is a fraction wherein the numerator is equal to the average tangential momentum flux transferred per collision, and the denominator is equal to the average tangential momentum flux possessed by the particle. This equation indicates that the specularity coefficient directly controls the relative amount of tangential momentum transferred from the discrete particle to the wall, and mathematically explains the reason for the variation of ϕ between zero (for perfectly specular collisions) and unity (for perfectly diffuse collisions) [22].

The relative tangential momentum flux of discrete particle i before collision is expressed according to the pre-tangential velocity $v_{t,iw}$ and pre-angular velocity ω_{iw}.

$$m_i v_{t,i} = m_i[v_{t,iw} + R_i(\omega_{iw} \cdot n_t)] \tag{3.68}$$

where n_t is the unit tangential vector directed from the boundary into the wall. Similarly, we can derive the equation for the relative tangential momentum flux of discrete particle i after collision according to the post-tangential velocity and post-angular velocity.

$$m_i v'_{t,i} = m_i[v'_{t,iw} + R_i(\omega'_{iw} \cdot n_t)] \tag{3.69}$$

The discrete particle specularity coefficient is defined to be the average fraction of relative tangential momentum flux transferred in a discrete particle–wall collision.

$$\phi_i = \frac{m_i v_{t,i} - m_i v'_{t,i}}{m_i v_{t,i}} = 1 - \frac{[v'_{t,iw} + R_i(\omega'_{iw} \cdot n_t)]}{[v_{t,iw} + R_i(\omega_{iw} \cdot n_t)]} \tag{3.70}$$

The granular phase specularity coefficient is obtained according to the discrete particle number N and discrete particle specularity coefficient in the grid.

$$\phi = \frac{1}{N}\sum_{i=1}^{N} \phi_i \tag{3.71}$$

Both Eq. (3.70) for discrete particles using DEM and Eq. (3.71) for granular phase used in KTGF link with the tangential momentum fluxes of Lagrangian discrete particles and granular phase through specularity coefficient.

These correlations suggest that the discrete particle restitution coefficient and specularity coefficient are determined according to the relative velocities of the before and after collision. The new value of discrete particle restitution coefficient will be applied to calculate the dashpot damping coefficient η_n according to Eq. (3.49). The new solution of the relative velocity after collisions can then be calculated, and the iterations of restitution coefficient and specularity coefficient can be repeated according to the calculation of residuals until the calculation attains a numerical convergence.

The discrete particles restitution coefficient and specularity coefficient are the ratio of the relative normal and tangential velocities after and before impact, indicating that the discrete particles restitution coefficient and specularity coefficient depend on impact velocity. They relate with impact velocity during the binary collisions. Hence, the dependence of the discrete particles restitution coefficient and specularity coefficient on the impacting relative velocity is a natural property assuming binary collisions. On the other hand, when the granular volume fractions are large in the grid, the multiple collisions are likely to become dominant because the binary collision might not be accurate to represent the momentum transfer through binary collisions. Indeed, the granular phase restitution coefficient and specularity coefficient relate with granular volume fractions. The probability of multiple collisions is relatively large at high granular volume fractions. When the inelastic collisions involving more than two discrete particles exist as frequently as the binary collisions, more kinetic energy will be dissipated, indicating that the granular phase restitution coefficient and specularity coefficient have to be determined from discrete particle restitution coefficient and specularity coefficient within the grids, linking between the KTGF and the DEM in multiphase flows.

3.3 Ghost Fluid

The CFD-KTGF-DEM model involves Euler fluid phase, Euler granular phase and Lagrangian discrete particles in the cell, as shown in Fig. 3.3 using MFIX code [28]. In MFIX code, the turbulence of fluid and granular phases is modelled using [36] or [3]. The discrete particles collisions are modelled using a linear spring dashpot or Hertzian collision models. The velocity and pressure of Euler fluid phase are coupled with the momentum balance equations of Euler granular phase and the motion equation of Lagrangian discrete particles. However, both Lagrangian discrete particles and Euler granular phase do not allowed to coexist at the same time in the control volume of CFD-KTGF-DEM approach. Hence, it is introduced the ghost fluid signified by *gh* in order to distinguish the Euler granular phase signified by *ep* and the Lagrangian discrete particles signified by *dp*. When the field of Euler granular phase is solved, the Lagrangian discrete particles is designated as the ghost phase at the same time in the cell during the stage of the computation [31]. Similarly, if the field of Lagrangian discrete particles is solved, the Euler granular phase acts as the ghost fluid. Therefore, the quantity ψ denoted by *sp* is expressed by

$$\psi_{sp} = \begin{cases} \psi_{ep}\big|_{gh=dp} \\ \psi_{dp}\big|_{gh=ep} \end{cases} \tag{3.72}$$

This equation gives a relation of the quantity between the Euler granular phase and the Lagrangian discrete particles in the cell. When the Lagrangian discrete particle is denoted as the ghost fluid, the Euler granular phase presents in the cell. On the other hand, the Lagrangian discrete particles present if the Euler granular phase acts as the ghost fluid. Thus, the ghost fluid communicates mass and momentum between the Euler granular phase and the Lagrangian discrete particles at the same time in the cell. The CFD-KTGF-DEM algorithm depends on ghost fluid to promote the parallel implementation of the calculations of inter-phase momentum transfer and impact velocity-dependent restitution coefficients. The ghost fluid acts as a communicator within Euler granular phase variables and Lagrangian discrete particles data at the same time in the cell.

The CFD-KTGF-DEM approach requires all processes both at the discrete particle scale such as drag force, contact force and impact velocity to be solved in the Lagrangian frame and at the macro-scale including fluid phase and granular phase velocities and volume fractions in the Eulerian grid. The time-splitting technique is used to determine granular restitution coefficient e_s in the cells. Two steps are required in the process of granular restitution coefficient calculations within the fluid phase time step Δt (i.e. granular phase time step).

$$\Delta t = \Delta t_a + \Delta t_d \tag{3.73}$$

$$\Delta t_a = c\Delta t = cN_a\Delta t_p \text{ and } \Delta t_d = (1-c)\Delta t = (1-c)N_d\Delta t_p \tag{3.74}$$

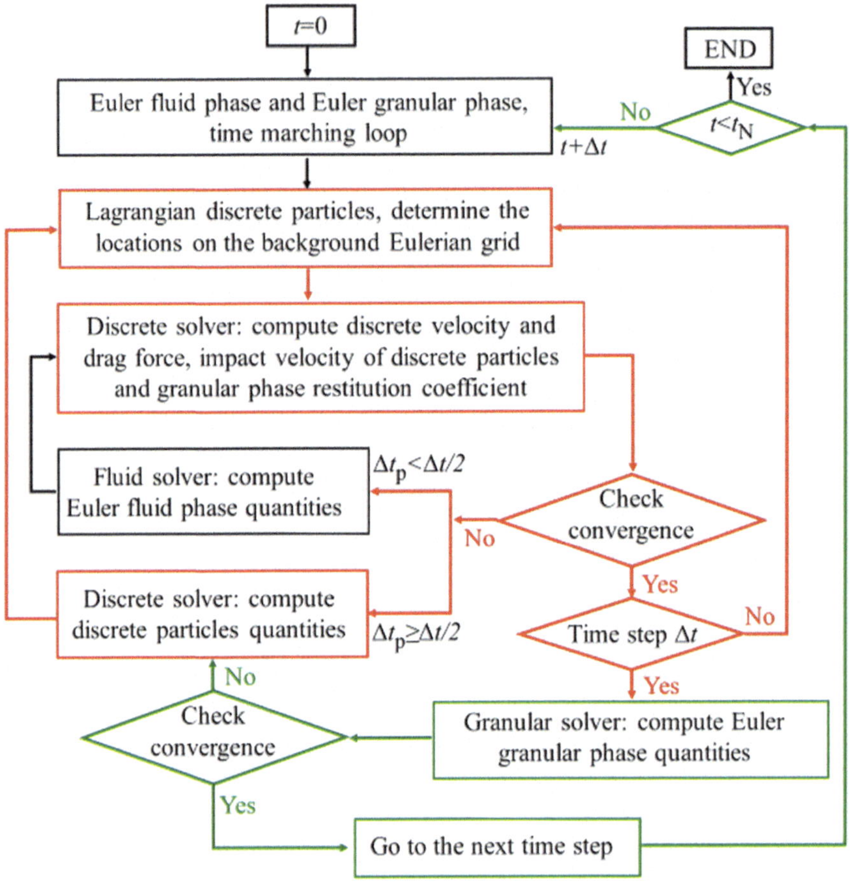

Fig. 3.3 Flow chart of CFD-KTGF-DEM approach

where c is the weighting parameter ($c = 0.5$). Δt_p is the particle time step. First, the contact interactions of discrete particles are calculated at the first time interval step Δt_a. The variables of discrete particles are updated at every cell. The particle restitution coefficient $e_{ij}(\Delta t_a)$ is computed, and the first granular restitution coefficient $e_s(\Delta t_a)$ is calculated. The granular restitution coefficient is given by:

$$e_s(\Delta t) = 0.5[e_s(\Delta t_a) + e_s(\Delta t_d)] \tag{3.75}$$

where $e_s(\Delta t_d)$ is the second granular restitution coefficient that is from the granular phase at the time interval Δt_d. The fluid phase and granular phase distributions are obtained according to the updated trajectories and velocities of Lagrangian discrete particles. Then, the mass and momentum equations of Euler granular phase and Euler fluid phase as well as the granular temperature equation of Euler granular phase are solved at the second time step Δt_d. Finally, the granular restitution coefficient

$e_s(\Delta t_d)$ is updated. This procedure is repeated until convergence is achieved. In the next fractional time step Δt_d, the trajectories and velocities of discrete particles are updated, and the granular restitution coefficient is calculated at the time interval Δt_d at every cell. The two steps are connected by the overall iteration procedure [32].

3.4 Dynamic Granular Restitution Coefficients

The CFD-KTGF-DEM approach consists of conservation equations of Euler fluid phase and Euler granular phase, and Lagrangian discrete particles motion equations in the cells. From Fig. 3.3, the computational algorithm is compiled through the discrete nodes applied for Euler fluid phase and Euler granular phase according to the mass and momentum conservation equations. The centroid method is applied for Lagrangian discrete particles according to the motion equations. The velocity variables of Euler granular phase and Euler fluid phase are calculated at the cell faces, and the fluid and granular volume fractions are calculated at the nodes. The velocities and center positions of Lagrangian discrete particles are computed within the cell. The discrete particles restitution coefficient $e_{n,ij}$ is calculated between two colliding particles i and j within the cell, and the dynamic granular restitution coefficient e_s is obtained according to the discrete particles restitution coefficient and the total number N of discrete particles in the cell.

Figure 3.4 shows the instantaneous granular and discrete particles velocities vectors and granular volume fraction contour at three times in the inlet fluidizing gas velocity of 0.6 m/s in a bubbling fluidized bed. Particle diameter and density are 2.5 mm and 2500 kg/m^3, respectively. The number of particles is 9240 in the initial bed height of 40 mm. The elastic modulus and Poisson's ratio of particles are 800 MPa and 0.33. The green color vectors represent the motion of discrete particles, showing the direction and variation of velocities of discrete particles within the bed. The white color vectors represent granular velocities at the cells. The volume fraction color contour indicates that the formation of bubbles exists as gas pass through the bed. As gas enters into the bed from the inlet, the small size bubbles are found near the inlet region. The size of bubbles is enlarged as gas passes through the bed due to coalescence. Furthermore, it is observed that both the discrete particles and granular phase flow upward through the center region and down-flow near the wall region, indicating that the circulation of bed materials exists within the bed [46].

Figure 3.5 shows instantaneous Euler granular velocities u_{gp} and Lagrangian discrete particles velocities u_{dp} in the cell. Both lateral and axial velocities fluctuate because of bubble motion and collisional interactions in the bed. The instantaneous axial and lateral velocities of Euler granular phase are very similar to those for Lagrangian discrete particles, having the same time-averaged lateral and axial velocities of granular phase and discrete particles. The positive lateral velocities mean both granular phase and discrete particles move from the wall to the center. Whereas the positive axial velocities indicate that both Euler granular phase and discrete particles

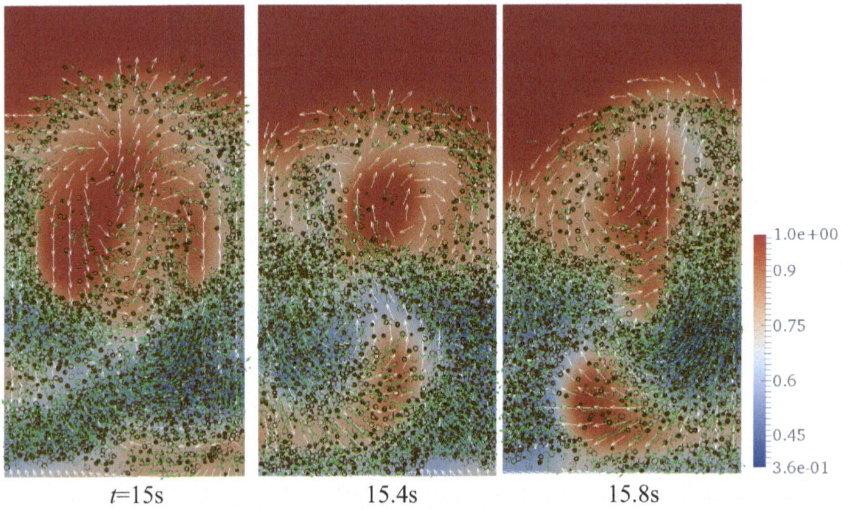

Fig. 3.4 Simulated instantaneous volume fractions and velocities (green and white color vectors represent velocities of Euler granular phase and discrete particles, color contour represents granular volume fraction)

flow from the inlet bottom to the bed top along height in the bubbling fluidized bed [47].

The discrete particles restitution coefficient $e_{n,ij}$ for a binary collision is simulated according to Eq. (3.59), and then the dynamic granular phase restitution coefficient e_s is calculated according to Eq. (3.61) in the cells. Figure 3.6 shows the instantaneous dynamic granular phase restitution coefficient e_s at two different cells. The simulated granular phase restitution coefficients fluctuate because of the motion of bubbles and

Fig. 3.5 Instantaneous velocities of Euler granular phase and Lagrangian discrete particles

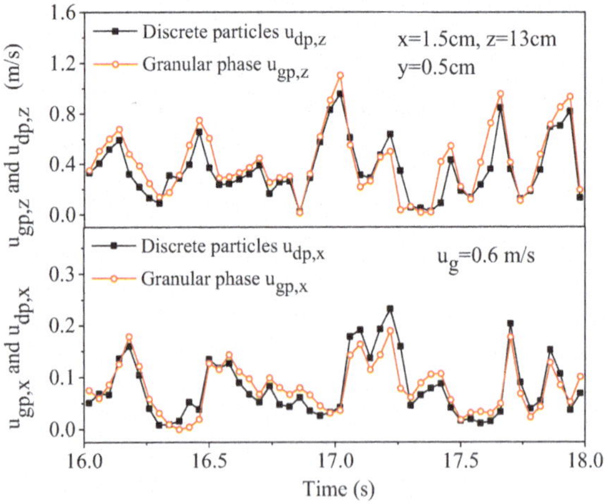

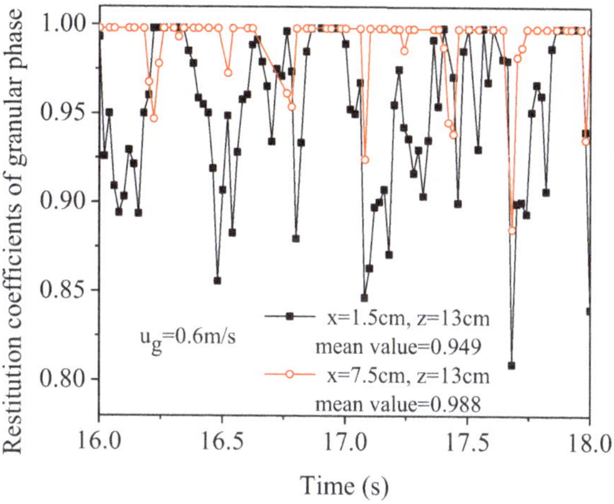

Fig. 3.6 Instantaneous dynamic coefficient of restitutions of granular phase

collisions of discrete particles within the bed. The time-averaged values of granular phase restitution coefficients are 0.949 and 0.988, indicating that the collisional energy dissipations of Euler granular phase change with time and position in the bubbling fluidized bed [47].

The KTGF models require several input parameters, and one of the vital parameters is granular restitution coefficients for the interactions of collisions of discrete particles. The discrete particle–particle restitution coefficient influences bubble formation due to the replacement of gas in voids beside the discrete particles. The discrete particle–wall restitution coefficient also has a significance on the hydrodynamic profile and heat-transfer ability at the wall since the contact time between a discrete particle and the wall affects the discrete particle–wall collosional interactions. The coefficient of restitution depends on impact velocity v_{ij} between the discrete particle i and j in the binary collision. [21] proposed a model that included plastic dissipation during impact and obtained an expression for the particle coefficient of restitution that was independent of the size of the spheres, but varied only with the impact velocity.

$$e_{ij} = 1.718\left[\frac{\sigma_{ys}^5}{\rho_s}\left(\frac{(1-\nu_i^2)E_j + (1-\nu_j^2)E_i}{E_iE_j}\right)^4\right]^{1/8} v_{ij}^{-1/4} \tag{3.76}$$

where σ_{ys} is the material yield stress of discrete particles. The kinetic energy dissipation can occur by two different mechanisms during the binary collisions of discrete particles. When the impact velocity v_{ij} is large between the discrete particle i and j, the colliding discrete particles deform fully plastically and $e_{ij} \propto {v_{ij}}^{-1/4}$. When the impact velocity v_{ij} is low, the deformations of discrete particles are elastic with mainly viscoelastic dissipation, and $1 - e_{ij} \propto {v_{ij}}^{1/5}$. The Lagrangian discrete particle impact-velocity-dependent restitution coefficient e_{ij} joins the two regimes of dissipation (viscoelastic and plastic) together, assuming that:

$$e_{ij} = \begin{cases} 1 - (1 - e_0)(v_{ij}/v_0)^{1/5} & v_{ij} \leq v_0 \\ e_0(v_{ij}/v_0)^{-1/4} & v_{ij} > v_0 \end{cases} \tag{3.77}$$

where $v_0 = 0.3$ m/s is chosen for stainless steel discrete particles, and e_0 is close to 0.95 [27].

The collision of particles is characterized by the compression (i.e., overlap) and by the compression rate. The compression gives rise to the elastic force, while the dissipative force appears due to the compression rate. Therefore, the restitution coefficient depends on the impact velocity. The impact-velocity-dependent restitution coefficient of Lagrangian discrete particles relates to dimensionless relative velocity v_r between the discrete particle i and j, and is calculated using the correlation proposed by [41]:

$$e_{ij} = \left[\left(\frac{6\sqrt{3}}{5}\right)\left(1 - \frac{1}{6}v_r^2\right)\right]^{0.5}\left[v_r\left(v_r + 2\sqrt{1.2 - 0.2v_r^2}\right)^{-1}\right]^{1/4} \tag{3.78}$$

$$v_r = \frac{3.194}{v_{ij}}\sqrt{\frac{(1.61\sigma_{ys})^5 m_i m_j}{m_i + m_j}\left(\frac{R_i R_j}{R_i + R_j}\right)^3\left[\frac{E_j(1 - v_i^2) + E_i(1 - v_j^2)}{E_i E_j}\right]^4} \tag{3.79}$$

From experiments as well as from theory it is well known that the discrete particle coefficient of normal restitution e_{ij} is not a constant but it depends sensitively on the impact velocity. The velocity-dependent restitution coefficient of discrete particles is expressed as a series in powers of the relative velocity before the collision of ${v_{ij}}^{1/5}$ [9, 25, 35].

$$\begin{aligned} e_{ij} = 1 &- 1.15344\left(\frac{3}{2}A\right)\left(\frac{\rho}{m_{eff}}\right)^{2/5} v_{ij}^{1/5} + 0.79826\left(\frac{3}{2}A\right)^2\left(\frac{\rho}{m_{eff}}\right)^{4/5} v_{ij}^{2/5} - \\ &0.483582\left(\frac{3}{2}A\right)^3\left(\frac{\rho}{m_{eff}}\right)^{6/5} v_{ij}^{3/5} + 0.285279\left(\frac{3}{2}A\right)^4\left(\frac{\rho}{m_{eff}}\right)^{8/5} v_{ij}^{4/5} \mp \cdots \end{aligned} \tag{3.80}$$

where ρ and A are the material constants depending on the Young modulus E and the Poisson ratio v of the particle material.

$$\rho = \frac{2E\sqrt{R_{eff}}}{3m_{eff}(1 - \nu^2)} \tag{3.81}$$

$$A = \frac{1}{3(1 - 2\nu)(1 + \nu)}\frac{[E(1 + \nu) - (1 - 2\nu)E]^2}{E(1 + \nu) + 2E(1 - 2\nu)}\left[\frac{(1 - \nu^2)(1 - 2\nu)}{E\nu^2}\right] \tag{3.82}$$

where R_{eff} and m_{eff} are the effective radius and mass of the particles, respectively. The discrete particle velocity-dependent restitution coefficient decreases with increasing impact velocity. Using the first four coefficients of this series expansion the approximation for restitution coefficient reproduces fairly well the tendency of energy losses for binary collisions of discrete particles.

When the discrete particle number density is large, the binary collision model of discrete particles might not be accurate, as multiple collisions are likely to occur. The multi-collisions affect the restitution coefficient of discrete particles. The granular phase restitution coefficient reveals energy dissipations due to the interactions of multiple collisions.

The instantaneous granular phase restitution coefficients are shown in Fig. 3.7 at three different times in the spouted fluidized bed [31]. In the bed, a high velocity gas stream flows through a spout located at the bottom of the bed, which results in generation of a local high velocity region in the center (referred to as spout) and a region with down flow of granular phase and discrete particles along the bed walls (referred to as annulus). Thus, in the spout discrete particles and granular phase move in a structured manner in the vertical direction accompanied by void (gas bubble) propagation. The instantaneous granular phase restitution coefficients are high in the spout region because of high velocity and low volume fraction of discrete particles. Simulations further illustrate that the instantaneous granular phase restitution coefficients are low in the annulus because of the higher volume fractions of discrete particles. The granular phase restitution coefficients are large at the bed surface because the gas volume fraction is high, indicating that the instantaneous dynamic granular phase restitution coefficients contribute to the impact velocity and gas volume fractions.

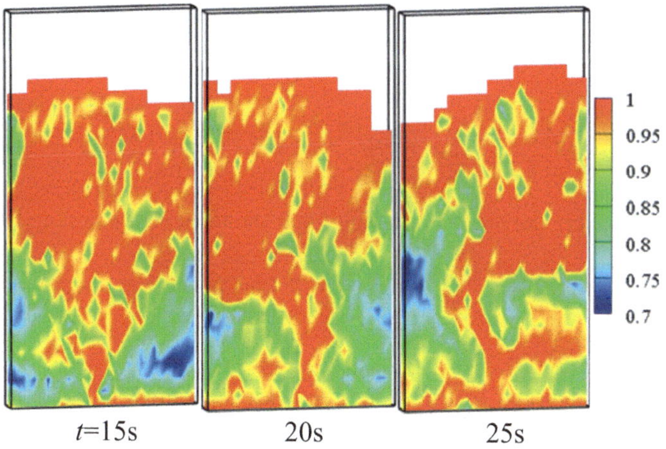

Fig. 3.7 Instantaneous granular phase restitution coefficient contours at three times

3.5 Energy Dissipations of Granular Phase and Discrete Particles

From simulated velocity and displacements of discrete particles, the energy dissipations of collision of discrete particles are obtained. The total mechanical energy of discrete particles is the sum of the kinetic energy, potential energy and energy losses due to the interactions of drag forces and collisions of discrete particles. The kinetic energy of Lagrangian discrete particles contains rotational and translational energy portions. The potential energy also includes elastic and gravitational energy portions. In the impact process of discrete particles in the bed, the kinetic energy is partially dissipated according to the motion of Newton's laws. One impact energy loss is due to the relative velocity at the contact point between discrete particles and between discrete particles and walls. The next one is the friction energy dissipation as a result of relative sliding between particles and between particles and walls. The last one is the rolling energy dissipation due to rolling friction arising from asymmetrical normal traction distribution on the contact area between particles. The collisional energy dissipation is the sum of the elastic potential energy ΔE_{cp}, impact energy dissipation ΔE_{dp} and friction energy dissipation ΔE_{sp} and rolling energy dissipation ΔE_{rp} between the discrete particles [15, 31].

$$\Delta E_{pp} = \Delta E_{cp} + \Delta E_{dp} + \Delta E_{sp} + \Delta E_{rp} \tag{3.83}$$

$$\Delta E_{cp} = \sum_{i=1}^{N} [(\vec{f}_{cn,ij} \cdot \vec{v}_{cn,ij} + \vec{f}_{ct,ij} \cdot \vec{v}_{ct,ij})]\Delta t \tag{3.84}$$

where f_{cn} and f_{ct} are the normal elastic force and tangential elastic force. v_{cn} and v_{ct} are the contact normal and tangential velocities.

$$\Delta E_{dp} = \sum_{i=1}^{N} [(\vec{f}_{dn,ij} \cdot \vec{v}_{dn,ij} + \vec{f}_{dt,ij} \cdot \vec{v}_{dt,ij})]\Delta t \tag{3.85}$$

where f_{dn} and f_{dt} are the normal damping force and tangential damping force.

$$\Delta E_{sp} = \sum_{i=1}^{N} [(\vec{f}_{cs,ij} \cdot \vec{v}_{t,ij})]\Delta t \tag{3.86}$$

where f_{cs} is the Coulomb friction forces between the particle i and j. The rolling energy dissipation between particles is given by

$$\Delta E_{rp} = \sum_{i=1}^{N} (\vec{T}_{r,ij} \cdot \vec{\omega}_{ij})\Delta t \tag{3.87}$$

where T_r is the rolling friction torque. The collisional energy dissipation between the discrete particles and the wall is

$$\Delta E_{pw} = \Delta E_{cw} + \Delta E_{dw} + \Delta E_{sw} + \Delta E_{rw}$$
$$= \sum_{i=1}^{N} [(\vec{f}_{cn,iw} \cdot \vec{v}_{cn,iw} + \vec{f}_{ct,iw} \cdot \vec{v}_{ct,iw}) + (\vec{f}_{dn,iw} \cdot \vec{v}_{dn,iw} + \vec{f}_{dt,iw} \cdot \vec{v}_{dt,iw})$$
$$+ (\vec{f}_{cs,iw} \cdot \vec{v}_{t,iw}) + (\vec{T}_{r,iw} \cdot \vec{\omega}_{iw})]\Delta t \tag{3.88}$$

The instantaneous energy dissipation portions of discrete particles are shown in Fig. 3.8 in the bed. The simulated energy dissipation portions of discrete particles indicate that the impact energy dissipation between discrete particles and between discrete particles and wall is the largest, and the friction energy dissipations between particles and between particles and wall are the smallest. The elastic potential energy between discrete particles and wall falls in between them.

The microscopic energy dissipation portions ΔE_{pp} and ΔE_{pw} are obtained from the discrete particle scale information for a given time interval Δt. From simulated energy dissipation portions, the total energy dissipations of collisions of discrete particles in the bed is

$$\Delta E_p = \Delta E_{pp} + \Delta E_{pw} \tag{3.89}$$

The variation of total energy dissipation is calculated from microscopic interactions of discrete particles. On the other hand, the variation of macroscopic energy dissipations of collisions is determined from granular phase. The rate of granular energy dissipation γ_s shows the energy loss of granular phase due to collisions between the particles. The value of collisional energy dissipation relates to Euler granular phase restitution coefficients.

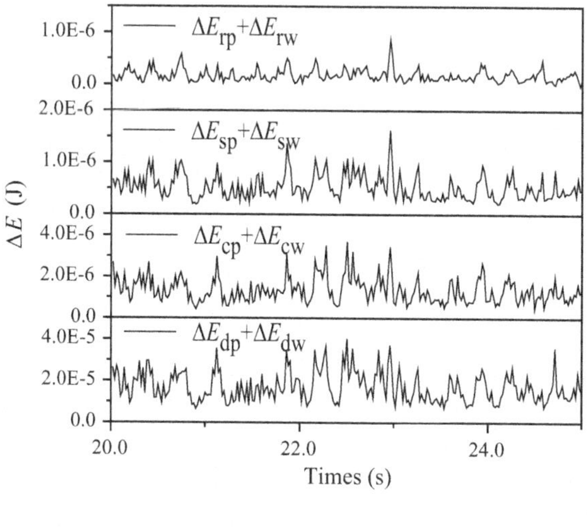

Fig. 3.8 Instantaneous energy dissipation portions of discrete particles

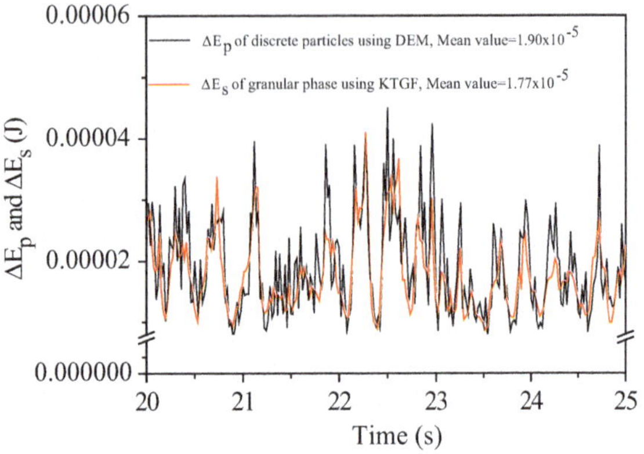

Fig. 3.9 Instantaneous collisional energy dissipations of granular phase and discrete particles

$$\Delta E_s = \sum_{i=1}^{N_{cell}} (\gamma_{s,i} \Delta V) \Delta t \tag{3.90}$$

where ΔV is the volume of the computational cell.

The instantaneous collisional energy dissipations of Euler granular phase and Lagrangian discrete particles are shown in Fig. 3.9 in the bed. The collisional energy dissipations of Euler granular phase have the same magnitude as the total energy dissipations of collisions of Lagrangian discrete particles. The mean values of collisional energy dissipations are 1.90×10^{-5} J and 1.77×10^{-5} J of Lagrangian discrete particles and Euler granular phase in the bed. These results show that the macroscopic variation of energy dissipation of granular phase using KTGF links to the microscopic information of Lagrangian discrete particles using DEM in the fluidized beds.

3.6 Summary

The CFD-KTGF-DEM approach, proposed by [31] and [46], applies the CFD for Euler fluid phase, kinetic theory of granular flow for Euler granular phase, and discrete element method for Lagrangian discrete particles to model fluid-particles fluidized beds and multiphase flows. The granular restitution coefficient is obtained from the collisions of discrete particles of DEM. The turbulence of discrete particles is based on the calculations of the turbulent interactions between the Euler fluid phase and the Euler granular phase. The CFD-KTGF-DEM approach provides the hydrodynamics of fluid phase, granular phase and discrete particles. A remarkable feature of the CFD-KTGF-DEM approach is that the model is fully free from model empirical constants. The numerical simulations are completely conducted according to material properties of discrete particles of Young's modulus and Poisson's ratio, allowing predictions to be made without extensive use of empirical constants. The CFD-KTGF-DEM method introduces a new state-of-the art tool for integrated multi-scale modeling of the discrete particles scale and the grid scale properties. Therefore, the CFD-KTGF-DEM model can apply for a great variety of properties of particles in multiphase flows and fluidized bed reactors.

The CFD-KTGF-DEM approach further illustrates that the simulations of hydrodynamics gas–solid fluidized beds need three momentum equations for Euler fluid phase, Euler granular phase and Lagrangian discrete particles instead of two momentum equations of Euler fluid phase and Euler granular phase in TFM-KTGF approach and Euler fluid phase and Lagrangian discrete particles in CFD-DEM approach. Thus, CFD-KTGF-DEM represents a novel approach coupling the control volume meso-scale and discrete particle micro-scale, and makes it possible to validate computational simulations in dense fluid–solid multiphase flows and fluidized beds.

Nomenclature

C	Constants
C_D	Single particle drag function
d_p	Diameter of particles, m
e_n	Normal coefficients of restitution of discrete particles
e_s	Granular coefficient of restitution
e_t	Tangential restitution coefficient of discrete particles
e_w	Wall-discrete particle coefficient of restitution
$\underline{E}_i$	Young's modulus of particle *i*, Pa
$\vec{f_c}$	Contact force between particles *i* and *j*, N
$\vec{f_d}$	Fluid–particle interaction force on particle *i*, N
$\vec{f_{fd}}$	Fluid phase–particle force, N
$\vec{f_n}$	Normal contact force between particles *i* and *j*, N
$\vec{f_t}$	Tangential contact force between particles *i* and *j*, N
$\vec{F}_{fd}$	Fluid phase-granular phase force, N
g	Gravitational acceleration vector, m/s^2
g_0	Radial distribution function at contact
$\vec{I}_{fs}$	Momentum transfer from fluid phase to solid phase, N/m^3
I_p	Moment of inertia of the particle
k_f	Fluid turbulent kinetic energy, m^2/s^2
k_{fs}	Fluid phase-granular phase velocity correlation tensor, m^2/s^2
k_n	Normal spring stiffness, N/m
k_s	Granular phase turbulent kinetic energy, m^2/s^2
k_t	Tangential spring stiffness, N/m
K_s	Granular conductivity, m^2/s
$K_{s,col}$	Collisional diffusivity, m^2/s
$K_{s,kin}$	Kinetic diffusivity, m^2/s
m	Mass of particle, kg

N	Total number of particles
p	Pressure in the fluid phase, Pa
p_s	Solids pressure, Pa
Re	Reynolds number
R_i	Radius of particle *i*, m
t	Time, s
$\vec{T}_p$	Sum of all torques acting on the particle, N·m
$\vec{u}_f$	Fluid velocity, m/s
$\vec{u}_s$	Granular phase velocity, m/s
$\vec{U}_r$	Fluid-particle relative velocity, m/s
v_r	Dimensionless relative velocity, m/s
$\vec{v}_t$	Translational velocity of particle *i*, m/s
$\vec{v}_{ij}$	Relative velocity at contact point, m/s
$\vec{v}_{rn}$	Normal relative velocity at contact point, m/s
$\vec{v}_{rt}$	Tangential relative velocity at contact point, m/s
V_p	Volume of a particle, m^3
$\vec{x}_i$	Center position vector of particle *i*, m

Greek symbols

γ_s	Granular energy dissipation due to collisions, J/m^3s
α_f	Volume fraction of fluid phase
α_s	Volume fraction of granular phase
θ	Granular temperature of granular phase, m^2/s^2
μ_{df}	Coefficient of dynamic friction
μ_f	Fluid phase viscosity, kg/m·s
μ_s	Granular phase viscosity, kg/m·s
$\mu_{s,co}$	Granular phase collisional viscosity, m^2/s
$\mu_{s,ki}$	Granular phase kinetic viscosity, m^2/s
ρ_f	Density of fluid phase, kg/m^3
ρ_s	Density of granular phase, kg/m^3
$\vec{\tau}_f$	Fluid phase stress tensor, Pa
$\vec{\tau}_s$	Granular phase stress tensor, Pa
τ_{12x}	Particle relaxation time, s
τ_{2c}	Particle-particle collision time, s
β	Momentum exchange coefficient, $kg/m^3 \cdot s$
β_s	Tangential restitution coefficient
δ_n	Normal overlap, m
δ_t	Tangential overlap, m
η_n	Normal damping coefficient
η_t	Tangential damping coefficient
υ_i	Poisson's ratio of particle *i*
$\vec{\omega}_i$	Rotational velocity of particle *i*, rad/s

$\vec{\omega}_{ij}$	Unit vector of relative angular velocity, rad/s
ϕ	Specularity coefficient
φ_w	Angle of internal friction
ΔE	Energy dissipation, J
ΔE_{dp}	Impact energy dissipation, J
ΔE_{cp}	Elastic potential energy dissipation, J
ΔE_{rp}	Rolling energy dissipation, J
ΔE_{sp}	Friction energy dissipation, J
Δt_p	Time step for motion of particle, s

Abbreviations

CFD	Computational fluid dynamics
DEM	Discrete element method
KTGF	Kinetic theory of granular flow
TFM	Two-fluid model

References

1. Arastoopour, H., Gidaspow, D., & Abbasi, E. (2017). *Computational transport phenomena of fluid-particle systems*. Springer International Publishing.
2. Balzer, G., Simonin, O., Boelle, A., Lavieville, J. (1996) A unifying modelling approach for the numerical prediction of dilute and dense gas-solid two phase flow. 5th International Conference on Circulating Fluidized Beds, Beijing, China, 1996
3. Cao, J., & Ahmadi, G. (1995). Gas-particle two-phase turbulent flow in a vertical duct. *International Journal of Multiphase Flow, 21*, 1203–1228.
4. Chapman, S., & Cowling, T. G. (1970). *The mathematical theory of non-uniform gases*. Cambridge Mathematical Library.
5. Crowe, C. T., Sommerfeld, M., & Tsuji, Y. (1998). *Multiphase flows with droplets and particles*. CRC Press.
6. Cundall, P. A., & Strack, O. D. L. (1979). A discrete numerical model for granular assemblies. *Geotechnique, 29*, 47–65.
7. Deen, N., Annaland, M. V. S., Van der Hoef, M., & Kuipers, J. (2007). Review of discrete particle modeling of fluidized beds. *Chemical Engineering Science, 62*, 28–44.
8. Ding, J., & Gidaspow, D. (1990). A bubbling fluidization model using kinetic theory of granular flow. *AIChE Journal, 36*, 523–538.
9. Dubey, A. K., Bodrova, A., Puri, S., & Brilliantov, N. (2013). Velocity distribution function and effective restitution coefficient for a granular gas of viscoelastic particles. *Physical Review E, 87*, 062202.
10. Elghobashi, S. E., & Abou-Arab, T. W. (1983). A two-equation turbulence model for two-phase flows. *Physics of Fluids, 26*, 931–938.
11. Fan, L. S., & Zhu, C. (1998). *Principles of gas-solid flows*. Cambridge University Press.
12. Feng, Y. Q., & Yu, A. B. (2004). Assessment of model formulations in the discrete particle simulation of gas-solid flow. *Industrial & Engineering Chemistry Research, 43*, 8378–8390.

13. Garg, R., Galvin, J., Li, T., & Pannala, S. (2012). *Documentation of open-source MFIX–DEM software for gas-solids flows*. National Energy Technology Laboratory, Morgantown, WV, USA.
14. Gidaspow, D. (1994). *Multiphase flow and fluidization: Continuum and kinetic theory descriptions*. Academic.
15. Hou, Q. F., Kuang, S. B., & Yu, A. B. (2017). A DEM-based approach for analyzing energy transitions in granular and particle-fluid flows. *Chemical Engineering Science, 161*, 67–79.
16. Hui, K., Haff, P. K., Ungar, J. E., & Jackson, R. (1984). Boundary conditions for high-shear grain flows. *Journal of Fluid Mechanics, 145*, 223–233.
17. Huilin, L., Gidaspow, D., Bouillard, J., & Wentie, L. (2003). Hydrodynamic simulation of gas–solid flow in a riser using kinetic theory of granular flow. *Chemical Engineering Journal, 95*, 1–13.
18. Huilin, L., Gidaspow, D., & Shuyan, W. (2021). *Computational fluid dynamics theory of fluidization: Kinetic theory of granular flow and applications*. Springer International Publishing.
19. Jenkins, J. T. (1992). Boundary conditions for rapid granular flow: Flat, frictional walls. *Journal of Applied Mechanics, 59*, 120–127.
20. Jenkins, J. T., & Richman, M. W. (1985). Grad's 13-moment system for dense gas of inelastic spheres. *Archive for Rational Mechanics and Analysis, 87*, 355–377.
21. Johnson, K. L. (1985). *Contact mechanics*. Cambridge University Press.
22. Johnson, P. C., & Jackson, R. (1987). Frictional-collisional constitutive relations for granular materials with application to plane shearing. *Journal of Fluid Mechanics, 176*, 67–93.
23. Johnson, P. C., Nott, P., & Jackson, R. (1990). Frictional–collisional equations of motion for participate flows and their application to chutes. *Journal of Fluid Mechanics, 210*, 501–535.
24. Louge, M. Y., Mastorakos, E., & Jenkins, J. T. (1991). The role of particle collisions in pneumatic transport. *Journal of Fluid Mechanics, 231*, 345–359.
25. Lun, C. K. K., & Savage, S. B. (1986). The effects of an impact velocity dependent coefficient of restitution on stresses developed by sheared granular materials. *Acta Meehanica, 63*, 15–44.
26. Lun, C. K. K., Savage, S. B., Jeffrey, D. J., & Chepurniy, N. (1984). Kinetic theories for granular flow: Inelastic particles in Couette flow and slightly inelastic particles in a general flow field. *Journal of Fluid Mechanics, 140*, 223–256.
27. McNamara, S., & Falcon, E. (2008). Simulations of dense granular gases without gravity with impact-velocity-dependent restitution coefficient. *Powder Technology, 182*, 232–240.
28. MFIX. (2020). MFiX-Exa documentation. MFS Development Group, National Energy Technology Laboratory.
29. Peirano, E., Delloume, V., Johnsson, F., Leckner, B., & Simonin, O. (2002). Numerical simulation of the fluid dynamics of a freely bubbling fluidized bed: Influence of the air supply system. *Powder Technology, 122*, 69–82.
30. Pope, S. B. (1983). A Lagrangian two-time probability density function equation for inhomogeneous turbulent flows. *Physics of Fluids, 26*, 3448–3450.
31. Qinghong, Z. (2020). Numerical simulation of fluidized bed using coupled Euler gas phase-Euler solid phase-Lagrangian discrete particles model. Ph. D. thesis, Harbin: Harbin Institute of Technology.
32. Qinghong, Z., Shuyan, W., Huilin, L., & Guodong, L. (2017). A coupled Eulerian fluid phase-Eulerian solids phase-Lagrangian discrete particles hybrid model applied to gas-solids bubbling fluidized beds. *Powder Technology, 315*, 385–397.
33. Qinghong, Z., Wenjian, C., Cailei, L., Gidaspow, D., & Huilin, L. (2019). Modified MFIX code to simulate hydrodynamics gas-solids bubbling fluidized beds: A model of coupled kinetic theory of granular flow and discrete element method. *Powder Technology, 357*, 417–427.
34. Reeks, M. W. (1993). On the constitutive relations for dispersed particles in non-uniform flows I: Dispersion in a simple shear flow. *Physics of Fluids, 5*, 750–761.
35. Rosa, R. A., Thorsten, P. E., Nikolai, V. B., & Thomas, S. (1999). Coefficient of restitution of colliding viscoelastic spheres. *Physical Review E, 60*, 4465–4472.

36. Simonin, O. (1996) Continuum modeling of dispersed two-phase flows. in Combustion and Turbulence in Two-Phase Flows, Von Karman Institute of Fluid Dynamics Lecture Series.
37. Soo, S. L. (1967). *Fluid dynamics of multiphase systems*. Blaisdell Publishing Company.
38. Squire, K. D., & Eaton, J. K. (1994). Effect of selective modification of turbulence on two-equation models for particle-laden turbulent flows. *ASME Journal of Fluids Engineering, 116*(778), 784.
39. Srivastava, A., & Sundaresan, S. (2003). Analysis of a frictional-kinetic model for gas-particle flow. *Powder Technology, 129*, 72–85.
40. Syamlal, M., Rogers, W. A., & O'Brien, T. J. (1993). *MFIX documentation. Department of Energy Office of Fossil Energy*, Morgantown Energy Technology Center, West Virginia, USA.
41. Thornton, C. (1997). Coefficient of restitution for collinear collisions of elastic perfectly plastic spheres. *ASME Journal of Applied Mechanics, 64*, 383–386.
42. Thornton, C., Cummins, S. J., & Cleary, P. W. (2013). An investigation of the comparative behavior of alternative contact force models during inelastic collisions. *Powder Technology, 233*, 30–46.
43. Tingwen, L., & Sofiane, B. (2012). Revisiting Johnson and Jackson boundary conditions for granular flows. *AIChE Journal, 58*, 2058–2068.
44. Tsuji, Y. (2007). Multi-scale modeling of dense phase gas–particle flow. *Chemical Engineering Science, 62*, 3410–3418.
45. Wu, C. Y. (2012). *Discrete element modelling of particulate media*. The Royal Society of Chemistry.
46. Xiaoxue, J. (2022). *Minimum energy dissipation rate drag coefficient model and numerical simulation on the active and passive regulation of cluster in riser*. Ph. D. Thesis, Harbin: Harbin Institute of Technology.
47. Xiaoxue, J., Shuyan, W., Baoli, S., & Huilin, L. (2021). Granular restitution coefficient-based kinetic theory computations of bubbling fluidized beds. *Powder Technology, 390*, 825–837.
48. Zhu, H. P., & Yu, A. B. (2003). The effects of wall and rolling resistance on the couple stress of granular materials in vertical flow. *Physica A: Statistical Mechanics and its Applications, 325*, 347–360.
49. Zhu, H. P., Zhou, Z., Yang, R., & Yu, A. B. (2007). Discrete particle simulation of particulate systems: Theoretical developments. *Chemical Engineering Science, 62*, 3378–3396.

Chapter 4
Low-Stokes-Number Kinetic Theory of Granular Flow

Abstract This chapter presents a novel and unique approach of low-Stokes-number kinetic theory of granular flow (KTGF). It is a well-known fact that the original KTGF has been successful in applications in which the effect of the interstitial fluid can be neglected in high-Stokes number fluid–solid flows. An incorporation of interstitial fluid effect in the derivation of the low-Stokes-number KTGF is presented with the Fokker–Planck operator. The formulated kinetic equation is solved with the Chapman-Enskog method as the function of Stokes numbers. The constitutive relationships are suitable for the study on low-Stokes-number fluid–solid multiphase flows.

The classical kinetic theory of granular flow (KTGF) is derived based on the inelastic particle assumptions [14, 17, 18, 23]. In the classical KTGF, both interactions between particles and fluid and those among particles are taken into account. A granular material is defined as a collection of a large number of discrete solid particles in a vacuum. Granular flow might thus be referred to as a flow of a powder in a vacuum in which the density ratio ρ_{sf} is very high, where $\rho_{sf} = \rho_s/\rho_f$, ρ_s and ρ_f are the particles and fluid densities. Thus, it is also known as high-density-ratio KTGF in which the equations are derived considering vacuum conditions during collisions of particles.

When two particles collide in the presence of an interstitial fluid, two different particle–particle interaction mechanisms occur. First, the particles can interact through direct collisions where momentum and energy are transferred through elastic–plastic deformations, which is similar to how particle interactions occur in a vacuum. This type of mechanism is referred to as the direct collision or the short-range potential interaction. Second, the particles can interact with each other through the interstitial fluid. The presence of a fluid velocity gradient and the pressure disturbance generated by the fluctuating motions of neighboring particles will cause a particle to experience a sequence of collisions. In other words, the Brownian-like motions of the neighboring particles result in momentum and energy transfers through the interstitial fluid to the particle. This type of mechanism is referred to as the noncontact collision or the long-range potential interaction. Lun and Savage [24]

H. Lu et al., *Computational Transport Phenomena of Multiphase Systems and Fluidization*, Fluid Mechanics and Its Applications 127,
https://doi.org/10.1007/978-981-96-0698-6_4

studied the effect of the interstitial fluid introducing a lubrication approximation for modeling interparticle collisions in the presence of an interstitial fluid. The model takes into account additional stresses exerted by the fluid on the particle through long-range potential interactions. An extensive incorporation of interstitial fluid effects in the derivation of the kinetic theory of granular flow was presented by Garzó et al. [12]. They modeled the instantaneous particle acceleration and incorporated it in the derivation. The solution is obtained using the method for states near the local homogeneous cooling state. Boelle et al. [4] extended [18] model to gas–solid flows by incorporating the drag force. Expressions for particle turbulent stresses, where fluid-particle velocity covariance is closed by a transport equation, are derived from the equation of the second moment of particle fluctuating velocity. Agrawal et al. [1] proposed modified closure relations for the equation of granular temperature. The relations are based on the findings of [4] and [21], who showed that the effect of the interstitial fluid is essential in fluid–solid flows.

The validation of classical KTGF was performed using direct numerical simulations [11]. The computational simulations using classical kinetic theory are compared to direct numerical simulations at different high and low density ratios ρ_{sf}. They found that the mean and fluctuating velocities agreed with both methods at high density ratios. On the contrary, the agreement deteriorated with decreasing density ratios. In order to check the agreement between the classical kinetic theory and direct numerical simulations, a criterion of the dimensionless number Φ was proposed using thermal Stokes number St_T.

$$\Phi = \frac{12}{\sqrt{\pi}} \frac{\alpha_s g_0}{F} St_T \geq 10^{1/2} \tag{4.1}$$

$$St_T = \frac{\rho_s}{\rho_f} \frac{Re_T}{9} = \frac{\rho_{sf} Re_T}{9} \tag{4.2}$$

where α_s denotes the solids volume fraction. g_0 is the radial distribution function at the contact. Re_T is the thermal Reynolds number. Re_T and St_T are measures of the fluid inertia to fluid viscous effect and the particle inertia to fluid viscous effect. F represents the Stokes deviation of the mean drag law as a function of solids volume fractions, and is given by

$$\begin{aligned} F = &\left[\frac{10\alpha_s}{(1-\alpha_s)^2} + (1-\alpha_s)^2(1+1.5\sqrt{\alpha_s})\right] \\ &+ \frac{0.413 Re_m}{24(1-\alpha_s)^2}\left[\frac{(1-\alpha_s)^{-1} + 3\alpha_s(1-\alpha_s) + 8.4 Re_m^{-0.343}}{1 + 10^{3\alpha_s} Re_m^{-(1+4\alpha_s)/2}}\right] \end{aligned} \tag{4.3}$$

$$Re_T = \frac{\rho_f d_p \sqrt{\theta}}{\mu_f} \tag{4.4}$$

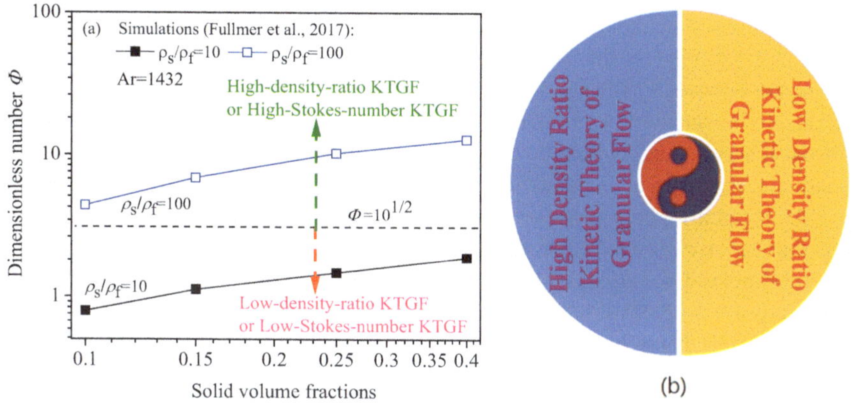

Fig. 4.1 Dimensionless numbers according to density ratios and solid volume fractions

where Re_m is the mean-flow Reynolds number. θ is the granular temperature. μ_f is the fluid dynamic viscosity. d_p is the particle diameter. Detailed investigations on this subject have been carried out by Fullme et al. [11].

The profile of dimensionless number Φ is illustrated in Fig. 4.1 at different density ratios as a function of solid volume fractions [31]. The horizontal dashed line represents a critical dimensionless number Φ, distinguishing the high-density-ratio kinetic theory of granular flow (i.e., classical KTGF) from the low-density-ratio kinetic theory of granular flow or low-Stokes-number KTGF. The predictions using the classical kinetic theory with two-fluid model are valid at large thermal Stokes number St_{T} with high density ratios, but invalid for small thermal Stokes number St_T with low density ratios. The effect of fluid–particle interaction on the particle fluctuating motion becomes important as a result of a non-linear increase in the dissipative fluctuating drag force. The fluid–particle interaction influence on the particle fluctuating motion is further intensified for low Stokes number flows, thereby the interstitial fluid effect on the particle collisional mechanism must be incorporated into the kinetic theory of granular flow formulation employed in the CFD two-fluid models [15]. Generally, the density ratio in air-particles fluidized beds is two to three orders of magnitude larger than that of water-particles fluidized beds. Hence, the high-Stokes-number KTGF has been accepted for numerical simulations of air-particles fluidized beds. However, this does not stand for liquid-particles fluidized beds.

Liquid–solid fluidization technology has been applied in many industries, such as coal and biomass supercritical fluid gasification, biocatalysis, mineral processing, electrolysis and water treatment. For the supercritical water gasification in a supercritical water fluidized bed reactor [22], the density ratio between the coal particles and the supercritical water fluid is in the range from 5.0 to 12.0 at different supercritical pressures and temperatures. Slurries are mixtures of solid particles in a carrier fluid. The carrier fluid may be water or some other Newtonian liquid, or it may be a non-Newtonian liquid. Typically coal-water slurries are highly loaded suspensions of fine coal in water, and they are utilized as a fuel to generate power. For slurry flows, the density ratio is less than 10. In the entrainment of bed sediment of solid–liquid two-phase flows [29], the sediment movement in open channels is of fundamental interest in the transport phenomena of rivers and estuaries. The density ratio of fluid-soil particles mixture is about 1.04–2.65 [30]. In the field of mineral processing, liquid-coal fluidized beds mainly utilize rising water flow to act on coal particles in the bed, separating coal particles according to the difference in settlement velocities. The density ratio between the coal and water is about 2.0–3.0 in the field of mineral fluidized bed processing. For liquid–solid fluidized bed with granular activated carbon for removal of copper ions from wastewater, the density ratio ranges from 1.2 to 1.5. These indicate that interstitial fluid plays a critical role during collisions of particles due to low density ratios. Hence, a more extensive incorporation of interstitial fluid effects in the derivation of the KTGF is essential in fluid-particles fluidization and multiphase flows.

4.1 Kinetic Transport

In kinetic modeling, the behavior of the particles is characterized by a kinetic equation of the particle probability density distribution function. The conservation equations are obtained through taking statistical moments of the kinetic equation, with the corresponding constitutive relations derived by integrating the probability density function over the whole velocity space.

The motion of single particle in fluid–solid flows is given by the following Lagrangian equations [9, 27].

$$\frac{dv_{pi}}{dt} = \frac{u_i - v_{pi}}{\tau_p} + F_i + W_i \tag{4.5}$$

$$\frac{dr_{pi}}{dt} = v_{pi} \tag{4.6}$$

where r_{pi} is the particle position, v_{pi} is the particle velocity, u_i is the fluid velocity at r_{pi}, F_i is the sum of the gravity and interphase forces except the drag, W_i is the interparticle collisional force, t is the time, and τ_p is the particle relaxation time which can be defined according to the fluid velocity $\boldsymbol{u}$ and solid velocity $\boldsymbol{v}$ [16]:

$$\tau_p = \frac{\alpha_s \rho_s}{\beta} \tag{4.7}$$

$$\beta = (1-\varphi)\left[150\frac{\alpha_s(1-\alpha_f)\mu_f}{\alpha_f d_p^2} + 1.75\frac{\alpha_s \rho_f |\boldsymbol{u}-\boldsymbol{v}|}{d_p}\right] + \varphi\frac{3}{4}C_d\frac{\alpha_s \alpha_f \rho_f}{d_p}|\boldsymbol{u}-\boldsymbol{v}|\alpha_f^{-2.65} \tag{4.8}$$

$$\varphi = \frac{\arctan[150 \times 1.75(\alpha_f - 0.8)]}{\pi} + 0.5 \tag{4.9}$$

where C_d is the drag coefficient. Usually, Reynolds numbers larger than 1000 are not often encountered in fluid–solid flows except at combinations of high velocity operation and large particles. The drag coefficient for a particle in the suspension is given as

$$C_d = \frac{24\mu_f}{\rho_f \alpha_f d_p |\boldsymbol{u} - \boldsymbol{v}|}\left[1 + 0.15\left(\frac{\rho_f \alpha_f d_p |\boldsymbol{u} - \boldsymbol{v}|}{\mu_f}\right)^{0.687}\right] \tag{4.10}$$

From Eq. (4.5), it describes the balance of forces acting on the moving particle. The first term on the right-hand side is the aerodynamic interphase force represented in the form of the Stokes drag law. But Eq. (4.7) accounts for the fact that the particle relaxation time is a function of the mean relative velocity of the fluid flow with respect to the moving particle. Equation (4.5) implies a linearization of the aerodynamic interphase force with respect to velocity fluctuations. Thus, the existence of the particle velocity fluctuations is caused by entrainment of particles into the fluid phase turbulent flow field; that is, by the interaction between particles and turbulent eddies.

From studies on the dynamics of a single particle in a fluid, two important mechanisms for the fluid–solid interaction force are considered: drag force caused by velocity differences between fluid and solid phases, and buoyancy caused by the fluid pressure gradient. Other forces, such as virtual mass effect caused by relative acceleration between phases, Magnus force caused by particle spin and Basset history forces are not considered due to the large density difference between fluid and solids. Thus, the force consists of the gravity, the drag forces with momentum transfer coefficient and the buoyancy effect which appears through the gradient of fluid pressure, and the external force per unit of mass acting on a sphere can be written as [14, 17]

$$\boldsymbol{F} = -\frac{1}{\rho_s}\nabla p + \frac{\beta}{\alpha_s \rho_s}(\boldsymbol{u} - \boldsymbol{v}) + \boldsymbol{g} \tag{4.11}$$

To transit from the Lagrangian equations to the Eulerian ones, a particle probability density distribution function (PDF) is introduced in terms of position and velocity in phase space (x_i, v_i):

$$f = \langle \delta(x_i - r_{pi})\delta(v_i - v_{pi}) \rangle \tag{4.12}$$

where $\langle\ \rangle$ denotes an averaging over an ensemble of all possible turbulence realizations, and δ is the Dirac delta function. The average volume fraction α_s and velocity $\langle v_i \rangle$ of the particles are defined as the following integrals taken over the whole velocity space, respectively:

$$\alpha_s = \int f d\boldsymbol{v} \tag{4.13}$$

$$\alpha_s \langle v_i \rangle = \int v_i f d\boldsymbol{v} \tag{4.14}$$

Using the hydrodynamic velocity of particles, the notation for the convective derivative follows

$$\frac{D}{Dt} = \frac{\partial}{\partial t} + \langle v_i \rangle \frac{\partial}{\partial x_i} \tag{4.15}$$

Differentiating Eq. (4.5) with respect to time t and changing the velocity coordinate from v_i to fluctuating velocity $v'_i = v_i - \langle v_i \rangle$, a kinetic equation of the probability density function is obtained:

$$\frac{Df}{Dt} + v'_i \frac{\partial f}{\partial x_i} - \left\{ \frac{D\langle v_i \rangle}{Dt} - \frac{\langle u_i \rangle - \langle v_i \rangle + \tau_p \langle F_i \rangle}{\tau_p} + v'_k \frac{\partial \langle v_i \rangle}{\partial x_k} \right\} \frac{\partial f}{\partial v'_i}$$
$$= -\frac{1}{\tau_p} \frac{\partial}{\partial v'_i} \left(\left\langle u'_i \delta(x_i - r_{pi}) \delta(v_i - v_{pi}) \right\rangle - v'_i f \right) - \frac{\partial}{\partial v'_i} \left\langle W_i \delta(x_i - r_{pi}) \delta(v_i - v_{pi}) \right\rangle \tag{4.16}$$

where $\langle u_i \rangle$ and u'_i are the average and fluctuating velocities of the fluid, respectively. $\langle v_i \rangle$ and v'_i denote the solid phase average and fluctuating velocities. The terms on the left-hand side describe the probability density function evolution in time and advection in phase space (x_i, v'_i). The terms on the right-hand side, respectively, characterize interactions of particles with turbulent eddies seen by the particles and interparticle collisions, respectively. From Eq. (4.16), the terms relate to particle-turbulence interactions and interparticle collisions should be modeled for constructing a closed kinetic equation. To quantify interphase interactions, [4] took a transport equation, while [33] applied an explicit Fokker–Planck diffusion operator. For practical purposes, a diffusion operator accounting for flow non-uniformity would be preferable.

$$-\left\langle u'_i \delta(x_i - r_{pi}) \delta(v_i - v_{pi}) \right\rangle = A_{ik} \frac{\partial f}{\partial x_k} + B_{ik} \frac{\partial f}{\partial x'_k} \tag{4.17}$$

where A_{ik} and B_{ik} are the functions that are related to fluid turbulent stresses and spatial gradients of mean fields. Both measure the response of particles to velocity fluctuations of the fluid and are related to the time scales of the fluid velocity viewed by a particle. They depend on the correlation functions of the fluid velocity determined along particle trajectories. Their simple forms for estimation of the gradient of particle averaged-velocity are as follows [2, 32, 34]:

$$A_{ik} = \left[1 - St\left(1 - \exp\left(-\frac{1}{St}\right)\right)\right]\frac{\tau_p}{St}\langle u_i' u_k' \rangle \tag{4.18}$$

$$B_{ik} = \frac{1}{1 - \exp(-\frac{1}{St})}\langle u_i' u_k' \rangle \tag{4.19}$$

The value $u_i{}'\ u_k{}'$ represents the fluid velocity tensor of fluctuations. Both functions are reduced with increasing Stokes number. The coefficients are obtained by assuming the frequently used exponential correlation function for fluid velocity determined along particle trajectories [32]. The particle Stokes number is defined by

$$St = \frac{\tau_p}{T_{Lp}} \tag{4.20}$$

where T_{Lp} is the fluid Lagrangian integral time seen by the particles. The time of interaction between particle motion and fluid phase fluctuations is defined according to fluid turbulent kinetic energy k_f and energy dissipation rate ε_f.

$$T_{Lp} = \frac{C_\mu k_f}{\varepsilon_f \sqrt{1 + 3C_\beta |(\boldsymbol{v} - \boldsymbol{u}) - \boldsymbol{u}_d|^2 / 2k_f}} \tag{4.21}$$

where the coefficient C_μ and C_β are constant. The former is 0.09, and the latter depends on the type of flow, but also on the direction. k_f and ε_f represent the fluid fluctuation kinetic energy and dissipation of fluid turbulent kinetic energy [25, 28]. $\boldsymbol{u}_d$ is the drift velocity and represents the correlation between the fluid phase fluctuating velocity and the spatial distribution of the particles. From Eqs. (4.7) and (4.21), the Stokes number is given by:

$$St = \frac{\varepsilon_f \alpha_s \rho_s \sqrt{1 + 1.5C_\beta |(\boldsymbol{v} - \boldsymbol{u}) - \boldsymbol{u}_d|^2 / k_f}}{C_\mu k_f \beta} \tag{4.22}$$

The coefficients A_{ik} and B_{ik} characterize an entrainment of particles into the fluid phase velocity fluctuations. With the help of a simple step-by-step approximation of the temporal correlation function of fluid velocity fluctuations along the particle trajectories, the following alternative expressions for the coefficients A_{ik} and B_{ik} have been proposed in the fluctuating motion of the fluid flow [8, 9, 33]:

$$A_{ik} = \frac{\tau_p}{St(1 + St)}\langle u_i' u_k' \rangle \tag{4.23}$$

$$B_{ik} = \frac{1}{1 + St}\langle u_i' u_k' \rangle \tag{4.24}$$

The calculations indicate that the coefficients A_{ik} and B_{ik} from the former Eqs. (4.23) and (4.24) are larger than that the latter Eqs. (4.18) and (4.19). The difference between them is not significant. Both formulations are simple in form, and consistent with the limit cases as Stokes number trends to zero and infinity, respectively.

In order to represent interparticle collisions, the Boltzmann operator is used without considering interrelations between two colliding particles resulting from interphase interactions [4, 33]. In fact, the Boltzmann operator only applies for nearly elastic collisions. In contrast, the revised Enskog theory (RET) provides a unique basis for analyzing systems with collisions across wide ranges of concentration and dissipation levels [6]. In the RET, the original Enskog equation was replaced by one in which a spatially dependent pair distribution function for the particles was used. Moreover, the interactions between two colliding particles may affect statistical characteristics of the collisions. For the sake of simplicity to facilitate the analytical study (while preserving the essential properties of RET), a modified operator is proposed as follows:

$$-\frac{\partial}{\partial v_i'}\left\langle W_i\delta(x_i-r_{pi})\delta(v_i-v_{pi})\right\rangle=-\frac{1+e}{2}\zeta(f-f_m)$$
$$-\frac{f_m}{\alpha_s\theta}\left[v_i'\frac{\partial p_{ik}^c}{\partial x_k}+\left(\frac{v_i'v_j'}{3\theta}-1\right)\left(\frac{\partial q_k^c}{\partial x_k}+p_{ik}^c\frac{\partial\langle v_k\rangle}{\partial x_i}+(1-e^2)w^c\right)\right] \tag{4.25}$$

$$f_m=\frac{\alpha_s}{(2\pi\theta)^{3/2}}\exp\left(-\frac{v_i'v_j'}{2\theta}\right) \tag{4.26}$$

where f_m is the Gaussian distribution function, ζ is the collision frequency; p_{ik}^c and q_k^c denote the collisional contributions to the particle stress and fluctuation energy flux, respectively. w^c is the fluctuation energy dissipation rate due to the interactions of collisions. All are functions of restitution coefficient e. The granular temperature θ is a measure of the velocity fluctuations in a fluid-particles flow process.

$$\theta=\frac{1}{3}\langle v_i'v_i'\rangle \tag{4.27}$$

The collision frequency ζ takes the effects of both fluid turbulence and particle crowding into account [32, 34]:

$$\zeta=\nu_0 g_0\sqrt{1-\frac{\langle u_i'v_i'\rangle^2}{\langle u_i'u_i'\rangle\langle v_i'v_i'\rangle}} \tag{4.28}$$

where ν_0 is the collision frequency in dilute rapid granular flows [6], and reflects the effect of fluid turbulence on collisions [28]. $v_i'\,v_i'$ and $u_i'u_i'$ represent the solid phase and fluid phase velocity tensor of fluctuations. The collision frequency is related to both the solid volume fraction and the kinetic granular temperature; it increases with increasing solid volume fraction and granular temperature.

$$\nu_0 = \frac{96}{5d_p}\alpha_s\sqrt{\frac{\theta}{\pi}} \tag{4.29}$$

The part $u_i' \, v_i'$ represents the fluid-particle velocity correlation tensor. If the fluid and particle velocities are uncorrelated with a fluid–particle velocity covariance of zero, the collision frequency of particles would be maximum.

Substituting Eqs. (4.17) and (4.25) into Eq. (4.16) leads to the kinetic equation:

$$\begin{aligned}
&\frac{Df}{Dt} + v_i'\frac{\partial f}{\partial x_i} - \frac{\partial f}{\partial v_i'}\left\{\frac{D\langle v_i\rangle}{Dt} - \frac{\langle u_i\rangle - \langle v_i\rangle + \tau_p\langle F_i\rangle}{\tau_p} + v_k'\frac{\partial\langle v_i\rangle}{\partial x_k}\right\} \\
&\qquad - \frac{1}{\tau_p}\left(A_{ik}\frac{\partial^2 f}{\partial x_i \partial v_k'} + B_{ik}\frac{\partial^2 f}{\partial v_i' \partial v_k'}\right) \\
&\qquad - \frac{1}{\tau_p}\left(\left(\theta + \tau_p\frac{\xi\theta}{2}\right)\delta_{ik}\frac{\partial^2 f}{\partial v_i' \partial v_k'}\right) \\
&- \frac{f_m}{\alpha_s\theta}\left[v_i'\frac{\partial p_{ik}^c}{\partial x_k} + (\frac{v_i'v_i'}{3\theta} - 1)(\frac{\partial q_k^c}{\partial x_k}\frac{\partial\langle v_k\rangle}{\partial x_i} + (1 - e^2)w^c)\right] \\
&= \left(\frac{\theta}{\tau_p} + \frac{\xi\theta}{2}\right)\frac{\partial^2 f}{\partial v_i' \partial v_i'} + \left(\frac{1}{\tau_p} + \frac{\xi}{2}\right)\frac{\partial v_i' f}{\partial v_i'} - \frac{1+e}{2}\nu_0(f - f_m)
\end{aligned} \tag{4.30}$$

δ_{ik} where ξ represents the cooling rate. It shows that an external energy input is required to compensate for the collisional cooling in order to reach a steady state. The cooling rate is a nonlinear functional of the distribution function f. As a consequence, a simple estimate is obtained from the local equilibrium distribution.

$$\xi = \frac{5\nu_0\left(1 - e^2\right)}{12} \tag{4.31}$$

For the first approximation, the collisions of particles are neglected. The transport equation of particle velocity distributions is expressed by means of two parts of *Pf* and *Qf* according to Eq. (4.30)

$$\begin{aligned}
Pf &= \frac{Df}{Dt} + v_i'\frac{\partial f}{\partial x_i} - \frac{\partial f}{\partial v_i'}\left\{\frac{D\langle v_i\rangle}{Dt} - \frac{\langle u_i\rangle - \langle v_i\rangle + \tau_p\langle F_i\rangle}{\tau_p} + v_k'\frac{\partial\langle v_i\rangle}{\partial x_k}\right\} \\
&- \frac{1}{\tau_p}\left(A_{ik}\frac{\partial}{\partial x_i}\frac{\partial f}{\partial v_k'} + B_{ik}\frac{\partial}{\partial v_i'}\frac{\partial f}{\partial v_k'} - \left(\theta + \frac{5\tau_p\nu_0\left(1 - e^2\right)\theta}{24}\right)\delta_{ik}\frac{\partial}{\partial v_i'}\frac{\partial f}{\partial v_k'}\right)
\end{aligned} \tag{4.32}$$

$$Qf = \left(\frac{\theta}{\tau_p} + \frac{5\nu_0\left(1 - e^2\right)\theta}{24}\right)\frac{\partial}{\partial v_i'}\frac{\partial f}{\partial v_i'} + \left(\frac{1}{\tau_p} + \frac{5\nu_0\left(1 - e^2\right)}{24}\right)\frac{\partial v_i' f}{\partial v_i'} - \frac{1+e}{2}\nu_0(f - f_m) \tag{4.33}$$

Multiplying with 1, v', and $v'v'$ and integrating over the whole velocity space, the conservation equations of mass, momentum, and fluctuation kinetic energy of particles are derived through statistical moments of Eq. (4.32) and (4.33).

$$\frac{D\alpha_s}{Dt} + \alpha_s \frac{\partial \langle v_k \rangle}{\partial x_k} = 0 \tag{4.34}$$

$$\begin{aligned} &\alpha_s \frac{D\langle v_k \rangle}{Dt} - \frac{\alpha_s}{\tau_p}[\langle u_i \rangle - \langle v_k \rangle + \tau_p \langle F_i \rangle] \\ &+ \frac{\partial}{\partial x_k}(\alpha_s \langle v_i' v_k' \rangle + p_{ik}^c) + \frac{A_{ik}}{\tau_p}\frac{\partial \alpha_s}{\partial x_k} = 0 \end{aligned} \tag{4.35}$$

$$\begin{aligned} &\alpha_s \frac{D(3/2)\theta}{Dt} + \frac{\partial}{\partial x_i}\left[\frac{1}{2}\alpha_s \langle v_k' v_k' v_i' \rangle + q_i^c\right] + \frac{\partial \langle v_i \rangle}{\partial x_k}\left[\alpha_s \langle v_i' v_k' \rangle + p_{ik}^c\right] \\ &+ (1 - e^2) w^c - \frac{\alpha_s}{\tau_p}(B_{ik} - 3\theta) = 0 \end{aligned} \tag{4.36}$$

In Eq. (4.35), the terms represent the average external force acting on the solid phase, the gradient of particulate pressure, and the momentum transfer resulting from particle diffusion. In the last term, A_{ik} constitutes one important part of particle diffusion tensor. The last two terms on the right-hand side of Eq. (4.36) represent the energy diffusion, work generation due to flow shear, dissipation due to in elastic collisions among particles and interphase transfer due to particle-turbulence interactions, respectively. Through Eqs. (4.18) and (4.19), the balance equations from Eqs. (4.34) to (4.36) become applicable for simulating fluid-particles flows due to the effect of Stokes number.

4.1.1 Solution of Kinetic Equation

With the Chapman-Enskog perturbation method, solution of Eq. (4.30) can be constructed as a series expansion in a formal parameter ε measuring the non-uniformity of the system [7].

$$f = f^0 + \varepsilon f^1 + \varepsilon^2 f^2 + \cdot\cdot \tag{4.37}$$

where each factor of ε means an implicit gradient of a hydrodynamic field. The uniformity parameter ε relates to the ratio of the mean free path of the particles to a length scale that characterizes the distance over which gradients in the hydrodynamic variables occur. In these situations, the Navier–Stokes approximation (first order in the expansion) only applies for nearly elastic systems. Here, we assume that the external forces (gravity and drag forces) do not induce any flux in the system and only modify the form of the transport coefficients.

For the first approximation, the velocity distribution function f^0 is expressed using Eq. (4.26) according to granular temperature.

$$\ln f^0 = \text{constant} + \ln \alpha_s - \ln \theta - \frac{v_i' v_i'}{2\theta} \tag{4.38}$$

The convective differentiation of Eq. (4.38) gives

$$\frac{\partial f^0}{\partial x_i} = 0 \tag{4.39}$$

$$\frac{\partial f^0}{\partial v_i'} = f^0 \frac{\partial \ln f^0}{\partial v_i'} = -f^0 \frac{v_i'}{\theta} \tag{4.40}$$

$$\frac{Df^0}{Dt} = \frac{\partial f^0}{\partial t} + \langle v_i \rangle \frac{\partial f^0}{\partial x_i} \tag{4.41}$$

The distribution function can be determined to a specified order provided that the function f still satisfies the moment relations defining the physical properties. From Eqs. (4.32) and (4.33), the functions are expressed using the Maxwellian velocity distribution function:

$$Pf^0 = Qf^0 \tag{4.42}$$

$$\begin{aligned} Pf^0 = {} & \frac{Df^0}{Dt} + v_i' \frac{\partial f^0}{\partial x_i} - \frac{\partial f^0}{\partial v_i'} \left\{ \frac{D\langle v_i \rangle}{Dt} - \frac{\langle u_i \rangle - \langle v_i \rangle + \tau_p \langle F_i \rangle}{\tau_p} + v_k' \frac{\partial \langle v_i \rangle}{\partial x_k} \right\} \\ & - \frac{1}{\tau_p} \left\{ A_{ik} \frac{\partial}{\partial x_i} \frac{\partial f^0}{\partial v_k'} + B_{ik} \frac{\partial}{\partial v_i'} \frac{\partial f^0}{\partial v_k'} - \left[\theta + \frac{\tau_p}{2} \frac{5\nu_0 (1 - e^2)\theta}{12} \right] \delta_{ik} \frac{\partial}{\partial v_i'} \frac{\partial f^0}{\partial v_k'} \right\} \end{aligned} \tag{4.43}$$

$$\begin{aligned} Qf^0 = {} & \left[\frac{\theta}{\tau_p} + \frac{5\nu_0 (1 - e^2)\theta}{24} \right] \frac{\partial}{\partial v_i'} \frac{\partial f^0}{\partial v_i'} \\ & + \left[\frac{1}{\tau_p} + \frac{5\nu_0 (1 - e^2)}{24} \right] \frac{\partial v_i' f^0}{\partial v_i'} - \frac{1 + e}{2} \nu_0 (f^0 - f_m) \end{aligned} \tag{4.44}$$

The first correction term f^1 represents a linear function of spatial gradients for multiple physical parameters of solid volume fraction α_s, velocity $\langle v_i \rangle$ and granular temperature θ. The distribution function f^1 satisfies the following normalization constraints:

$$\int f^1 d\boldsymbol{v} = \int v_i' f^1 d\boldsymbol{v} = \int v_i' v_i' f^1 d\boldsymbol{v} = 0 \tag{4.45}$$

The zero-order solution f^0 satisfies:

$$\frac{\theta}{\tau_p}\frac{\partial^2 f^0}{\partial v_i'\partial v_i'}+\frac{1}{\tau_p}\frac{\partial v_i' f^0}{\partial v_i'}-\frac{1+e}{2}\nu_0(f^0-f_m)=0 \tag{4.46}$$

The corresponding solution is $f^0=f_m$, which represents a local equilibrium state without spatial gradients. Keeping in mind that heat flux is zero at the Maxwellian distribution.

$$q_k^0=0 \tag{4.47}$$

Closing Eqs. (4.34) to (4.36) with f^0 results in:

$$\frac{D\alpha_s}{Dt}=-\alpha_s\frac{\partial\langle v_k\rangle}{\partial x_k} \tag{4.48}$$

$$\frac{D\langle v_k\rangle}{Dt}=\frac{\langle u_i\rangle-\langle v_k\rangle}{\tau_p}+\langle F_i\rangle-\frac{\partial\theta}{\partial x_i}-\frac{1}{\alpha_s}\frac{\partial p_{ik}^c}{\partial x_k}-\frac{1}{\alpha_s}\left(\theta\delta_{ik}+\frac{A_{ik}}{\tau_p}\right)\frac{\partial\alpha_s}{\partial x_k} \tag{4.49}$$

$$\frac{D\theta}{Dt}=-\frac{2\theta}{3}\frac{\partial\langle v_i\rangle}{\partial x_i}-\frac{2}{3\alpha_s}\frac{\partial\langle v_i\rangle}{\partial x_k}p_{ik}^c-\frac{2(1-e^2)w_0^c}{3\alpha_s}-\frac{2(3\theta-B_{ik})}{3\tau_p} \tag{4.50}$$

where p_{ik}^c and w_0^c are the collisional quantities corresponding to f^0 and retaining no spatial gradients. The equation for f^1 is then given by

$$\begin{aligned}&\frac{Df^0}{Dt}+v_i'\frac{\partial f^0}{\partial x_i}-\frac{\partial f^0}{\partial v_i'}\left\{\frac{D\langle v_i\rangle}{Dt}-\frac{\langle u_i\rangle-\langle v_i\rangle+\tau_p\langle F_i\rangle}{\tau_p}+v_k'\frac{\partial\langle v_i\rangle}{\partial x_k}\right\}\\&-\frac{A_{ik}}{\tau_p}\frac{\partial^2 f^0}{\partial x_i\partial v_k'}-\frac{B_{ik}-\theta\delta_{ik}}{\tau_p}\frac{\partial^2 f^0}{\partial v_i'\partial v_k'}\\&-\frac{f_m}{\alpha_s\theta}\left[v_i'\frac{\partial p_{ik}^c}{\partial x_k}+\left(\frac{v_i'v_i'}{3\theta}-1\right)\left(p_{ik}^c\frac{\partial\langle v_k\rangle}{\partial x_i}+(1-e^2)w_0^c\right)\right]\\&=\frac{\theta}{\tau_p}\frac{\partial^2 f^1}{\partial v_i'\partial v_i'}+\frac{1}{\tau_p}\frac{\partial v_i' f^1}{\partial v_i'}-\zeta f^1\end{aligned} \tag{4.51}$$

From Eqs. (4.43) and (4.44), Eq. (4.51) is expressed by

$$\begin{aligned}&\left(\frac{\theta}{\tau_p}+\frac{5\nu_0(1-e^2)\theta}{24}\right)\frac{\partial}{\partial v_i'}\frac{\partial f^1}{\partial v_i'}+\left(\frac{1}{\tau_p}+\frac{5\nu_0(1-e^2)}{24}\right)\frac{\partial v_i' f^1}{\partial v_i'}-\frac{1+e}{2}\nu_0 f^1\\&=f_m\left(\frac{\partial\langle v_i\rangle}{\partial x_k}-\frac{B_{ik}}{\tau_p\theta}\right)\left(\frac{v_i'v_k'}{\theta}-\frac{v_l'v_l'}{3\theta}\delta_{ik}\right)+f_m\left(\delta_{ik}+\frac{A_{ik}}{\tau_p\theta}\right)\frac{\partial\ln\theta}{\partial x_i}\left(\frac{v_l'v_l'}{2\theta}-\frac{5}{2}\right)v_k'\end{aligned} \tag{4.52}$$

The solution for the probability density function is derived in terms of the spatial gradients of parameters as follows:

$$
\begin{aligned}
f^1 &= f_m - \frac{\tau_p f_m}{2+\zeta\tau_p}\left(\frac{\partial\langle v_k\rangle}{\partial x_k} - \frac{B_{ik}}{\tau_p\theta}\right)\left(\frac{v'_i v'_k}{\theta} - \frac{v'_i v'_i}{3\theta}\delta_{ik}\right) \\
&\quad - \frac{\tau_p f_m}{3+\zeta\tau_p}\left(\delta_{ik} + \frac{A_{ik}}{\tau_p\theta}\right)\frac{\partial\theta}{\partial x_i}\left(\frac{v'_i v'_i}{2\theta} - \frac{5}{2}\right)\frac{v'_k}{\theta}
\end{aligned} \tag{4.53}
$$

Using Eqs. (4.26) and (4.53) at the local equilibrium ($f^0 = f_m$), the velocity distribution function f^1 can be written as:

$$
\begin{aligned}
f &= f^0 - f^0 \frac{\tau_p\left[\frac{\partial\langle v_i\rangle}{\partial x_k} - \frac{B_{ik}}{\tau_p\theta}\right]\left[\frac{v'_i v'_k}{\theta} - \frac{v'_l v'_l}{3\theta}\delta_{ik}\right]}{2+\left[\frac{1+e}{2}\nu_0 + \frac{5\nu_0(1-e^2)}{12}\right]\tau_p} \\
&\quad - f^0 \frac{\tau_p\left[\delta_{ik} + \frac{A_{ik}}{\tau_p\theta}\right]\frac{\partial\ln\theta}{\partial x_i}\left[\frac{v'_l v'_l}{2\theta} - \frac{5}{2}\right]v'_k}{3+\left[\frac{1+e}{2}\nu_0 + \frac{3}{2}\frac{5\nu_0(1-e^2)}{12}\right]\tau_p}
\end{aligned} \tag{4.54}
$$

It can be seen that the solution of particle velocity probability distribution function represents the effects of both particle–particle collisions and particle–fluid turbulence interactions. Grad [13] wrote the particle velocity probability density function as a series of Hermite polynomials, where the Hermite polynomials are defined in terms of the derivatives of the Maxwellian distribution function. Hence, the second and third terms on the right-hand side represent first and second order tensor valued functions of particles.

4.1.2 Viscous Kinetic Stress Tensor

From Eq. (4.54), the second order moment of the particle fluctuating velocity is obtained through integrals of the probability density function over the whole velocity space:

$$
\begin{aligned}
\langle v'_i v'_k\rangle &= \frac{1}{\alpha_s}\int v'_i v'_k f\,\mathrm{d}\boldsymbol{v} \\
&= \frac{1}{\alpha_s}\int v'_i v'_k f^0\mathrm{d}\boldsymbol{v} - \frac{1}{\alpha_s}\int \frac{v'_i v'_k f^0\tau_p}{2+\left[\frac{1+e}{2}\nu_0 + \frac{5\nu_0(1-e^2)}{12}\right]\tau_p}\frac{\partial\langle v_i\rangle}{\partial x_k}\left(\frac{v'_i v'_k}{\theta} - \frac{v'_l v'_l}{3\theta}\delta_{ik}\right)\mathrm{d}\boldsymbol{v} \\
&\quad + \frac{1}{\alpha_s}\int \frac{v'_i v'_k f^0}{2+\left[\frac{1+e}{2}\nu_0 + \frac{5\nu_0(1-e^2)}{12}\right]\tau_p}\frac{B_{ik}}{\theta}\left(\frac{v'_i v'_k}{\theta} - \frac{v'_l v'_l}{3\theta}\delta_{ik}\right)\mathrm{d}\boldsymbol{v}
\end{aligned} \tag{4.55}
$$

Using Eq. (4.26), all terms on the right-hand side of Eq. (4.55) are integrated.

$$
\frac{1}{\alpha_s}\int v'_i v'_k f^0\mathrm{d}\boldsymbol{v} = \theta\delta_{ik} \tag{4.56}
$$

$$\frac{1}{\alpha_s}\int v_i'v_k'f^0\frac{\tau_p\frac{\partial\langle v_i\rangle}{\partial x_k}\left(\frac{v_i'v_k'}{\theta}-\frac{v_l'v_l'}{3\theta}\delta_{ik}\right)}{2+[\frac{1+e}{2}\nu_0+\frac{5\nu_0(1-e^2)}{12}]\tau_p}\mathrm{d}\boldsymbol{v} = \frac{\tau_p\theta\left(\frac{\partial\langle v_i\rangle}{\partial x_k}+\frac{\partial\langle v_k\rangle}{\partial x_i}-\frac{2}{3}\frac{\partial\langle v_l\rangle}{\partial x_l}\delta_{ik}\right)}{2+\left[\frac{1+e}{2}\nu_0+\frac{5\nu_0(1-e^2)}{12}\right]\tau_p} \tag{4.57}$$

$$\frac{1}{\alpha_s}\int v_i'v_k'f^0\frac{\frac{B_{ik}}{\theta}\left(\frac{v_i'v_k'}{\theta}-\frac{v_i'v_l'}{3\theta}\delta_{ik}\right)}{2+\left[\frac{1+e}{2}\nu_0+\frac{5\nu_0(1-e^2)}{12}\right]\tau_p}\mathrm{d}\boldsymbol{v}=\frac{\left(B_{ik}+B_{ki}-\frac{2}{3}B_{ll}\delta_{ik}\right)}{2+\left[\frac{1+e}{2}\nu_0+\frac{5\nu_0(1-e^2)}{12}\right]\tau_p} \tag{4.58}$$

Using Eq. (4.19), the second order moment of the particle fluctuating velocity is correlated to granular temperature and Stokes number.

$$\left\langle v_i'v_k'\right\rangle=\theta\delta_{ik}-\frac{\tau_p\theta}{2+\varphi(e)\nu_0\tau_p}\left(\frac{\partial\langle v_i\rangle}{\partial x_k}+\frac{\partial\langle v_k\rangle}{\partial x_i}-\frac{2}{3}\frac{\partial\langle v_l\rangle}{\partial x_l}\delta_{ik}\right)+ \frac{2}{2+\varphi(e)\nu_0\tau_p}\frac{1}{1-\exp(-St^{-1})}\left(\left\langle u_i'u_k'\right\rangle-\frac{1}{3}\left\langle u_i'u_k'\right\rangle\delta_{ik}\right) \tag{4.59}$$

$$\varphi(e)=\frac{1+e}{2}\left[1+\frac{5}{6}(1-e)\right] \tag{4.60}$$

Adopting the Boussinesq turbulent viscosity concept for shear viscosity of fluid phase, the turbulent closures are based on the gradient hypothesis and the coefficient of fluid eddy viscosity μ_t.

$$\left\langle u_i'u_k'\right\rangle=-\mu_t\left(\frac{\partial\langle u_i\rangle}{\partial x_k}+\frac{\partial\langle u_k\rangle}{\partial x_i}-\frac{2}{3}\frac{\partial\langle u_l\rangle}{\partial x_l}\delta_{ik}\right)+\frac{1}{3}\left\langle u_i'u_k'\right\rangle\delta_{ik} \tag{4.61}$$

The second order velocity moment of particles is expressed as follows:

$$\left\langle v_i'v_k'\right\rangle=\theta\delta_{ik}-\frac{\tau_p\theta}{2+\varphi(e)\nu_0\tau_p}\left(\frac{\partial\langle v_i\rangle}{\partial x_k}+\frac{\partial\langle v_k\rangle}{\partial x_i}-\frac{2}{3}\frac{\partial\langle v_l\rangle}{\partial x_l}\delta_{ik}\right) -\frac{2}{2+\varphi(e)\nu_0\tau_p}\frac{\mu_t}{1-\exp(-St^{-1})}\left(\frac{\partial\langle u_i\rangle}{\partial x_k}+\frac{\partial\langle u_k\rangle}{\partial x_i}-\frac{2}{3}\frac{\partial\langle u_l\rangle}{\partial x_l}\delta_{ik}\right) \tag{4.62}$$

The second order velocity moment is related to granular temperatures, particle relaxation time and Stokes number. Two additional terms appear on the right-hand side. The first term in the denominator represents the contribution of particle inertia. The latter term is important when particle motion is controlled by the aerodynamic forces due to the rate of shear tensors of fluid and particles. This term exists on the presence of interstitial fluid. It is also inverse proportional to particle Stokes numbers. This term is negligible at high Stokes numbers. The second-order velocity moments of particles relate to rates of shear tensors of fluid and solid phases. Hence, the interstitial fluid influences on the momentum transfer of particles.

From Eq. (4.54), the third order moments of the particle fluctuating velocity is obtained through integrals of the probability density function over the whole velocity space.

$$\left\langle v_i'v_k'v_k'\right\rangle = \frac{1}{\alpha_s}\int v_i'v_k'v_k' f\,d\boldsymbol{v} =$$

$$\frac{1}{\alpha_s}\int v_i'v_k'v_k' f^0\{1 - \frac{\tau_p}{2+\left[\frac{1+e}{2}\nu_0 + \frac{5\nu_0(1-e^2)}{12}\right]\tau_p}\left(\frac{\partial\langle v_i\rangle}{\partial x_k} - \frac{B_{ik}}{\tau_p\theta}\right)\left(\frac{v_i'v_k'}{\theta} - \frac{v_l'v_l'}{3\theta}\delta_{ik}\right)$$
$$-\frac{\tau_p}{3+\left(\frac{1+e}{2}\nu_0 + \frac{3}{2}\frac{5\nu_0(1-e^2)}{12}\right)\tau_p}\left(\delta_{ik} + \frac{A_{ik}}{\tau_p\theta}\right)\frac{\partial \ln\theta}{\partial x_i}\left(\frac{v_l'v_l'}{2\theta} - \frac{5}{2}\right)v_k'\}\,d\boldsymbol{v} \tag{4.63}$$

These integrals can be evaluated analytically by means of known integrals from [7]. Note that the first part on the right-hand side is vanished using the Maxwellian distribution function. Thus, the final expression for third order moment becomes as follows:

$$\begin{aligned}\langle v'_i v'_k v'_k\rangle &= -\frac{1}{\alpha_s}\int v'_i v'_k v'_k f^0 \frac{\tau_p}{3+[\frac{1+e}{2}\nu_0 + \frac{3}{2}\frac{5\nu_0(1-e^2)}{12}]\tau_p}\left(\delta_{ik} + \frac{A_{ik}}{\tau_p\theta}\right)\frac{\partial\ln\theta}{\partial x_i}\left(\frac{v'_l v'_l}{2\theta} - \frac{5}{2}\right)v'_k d\boldsymbol{v}\\ &= -\frac{5\tau_p\theta}{3+[\frac{1+e}{2}\nu_0 + \frac{3}{2}\frac{5\nu_0(1-e^2)}{12}]\tau_p}\frac{\partial\theta}{\partial x_i} - \frac{5A_{ik}}{3+[\frac{1+e}{2}\nu_0 + \frac{3}{2}\frac{5\nu_0(1-e^2)}{12}]\tau_p}\frac{\partial\theta}{\partial x_k}\end{aligned} \tag{4.64}$$

Using Eq. (4.18), the second term on the right-hand side is substituted and given by

$$\begin{aligned}\left\langle v_i'v_k'v_k'\right\rangle = &-\frac{5\tau_p\theta}{3+\left[\frac{1+e}{2}\nu_0 + \frac{3}{2}\frac{5\nu_0(1-e^2)}{12}\right]\tau_p}\frac{\partial\theta}{\partial x_i}\\ &-\frac{5\left\{1-St\left[1-\exp\left(-\frac{1}{St}\right)\right]\right\}\tau_{Lp}\left[-\mu_t\left(\frac{\partial\langle u_i\rangle}{\partial x_k} + \frac{\partial\langle u_k\rangle}{\partial x_i}\right) + \mu_t\frac{2}{3}\frac{\partial\langle u_l\rangle}{\partial x_l}\delta_{ik} + \frac{1}{3}\left(-2\mu_t\frac{\partial\langle u_l\rangle}{\partial x_l}\right)\delta_{ik}\right]}{3+\left[\frac{1+e}{2}\nu_0 + \frac{3}{2}\frac{5\nu_0(1-e^2)}{12}\right]\tau_p}\frac{\partial\theta}{\partial x_k}\end{aligned} \tag{4.65}$$

Substituting Eq. (4.61) into the complete solution, simplifying and rearranging yields the final solution for the granular kinetic energy equation as

$$\left\langle v_i'v_k'v_k'\right\rangle = -\frac{5\theta}{3+\phi(e)\nu_0\tau_p}\left\{\tau_p\delta_{ik} - \left[1-St\left(1-\exp\left(-\frac{1}{St}\right)\right)\right]\frac{\tau_{Lp}\mu_t}{\theta}\left(\frac{\partial\langle u_i\rangle}{\partial x_k} + \frac{\partial\langle u_k\rangle}{\partial x_i}\right)\right\}\frac{\partial\theta}{\partial x_k} \tag{4.66}$$

$$\phi(e) = \frac{1+e}{2}\left[1 + \frac{5}{4}(1-e)\right] \tag{4.67}$$

Both Eq. (4.62) for the second order moment and Eq. (4.66) for the third order moment of velocity fluctuations explicitly include contributions of fluid turbulence. That is, fluid turbulence affects particle motion through the balance equations as well as these constitutive relations. Such effects will be crucial in determining particle motion. In addition to considering inelastic collisions and particle–fluid turbulence interactions, the above constitutive relations cover wide ranges of solid volume fractions and particle inertia and become applicable to dilute and moderately solid volume fraction flows. Both Stokes number and solid volume fractions determine the importance of the collisional components relative to the kinetic ones in particle stresses.

4.2 Dense Transport Theorem

As the solids volume fraction increases, the interactions between particles are significant. Effects of fluid turbulence and direct collisions between particles on granular flow should be taken into account comprehensively, while the latter is often evaluated for collisional transport through pair-distribution function [7, 19]. The transport equation for a quantity Ψ is expressed according to the Boltzmann equation integrating peculiar velocity $\boldsymbol{C}$ [3, 14].

$$\frac{Dn\langle\Psi\rangle}{Dt} + n\langle\Psi\rangle\frac{\partial \boldsymbol{v}}{\partial \boldsymbol{x}} + \frac{\partial}{\partial \boldsymbol{x}}n\langle\Psi\boldsymbol{C}\rangle - n\left(F - \frac{D\boldsymbol{v}}{Dt}\right)\left\langle\frac{\partial\Psi}{\partial\boldsymbol{C}}\right\rangle - n\left\langle\frac{\partial\Psi}{\partial\boldsymbol{C}}\boldsymbol{C}\right\rangle : \frac{\partial\boldsymbol{v}}{\partial\boldsymbol{x}} = n\langle\Psi_c\rangle \tag{4.68}$$

where $\boldsymbol{F}$ represents the forces per unit mass. The left-hand side of the balance is an accumulation plus a divergence and net rate of outflow of Ψ with respect to $\boldsymbol{x}$ and $\boldsymbol{C}$. The right-hand side denotes the rate of change of Ψ per unit volume by collisions. The mean change of the collisional property is expressed according to the Maxwell transport equation [14]).

$$\langle\psi_c\rangle = \frac{1}{2}\int\limits_{\boldsymbol{c}_{12}\cdot\boldsymbol{k}>0}[(\psi_2' + \psi_1') - (\psi_2 + \psi_1)]f^2(\boldsymbol{x}_1, \boldsymbol{x}_2, \boldsymbol{c}_1, \boldsymbol{c}_2, t)d_p^2(\boldsymbol{c}_{12}\cdot\boldsymbol{k})d\boldsymbol{k}d\boldsymbol{c}_1d\boldsymbol{c}_2 \tag{4.69}$$

where $\boldsymbol{c}_1$ and $\boldsymbol{c}_2$ denote the velocity of a pair of particles. $\boldsymbol{k}$ is a unit vector directed from the center of the particle 1 to that of particle 2 at contact. $\boldsymbol{c}_{12}$ is the relative velocity of the particles. It represents the loss of Ψ caused by inelastic collisions of particles.

The collisional pair distribution function can be expressed as the product of the two particle velocity probability density functions $f_1(\boldsymbol{x}_1, \boldsymbol{c}_1, t)$ and $f_2(\boldsymbol{x}_2, \boldsymbol{c}_2, t)$, and is given [3, 14, 17, 26]

$$f^{(2)}\left(\boldsymbol{c}_1, \boldsymbol{x}_1 - \frac{d_p}{2}, \boldsymbol{c}_2, \boldsymbol{x}_2 + \frac{d_p}{2}, t\right) = g_0 f_1\left(\boldsymbol{c}_1, \boldsymbol{x}_1 - \frac{d_p}{2}, t\right) f_2\left(\boldsymbol{c}_2, \boldsymbol{x}_2 + \frac{d_p}{2}, t\right) \tag{4.70}$$

The radial distribution function g_0 trends to be unity at low solid volume fractions, and increases with increasing solid volume fractions, becoming infinite as particles approach the state in which the particles are packed so closely that motion is impossible. When the single particle velocity distribution function f takes the first order approximation to the Maxwellian distribution and using Eq. (4.70), Eq. (4.69) is expressed as follows [7, 14]:

$$\langle \psi_c \rangle = -\nabla \cdot \boldsymbol{P}_c + N_c \tag{4.71}$$

where the collisional part of the vector of the flow Ψ is

$$\boldsymbol{P}_c = -\frac{1}{2} g_0 d_p^3 \iiint\limits_{\boldsymbol{c}_{12} \cdot \boldsymbol{k} > 0} (\psi_1' - \psi_1) f_1 f_2 \boldsymbol{k} (\boldsymbol{c}_{12} \cdot \boldsymbol{k}) d\boldsymbol{k} d\boldsymbol{c}_1 d\boldsymbol{c}_2$$
$$-\frac{1}{4} g_0 d_p^4 \iiint\limits_{\boldsymbol{c}_{12} \cdot \boldsymbol{k} > 0} (\psi_1' - \psi_1) f_1 f_2 \boldsymbol{k} \nabla \ln \frac{f_2}{f_1} \boldsymbol{k} (\boldsymbol{c}_{12} \cdot \boldsymbol{k}) d\boldsymbol{k} d\boldsymbol{c}_1 d\boldsymbol{c}_2 \tag{4.72}$$

and the source-like contribution is

$$N_c = \frac{g_0 d_p^2}{2} \iiint\limits_{\boldsymbol{c}_{12} \cdot \boldsymbol{k} > 0} (\psi_2' + \psi_1' - \psi_2 - \psi_1) f_1 f_2 \boldsymbol{k} (\boldsymbol{c}_{12} \cdot \boldsymbol{k}) d\boldsymbol{k} d\boldsymbol{c}_1 d\boldsymbol{c}_2 \tag{4.73}$$

Let $\Psi = mC$ in order to obtain collisional momentum transfer, the relative velocity of the particles relate to its hydrodynamic velocity. Equation (4.72) represents the collisional stress tensor, and it becomes the sum of two integrals [10, 14].

$$\boldsymbol{P}_c = \frac{1}{4} g_0 d_p^3 m(1+e) \iiint\limits_{\boldsymbol{c}_{12} \cdot \boldsymbol{k} > 0} (\boldsymbol{k} \cdot \boldsymbol{c}_{12})^2 \boldsymbol{k}\boldsymbol{k} f_1 f_2 d\boldsymbol{k} d\boldsymbol{c}_1 d\boldsymbol{c}_2 +$$
$$\frac{1}{8} g_0 d_p^4 m(1+e) \iiint\limits_{\boldsymbol{c}_{12} \cdot \boldsymbol{k} > 0} (\boldsymbol{k} \cdot \boldsymbol{c}_{12})^2 \boldsymbol{k}\boldsymbol{k} f_1^{(0)} f_2^{(0)} \boldsymbol{k} \cdot \nabla \ln \frac{f_2^{(0)}}{f_1^{(0)}} d\boldsymbol{k} d\boldsymbol{c}_1 d\boldsymbol{c}_2 = \boldsymbol{\tau}_{ik,1}^c + \boldsymbol{\tau}_{ik,2}^c \tag{4.74}$$

Using Maxwellian distributions, the first and second parts of the collisional integral can be written

$$\tau_{ik,1}^{c} = -\frac{4}{5}(1+e)\rho_s\alpha_s^2 g_0\left[\langle v_i' v_k'\rangle + \frac{3}{2}\theta\delta_{ik}\right] \tag{4.75}$$

$$\tau_{ik2}^{c} = \frac{4}{5\sqrt{\pi}}(1+e)\rho_s\alpha_s^2 g_0 d_p\sqrt{\theta}\left(\frac{\partial\langle v_i\rangle}{\partial x_k} + \frac{\partial\langle v_k\rangle}{\partial x_i} + \frac{\partial\langle v_l\rangle}{\partial x_l}\delta_{ik}\right) \tag{4.76}$$

Then, the value of the collisional stress tensor is the sum of that given by Eqs. (4.75) and (4.76), and expressed as follows:

$$\tau_{ik}^{c} = -\frac{4}{5}(1+e)\rho_s\alpha_s^2 g_0\left[\langle v_i' v_k'\rangle + \frac{3}{2}\theta\delta_{ik} - d_p\sqrt{\frac{\theta}{\pi}}\left(\frac{\partial\langle v_i\rangle}{\partial x_k} + \frac{\partial\langle v_k\rangle}{\partial x_i} + \frac{\partial\langle v_l\rangle}{\partial x_l}\delta_{ik}\right)\right] \tag{4.77}$$

Substituting Eq. (4.62) into Eq. (4.77), the expressions for the solid collisional stress in fluid–solid two phase flows are expressed as follows:

$$\begin{aligned}\tau_{ik}^{c} = &-\frac{4}{5}(1+e)\rho_s\alpha_s^2 g_0 \\ &\left\{\frac{5}{2}\theta\delta_{ik} - \frac{2}{2+\nu_0\tau_p\varphi(e)}\frac{\mu_t}{1-\exp(-St^{-1})}\left(\left(\frac{\partial\langle u_i\rangle}{\partial x_k} + \frac{\partial\langle u_k\rangle}{\partial x_i}\right) - \frac{2}{3}\frac{\partial\langle u_l\rangle}{\partial x_l}\delta_{ik}\right)\right. \\ &- \frac{\theta\tau_p}{2+\nu_0\tau_p\varphi(e)}\left(\left(\frac{\partial\langle v_i\rangle}{\partial x_k} + \frac{\partial\langle v_k\rangle}{\partial x_i}\right) - \frac{2}{3}\frac{\partial\langle v_l\rangle}{\partial x_l}\delta_{ik}\right) \\ &\left.-d_p\sqrt{\frac{\theta}{\pi}}\left(\frac{\partial\langle v_i\rangle}{\partial x_k} + \frac{\partial\langle v_k\rangle}{\partial x_i} + \frac{\partial\langle v_l\rangle}{\partial x_l}\delta_{ik}\right)\right\}\end{aligned} \tag{4.78}$$

It is clear that the granular collisional stress relates to Stokes number and rates of shear tensors of fluid phase and solid phase.

The granular energy associated with granular temperature of particles is assumed characterized by scales comparable to the particle size being considerable smaller than the microscopic length scales of the flow. Thus, the granular temperature is associated with the turbulent kinetic energy which is a continuum fluid flow phenomenon. The granular kinetic energy flow vector $\boldsymbol{q}$ is computed by means of Eq. (4.72) as the quantity Ψ is $m\boldsymbol{CC}/2$. It consists of two collisional parts.

$$\begin{aligned}\boldsymbol{q}_c = &\frac{1}{4}g_0 d_p^3 m \iiint_{\boldsymbol{c}_{12}\cdot\boldsymbol{k}>0}(\boldsymbol{C}_1'^2 - \boldsymbol{C}_1^2)\boldsymbol{k}(\boldsymbol{c}_{12}\cdot\boldsymbol{k})^2 f_1 f_2 d\boldsymbol{k}d\boldsymbol{c}_1 d\boldsymbol{c}_2 + \\ &\frac{1}{8}g_0 d_p^4 m \iiint_{\boldsymbol{c}_{12}\cdot\boldsymbol{k}>0}(\boldsymbol{C}_1'^2 - \boldsymbol{C}_1^2)f_1 f_2 \boldsymbol{k}\nabla\ln\frac{f_2^{(0)}}{f_1^{(0)}}\boldsymbol{k}(\boldsymbol{c}_{12}\cdot\boldsymbol{k})d\boldsymbol{k}d\boldsymbol{c}_1 d\boldsymbol{c}_2 = \boldsymbol{q}_1^c + \boldsymbol{q}_2^c\end{aligned} \tag{4.79}$$

where m is the mass of a particle. Introducing the gradient of granular temperature, the first and second integrals are evaluated [10, 23].

$$q_{ik,1}^{c} = \frac{3}{5}\alpha_s^2 \rho_s g_0 (1+e)\langle v_i' v_k' v_k' \rangle \tag{4.80}$$

$$q_{ik,2}^{c} = -2\alpha_s^2 \rho_s g_0 (1+e) d_p \sqrt{\frac{\theta}{\pi}} \frac{\partial \theta}{\partial x_i} \tag{4.81}$$

Both parts relate with coefficient of restitution of particles. The total kinetic energy flux caused by collisions of particles is given by the sum of q_{ik}^{c}, 1 and $q_{ik,2}^{c}$ 2 [10, 14].

$$q_{ik}^{c} = \alpha_s^2 \rho_s g_0 (1+e)\left(\frac{3}{5}\langle v_i' v_k' v_k' \rangle - 2d_p \sqrt{\frac{\theta}{\pi}} \frac{\partial \theta}{\partial x_i}\right) \tag{4.82}$$

Substituting Eq. (4.66) into Eq. (4.82), the expressions for the fluctuation energy flux in fluid–solid two-phase flow is obtained:

$$\begin{aligned} q_{ik}^{c} &= \alpha_s^2 \rho_s g_0 (1+e) \\ &\times \left\{ -\frac{3\theta}{3+\nu_0 \tau_p \phi(e)} \left\{ \tau_p \delta_{ik} - \left[1 - St\left(1 - \exp\left(-\frac{1}{St}\right)\right)\right] \right.\right. \\ &\left.\left. \frac{\tau_{Lp}\mu_t}{\theta}\left(\frac{\partial \langle u_i \rangle}{\partial x_k} + \frac{\partial \langle u_k \rangle}{\partial x_i}\right)\right\} \frac{\partial \theta}{\partial x_k} - 2d_p \sqrt{\frac{\theta}{\pi}} \frac{\partial \theta}{\partial x_i} \right\} \end{aligned} \tag{4.83}$$

It is clear that the granular heat flux relates to Stokes number and rates of shear tensors of fluid phase and solid phase.

The last term on the right-hand side of Eq. (4.71) represents the loss of energy due to collisions. Using Eq. (4.70) with the quantity Ψ is $mc^2/2$, the dissipation rate of energy due to collisions are the first to evaluate such an integral using the methods of dense phase kinetic theory [10].

$$N_c\left(\frac{m}{2}c^2\right) = 3(1-e^2)\alpha_s^2 \rho_s g_0 \theta \left(\frac{4}{d_p}\sqrt{\frac{\theta}{\pi}} - \nabla \cdot v\right) \tag{4.84}$$

4.3 Granular Constitutive Relations

The rigorous procedure deriving the governing equations of solid and fluid dynamics from first principles is established. It is stressed that in classical solid and fluid dynamics the continuum hypothesis is used so that the governing equations of change are derived directly on the macroscopic scales applying the balance principle over a suitable control volume. Thereafter, closure laws are required relating the internal stresses and diffusive transport fluxes of mass and energy to the gradients of the macroscopic quantities (e.g., Newtons law of viscosity). Using these constitutive relations, we also introduce several unknown parameters, the so-called transport coefficients of viscosity and thermal conductivity into the governing equations. Unfortunately, dynamics do neither provide any guidelines on how to determine the mathematical form of the constitutive relations nor the values of transport coefficients.

A unique application of the kinetic theory of granular flow provides that a solution of the Boltzmann equation is available. In this case particular moments of the distribution function are obtained multiplying the Boltzmann equation by an invariant function (i.e., successively representing mass, momentum and energy), and thereafter integrating the resulting equation over the whole velocity space. The resulting transport equations are recognized as being analogue to the conservation equations of continuum mechanics. In addition, the ensemble averaged equations obtained from the kinetic theory of granular flow do not contain any unknown coefficients. The set of closure laws is established.

The granular momentum balance is obtained according to Eq. (4.35). The forces consist of the externally force and the forces of interaction between phases. The latter forces are expressed by the drag coefficient multiplied by the slip velocity.

$$\frac{\partial(\alpha_s\rho_s\langle v_i\rangle)}{\partial t}+\frac{\partial(\alpha_s\rho_s\langle v_i\rangle\langle v_k\rangle)}{\partial x_k}=-\alpha_s\frac{\partial p}{\partial x_i}+\frac{\partial\tau_{sik}}{\partial x_k}+\beta(\langle u_i\rangle-\langle v_i\rangle)+\alpha_s\rho_s g_i \tag{4.85}$$

The total pressure tensor contains both the kinetic and collisional contributions. The kinetic pressure tensor accounts for the transport phenomena due to the translational particle movement, and the collisional pressure tensor accounts for the transport phenomena due to particle collisions. The total solid stress tensor can be written as:

$$\begin{aligned}
\tau_{sik} &= \tau_{ik}^k+\tau_{ik}^c=\{-\rho_s\alpha_s(1+2(1+e)\alpha_s g_0)\theta \\
&-\frac{(1+\frac{4}{5}(1+e)\alpha_s g_0)}{2+\nu_0\tau_p\varphi(e)}\frac{2\rho_s\alpha_s\mu_t}{1-\exp(-St^{-1})}\frac{2}{3}\frac{\partial\langle u_l\rangle}{\partial x_l} \\
&+\left[\frac{4}{5}(1+e)\rho_s\alpha_s^2 g_0 d_p\sqrt{\frac{\theta}{\pi}}-\frac{2}{3}\left(1+\frac{4}{5}(1+e)\alpha_s g_0\right)\frac{\theta\tau_p\rho_s\alpha_s}{2+\nu_0\tau_p\varphi(e)}\right]\frac{\partial\langle v_l\rangle}{\partial x_l}\Bigg\}\delta_{ik} \\
&+\Bigg\{\left(1+\frac{4}{5}(1+e)\alpha_s g_0\right)\frac{\rho_s\alpha_s\theta\tau_p}{2+\nu_0\tau_p\varphi(e)}+\frac{4}{5}(1+e)\rho_s\alpha_s^2 g_0 d_p\sqrt{\frac{\theta}{\pi}} \\
&+\left(1+\frac{4}{5}(1+e)\alpha_s g_0\right)\frac{2\rho_s\alpha_s}{2+\nu_0\tau_p\varphi(e)}\frac{\mu_t}{1-\exp(-St^{-1})}\frac{\left(\frac{\partial\langle u_i\rangle}{\partial x_k}+\frac{\partial\langle u_k\rangle}{\partial x_i}\right)}{\left(\frac{\partial\langle v_i\rangle}{\partial x_k}+\frac{\partial\langle v_k\rangle}{\partial x_i}\right)}\Bigg\} \\
&\left(\frac{\partial\langle v_i\rangle}{\partial x_k}+\frac{\partial\langle v_k\rangle}{\partial x_i}\right)
\end{aligned} \tag{4.86}$$

Analogous to the viscous fluid mechanics constitutive equation, the constitutive relationship for the solid stress tensor can be written as in terms of solid pressure p_s and shear viscosity μ_s:

$$\tau_{sik} = \left(-p_s + \mu_b \frac{\partial \langle v_l \rangle}{\partial x_l}\right)\delta_{ik} + 2\mu_s\left(\frac{1}{2}\left(\frac{\partial \langle v_i \rangle}{\partial x_k} + \frac{\partial \langle v_k \rangle}{\partial x_i}\right) - \frac{1}{3}\frac{\partial \langle v_l \rangle}{\partial x_l}\delta_{ik}\right) \quad (4.87)$$

where μ_b represents the solid bulk viscosity.

From Eq. (4.86), the solid phase pressure is

$$p_s = \rho_s \alpha_s \theta + 2(1+e)\alpha_s^2 g_0 \rho_s \theta \quad (4.88)$$

The first part on the right-hand side represents kinetic contribution, and the second part denotes collisional contribution.

The solid shear viscosity can be written as:

$$\begin{aligned} \mu_s = &\left(1 + \frac{4}{5}(1+e)\alpha_s g_0\right)\frac{\rho_s \alpha_s \theta \tau_p}{2 + \nu_0 \tau_p \varphi(e)} + \frac{4}{5}(1+e)\rho_s \alpha_s^2 g_0 d_p \sqrt{\frac{\theta}{\pi}} \\ &+ \left(1 + \frac{4}{5}(1+e)\alpha_s g_0\right)\frac{2\rho_s \alpha_s}{2 + \nu_0 \tau_p \varphi(e)} \frac{\mu_t}{1 - \exp(-St^{-1})} \frac{S_f}{S_p} \end{aligned} \quad (4.89)$$

where S_f and S_p represent the strain rates of fluid and solid phases. The first term on the right-hand side in the granular shear viscosity denotes the kinetic contribution and dominates in the dilute regime with low solid volume fractions. The second term denotes the collisional contribution and dominates in the dense flow regime at high solid volume fractions. The last term represents interstitial fluid effects. This term is inversely proportional to Stokes number, and it vanishes with the increase of Stokes numbers. Thus, the solid shear viscosity represents a kinetic part and collisional part.

$$\mu_s = \left(\mu_{k,s} + \mu_{k,f}\right) + \left(\mu_{c,s} + \mu_{c,f}\right) \quad (4.90)$$

The granular kinetic viscosity component consists of particles diffusion part $\mu_{k,s}$ and fluid turbulence contribution $\mu_{k,f}$.

$$\mu_{k,s} = \frac{\rho_s \alpha_s \theta \tau_p}{2 + \nu_0 \tau_p \varphi(e)} \quad (4.91)$$

$$\mu_{k,f} = \frac{2\rho_s \alpha_s}{2 + \nu_0 \tau_p \varphi(e)} \frac{\mu_t}{1 - \exp(-St^{-1})} \frac{S_f}{S_p} \quad (4.92)$$

Similarly, the collisional part includes collisions of particles and fluid turbulence interaction effects.

$$\mu_{c,s} = \frac{4}{5}(1+e)\alpha_s g_0 \frac{\rho_s \alpha_s \theta \tau_p}{2 + \nu_0 \tau_p \varphi(e)} + \frac{4}{5}(1+e)\rho_s \alpha_s^2 g_0 d_p \sqrt{\frac{\theta}{\pi}} \quad (4.93)$$

$$\mu_{c,f} = \frac{4}{5}\frac{2\rho_s\alpha_s}{2 + \nu_0\tau_p\varphi(e)}\frac{(1+e)\alpha_s g_0\mu_t}{1-\exp(-St^{-1})}\frac{S_f}{S_p} \tag{4.94}$$

Increasing Stokes number, the fluid turbulent interaction effect will be vanished. The granular shear viscosity is reduced as the Stokes number trends to infinite. The particle shear viscosity part $\mu_{k,f}$ and $\mu_{c,f}$ correlate with solid volume fractions. It also depends on the ratio of the fluid deformation tensor to the solid phase deformation tensor (i.e., S_f/S_p). As can be observed, they can both decrease and increase with changing strain-rate tensors of fluid and solid phases. The shear viscosity of particles can be expressed as the Stokes number trends to infinite.

$$\mu_{s,\infty} = \frac{10\rho_s d_p\sqrt{\pi\theta}}{96g_0(1+e)\left[1+\frac{5}{6}(1-e)\right]}\left(1+\frac{4}{5}(1+e)\alpha_s g_0\right) + \frac{4}{5}(1+e)\rho_s\alpha_s^2 g_0 d_p\sqrt{\frac{\theta}{\pi}} \tag{4.95}$$

The first term on the right-hand side represents kinetic contribution, and the second term represents collisional contribution. On the other hand, the granular shear viscosity in terms of the classical kinetic theory of granular flow is expressed by both kinetic part and collisional part [14].

$$\mu_{s,G} = \frac{10\rho_s d_p\sqrt{\pi\theta}}{96g_0(1+e)}\left(1+\frac{4}{5}(1+e)\alpha_s g_0\right)^2 + \frac{4}{5}(1+e)\rho_s\alpha_s^2 g_0 d_p\sqrt{\frac{\theta}{\pi}} \tag{4.96}$$

For dilute flow with low solid volume fractions, the ratio is

$$\frac{\mu_{s,G}}{\mu_{s,\infty}} = \left(1+\frac{4}{5}(1+e)\alpha_s g_0\right)\left[1+\frac{5}{6}(1-e)\right] \tag{4.97}$$

The ratio is close to unity at low solid volume fractions.

The normal component of the collisional deviatoric pressure tensor resembles the bulk viscosity term. Thus, the solid bulk viscosity can be written as follows:

$$\mu_b = \frac{4}{3}(1+e)\rho_s\alpha_s^2 g_0 d_p\sqrt{\frac{\theta}{\pi}} + \frac{2}{3}\left(1+\frac{4}{5}(1+e)\alpha_s g_0\right)$$
$$\frac{2\rho_s\alpha_s}{2+\nu_0\tau_p\varphi(e)}\frac{\mu_t}{1-\exp(-St^{-1})}\frac{S_f}{S_p} \tag{4.98}$$

The first term on the right-hand side represents the collisional contribution and dominates at high solid volume fractions. The second term denotes the contribution of interstitial fluid effects. This term is inversely proportional to Stokes number, and it decreases with increasing Stokes numbers. As the Stokes number trends to infinite, the solid bulk viscosity is expressed

$$\mu_{b,\alpha} = \frac{4}{3}(1+e)\rho_s\alpha_s^2 g_0 d_p\sqrt{\frac{\theta}{\pi}} \tag{4.99}$$

Both shear viscosity and bulk viscosity explicitly include contributions of interstitial fluid effects. That is, fluid turbulent interaction affects particle motion through the balance equations and these constitutive relations. For dilute flow with low solid volume fractions, such effects will be crucial in determining particle motion. In addition to considering inelastic collisions and particle–fluid interactions, these constitutive relations cover wide ranges of solid volume fractions and particle inertia, and become applicable to dilute and dense fluid-particles flows.

4.4 Granular Conductivity

The total granular fluctuation energy flux consists of a kinetic part and collisional part, and it can be written as:

$$q_{ik} = \frac{1}{2}\rho_s\alpha_s\langle v_i' v_k' v_k'\rangle + q_{ik}^c \tag{4.100}$$

In the zero order solution ($f = f^0$), there is no heat flow. Using Eqs. (4.66) and (4.83), the granular fluctuation energy flux is expressed as follows:

$$\begin{aligned} q_{ik} = &-\rho_s\alpha_s\left\{\frac{5}{2}\frac{\theta\tau_p\delta_{ik}}{3+\nu_0\tau_p\phi(e)} - \frac{5}{2}\frac{\theta[1-St(1-\exp(-St^{-1}))]}{3+\nu_0\tau_p\phi(e)}\frac{\tau_{Lp}\mu_t}{\theta}\right.\\ &\left.\left(\frac{\partial\langle u_i\rangle}{\partial x_k}+\frac{\partial\langle u_k\rangle}{\partial x_i}\right)\right\}\frac{\partial\theta}{\partial x_k}\\ &-\alpha_s\rho_s\left\{\left(\alpha_s g_0(1+e)\frac{3\theta\tau_p}{3+\nu_0\tau_p\phi(e)} + \alpha_s g_0(1+e)2d_p\sqrt{\frac{\theta}{\pi}}\right)\delta_{ik}\right.\\ &\left.-\alpha_s g_0(1+e)\frac{3\theta[1-St(1-\exp(-St^{-1}))]}{3+\nu_0\tau_p\phi(e)}\frac{\tau_{Lp}\mu_t}{\theta}\left(\frac{\partial\langle u_i\rangle}{\partial x_k}+\frac{\partial\langle u_k\rangle}{\partial x_i}\right)\right\}\frac{\partial\theta}{\partial x_k} \end{aligned} \tag{4.101}$$

The total granular heat flux vectors can be expressed using granular conductivity, and is given by:

$$q_{ik} = -\kappa_s\frac{\partial\theta}{\partial x_i} \tag{4.102}$$

The solid phase conductivity of the particle fluctuating energy can be expressed as:

$$\kappa_s = \frac{5\rho_s\alpha_s\theta}{3+\nu_0\tau_p\phi(e)}\left(\frac{1}{2}+\frac{3}{5}\alpha_s g_0(1+e)\right)\tau_p + \rho_s\alpha_s^2 g_0(1+e)2d_p\sqrt{\frac{\theta}{\pi}}$$
$$-\left(\frac{1}{2}+\frac{3}{5}\alpha_s g_0(1+e)\right)[1-St(1-\exp(-St^{-1}))]\frac{10\rho_s\alpha_s\tau_{Lp}\mu_t}{3+\nu_0\tau_p\phi(e)}\left(\frac{\partial\langle u_i\rangle}{\partial x_k}+\frac{\partial\langle u_k\rangle}{\partial x_i}\right) \tag{4.103}$$

The first term on the right-hand side represents the kinetic contribution and dominates in the dilute regime at low solid volume fractions. The second term reflects the contribution of collisions of particles and dominates in the dense flow regime at high solid volume fractions. The last term denotes interstitial fluid effects, and is inversely proportional to Stokes number. It vanishes at large Stokes numbers.

As the Stokes number trends to infinite, the third term is vanished, and the granular conductivity is expressed

$$\kappa_{s,\infty} = \frac{25\rho_s d_p\sqrt{\pi\theta}}{96 g_0(1+e)[1+\frac{5}{4}(1-e)]}\left(1+\frac{6}{5}\alpha_s g_0(1+e)\right) + 2d_p\rho_s\alpha_s^2 g_0(1+e)\sqrt{\frac{\theta}{\pi}} \tag{4.104}$$

On the other hand, the granular conductivity from the classical kinetic theory of granular flow includes kinetic part and collisional part [14].

$$\kappa_{s,G} = \frac{25\rho_s d_p\sqrt{\pi\theta}}{64 g_0(1+e)}\left(1+\frac{6}{5}\alpha_s g_0(1+e)\right)^2 + 2d_p\rho_s\alpha_s^2 g_0(1+e)\sqrt{\frac{\theta}{\pi}} \tag{4.105}$$

Both second terms of the collisional part are the same, indicating both models predict the same granular thermal conductivity at high solid volume fractions. For the first term of kinetic part, the ratio is

$$\frac{\kappa_{s,G}}{\kappa_{s,\infty}} = \frac{3}{2}\left(1+\frac{5}{4}(1-e)\right)\left(1+\frac{6}{5}\alpha_s g_0(1+e)\right) \tag{4.106}$$

Both granular shear viscosity and granular conductivity determine the importance of the collisional components relative to the kinetic ones in granular stresses and heat fluxes. As solid volume fraction approaches zero, the collisional part is on the order of ${\alpha_s}^2$, while the kinetic part is on the order of α_s. Consequently, the kinetic components play a dominant role at low solid volume fractions. At high solid volume fractions, the collisional contributions become significant due to large values of g_0. In comparison, particle inertia affects particle–fluid turbulence interactions. If Stokes number approaches infinite, then both A_{ik} and B_{ik} of Eqs. (4.18) and (4.19) approach zero, and the derived constitutive relations become identical to those proposed by [14, 23] . As Stokes number trends to zero, the coefficient A_{ik} approaches toward T_{Lp} $\langle u_i' u_k' \rangle$, and the coefficient B_{ik} trends to $\langle u_i' u_k' \rangle$, indicating that fluid turbulence dominates particles motion.

4.5 Constitutive Relations of Low-Stokes-Number KTGF

For flow of fluid and particles with high Stokes number and low Reynolds number (e.g. gas-particle settling), the particle fluctuating motion is not affected by the fluctuations in the small-scale fluid turbulent eddies as a result of its high inertia, such that the particle travels with a minimal velocity change between successive interparticle collisions. For such flows, the particle fluctuating motion is governed by interparticle collisions and not significantly affected by fluid–particle interactions. For flow at the low Stokes number and high Reynolds number, however, the fluid–particle interaction influence on the particle fluctuating motion becomes important as a result of a non-linear increase in the dissipative fluctuating drag force. Thereby, the interstitial fluid effect on the particle collisional mechanism must be incorporated into the kinetic theory of granular flow.

From the low-Stokes-number kinetic theory of granular flow, the solid pressure is modelled in terms of solid volume fraction and granular temperature.

$$p_s = \rho_s\alpha_s(1 + 2(1 + e)\alpha_s g_0)\theta \tag{4.107}$$

The granular shear viscosity relates to Stokes number, frequency of collisions and strain rate of liquid and solid phases except that solid volume fraction and granular temperature.

$$\begin{aligned}\mu_s = {} & \frac{\rho_s\alpha_s\theta\tau_p}{2 + \nu_0\tau_p\varphi(e)} + \frac{4}{5}(1 + e)\alpha_s g_0\frac{\rho_s\alpha_s\theta\tau_p}{2 + \nu_0\tau_p\varphi(e)} + \frac{4}{5}(1 + e)\rho_s\alpha_s^2 g_0 d_p\sqrt{\frac{\theta}{\pi}} \\ & + \left(1 + \frac{4}{5}(1 + e)\alpha_s g_0\right)\frac{2\rho_s\alpha_s}{2 + \nu_0\tau_p\varphi(e)}\frac{\mu_t}{1 - \exp(-St^{-1})}\frac{S_{fik}}{S_{pik}}\end{aligned} \tag{4.108}$$

Similarly, we can obtain the constitutive equations for the solid phase shear viscosity using Eqs. (4.23) and (4.24) [5, 31]. These equations are similar to the second one-point one-time moments of the fluid velocity fluctuations in the Lagrangian equations for individual particle motion. With assistance of the formula (4.58) an alternative expression representing solid phase shear viscosity was derived and expressed as the function of Stokes numbers [5, 31].

The Stokes number is calculated according to the particle relaxation time and the time of interaction between particle motion and fluid fluctuations.

$$St = \frac{2\varepsilon_f\alpha_s\rho_s\sqrt{1 + 1.5C_\beta\left|\boldsymbol{v}_s - \boldsymbol{u}_f\right|^2/k_f}}{3C_\mu k_f\beta} \tag{4.109}$$

The rate of strain tensor of fluid phase and solid phase is of the following form according to fluid velocity u_f and solid phase velocity v_s.

$$S_{fik} = \frac{1}{2}\left(\frac{\partial u_{fi}}{\partial x_k} + \frac{\partial u_{fk}}{\partial x_i}\right) \tag{4.110}$$

and

$$S_{pik} = \frac{1}{2}\left(\frac{\partial v_{si}}{\partial x_k} + \frac{\partial v_{sk}}{\partial x_i}\right) \tag{4.111}$$

They represent the symmetric part of the deformation tensor of fluid and solid phases, and describes the effect of fluid and solid phases deformation tensor on solid shear viscosities.

The granular bulk viscosity and conductivity coefficient are given by according to the rates of strain tensor of fluid and solid phases and Stokes number.

$$\begin{aligned} \mu_b = \frac{4}{3}(1+e)\rho_s\alpha_s^2 g_0 d_p\sqrt{\frac{\theta}{\pi}} + \frac{2}{3}\left[1+\frac{4}{5}(1+e)\alpha_s g_0\right] \\ \frac{2\rho_s\alpha_s}{2+\nu_0\tau_p\varphi(e)}\frac{\mu_t}{1-\exp(-St^{-1})}\left(\frac{S_{fik}}{S_{pik}} - \frac{S_{fii}}{S_{pii}}\right) \end{aligned} \tag{4.112}$$

$$\begin{aligned} \kappa_s = \rho_s\alpha_s\left\{\frac{5\theta}{3+\nu_0\tau_p\phi(e)}\left[\frac{1}{2}+\frac{3}{5}\alpha_s g_0(1+e)\right]\tau_p + \alpha_s g_0(1+e)2d_p\sqrt{\frac{\theta}{\pi}}\right\} \\ -\frac{10\rho_s\alpha_s}{3+\nu_0\tau_p\phi(e)}\left[\frac{1}{2}+\frac{3}{5}\alpha_s g_0(1+e)\right][1-St(1-\exp(-St^{-1}))]\tau_{Lp}\mu_t S_{fik} \end{aligned} \tag{4.113}$$

Using Eqs. (4.23) and (4.24), the alternative expressions of solid phase bulk viscosity and granular conductivity can be derived, and they were expressed in Table (6.2) as the function of Stokes number and the rate of strain tensor of fluid and solid phases [5, 31].

The solid viscosity and conductivity relate to Stokes numbers. The Stokes number can be used to describe the behavior of particles in fluid flows based on their size, density and the fluid properties. A low Stokes number shows that the particle will adhere to the fluid flow closely, while a high Stokes number indicates that the particle will be less affected by the fluid flow and move more independently. For gas-particles flow, particle inertia could be significant even when the Reynolds number is small because the particle density is usually three orders of magnitude larger than the gas density. In this case, particles interact due to collisions or due to hydrodynamic interactions. On the contrary, for liquid-particle flow, particles with low Stokes number are characterized by low inertia and follow fluid streamlines closely. These indicate that the effect of the interstitial fluid is taken into account by incorporating the Stokes number in the solid viscosity and solid conductivity models.

An accurate description of the boundary conditions is necessary. The boundary conditions have to be specified at the wall for the fluid volume fraction, fluid velocity, and fluid phase turbulent kinetic energy and dissipation rate.

$$\frac{\partial \alpha_s}{\partial \boldsymbol{n}} = \frac{\partial \boldsymbol{u}_f}{\partial \boldsymbol{n}} = \frac{\partial k_f}{\partial \boldsymbol{n}} = \frac{\partial \varepsilon_f}{\partial \boldsymbol{n}} = 0 \tag{4.115}$$

where $\boldsymbol{n}$ is the direction normal to the wall.

For the solid phase mean velocity and granular temperature, the solid phase boundary conditions are expressed according to specularity coefficient ϕ [20]:

$$v_{sw} = -\frac{6\mu_{sw}\alpha_{s,max}}{\sqrt{3}\pi\phi\rho_s\alpha_s g_0\sqrt{\theta}} \left.\frac{\partial \boldsymbol{v}_s}{\partial \boldsymbol{n}}\right|_w \tag{4.116}$$

$$\theta_w = -\frac{k_{sw}\theta}{\gamma_w} \left.\frac{\partial \theta}{\partial \boldsymbol{n}}\right|_w + \frac{\sqrt{3}\pi\phi\rho_s\alpha_s v_{slip}^2 g_0 \theta^{3/2}}{6\alpha_{s,max}\gamma_w} \tag{4.117}$$

where v_{slip} is the slip velocity parallel to the wall. ϕ is the specularity coefficient between the particle and the wall. The expressions of granular thermal conductivity k_{sw} and viscosity μ_{sw} are deduced by means of Eqs. (4.108) and (4.113) in terms of particle–wall restitution coefficient e_w. The energy dissipation due to inelastic collision between the wall and particles is given by

$$\gamma_w = \frac{12(1 - e_w^2)\alpha_s^2\rho_s g_0 \theta}{d_p}\sqrt{\frac{\theta}{\pi}} \tag{4.118}$$

The dissipation of solids turbulent kinetic energy by collisions with the wall is specified by the particle–wall restitution coefficient e_w. A value of e_w close to unity implies low dissipation of granular energy at the wall. It is expected that the specularity coefficient and the particle–wall restitution coefficient need to be calibrated for a given gas-particle flow system.

4.6 Summary

The low-Stokes-number KTGF or low-density-ratio KTGF takes the effect of interstitial fluid on interparticle collisions into account because the interstitial fluid effect does not fully reveal in classical KTGF. The low-Stokes-number KTGF model is developed for inelastic particle–particle collisions using the Fokker–Planck operator for fluid turbulence-particle interactions. The Chapman-Enskog's solution is derived from the model of kinetic equation. The granular constitutive relations are expressed as a function of Stokes number. It has been generalized to granular flow taking interstitial fluid effect into account. Hence, a remarkable feature of the low-Stokes-number KTGF links to TFM for computational simulations of low density ratio fluid-particle multiphase flows.

Nomenclature

A, B	Function
$\boldsymbol{c}$	Instantaneous velocity of particles, m/s
$\boldsymbol{C}$	Fluctuation velocity of a particle, m/s
C_D	Drag coefficient
$C\mu$	Constant in k-ε model
d_p	Particle diameter, m
e	Coefficient of restitution
e_w	Coefficient of restitution at wall
f	Velocity distribution function of particles
$f^{(1)}$	Single distribution function
$f^{(2)}$	Pair distribution function
f^0	Normal distribution function
f^1	Distribution function for particle 1
f^2	Distribution function for particle 2
f_m	Gaussian distribution function
$\boldsymbol{F}$	Force, N
g	Gravitational acceleration, m/s^2
g_0	Radial distribution function
k	Turbulence kinetic energy, m^2/s^2
k_{gs}	Fluid-particle fluctuation kinetic energy, m^2/s^2
m	Mass, kg
p	Pressure, Pa
p_c	Collisional part of solid pressure, Pa
p_k	Kinetic part of solid pressure, Pa
q	Granular heat flux, kg/s^3
q_c	Collisional part of the flux of fluctuation energy, kg/s^3
q_k	Kinetic part of the flux of fluctuation energy, kg/s^3
Re	Reynolds number
Re_p	Particle Reynolds number
Re_T	Thermal Reynolds number
$\boldsymbol{r}$	Radial coordinate, m
$\boldsymbol{S}$	Strain rate tensor of fluid and solid phases, 1/s
St	Stokes number
St_T	Thermal Stokes number
t	Time, s
T_{Lp}	Fluid integral time scale along particle trajectories, s
$\boldsymbol{u}$, $\boldsymbol{v}$	Fluid and solid velocities, m/s
$\boldsymbol{u}'$, $\boldsymbol{v}'$	Fluid and solids velocity fluctuation, m/s
$u'_i v'_i$	Variance of solid and fluid velocity fluctuations, m^2/s^2
V	Volume, m^3
v_0	Collisional frequency, 1/s
$\boldsymbol{w}_c$	Collisional force, N/m^2

x, y, z Cartesian coordinates

Greek Symbols

θ	Granular temperature (random kinetic energy per unit mass), m^2/s^2
α_f	Fluid phase volume fraction
α_s	Solid phase volume fraction
β	Inter-phase drag function, kg/m^3s
γ	Dissipation rate of granular temperature due to collision, $kg/m{\cdot}s^3$
ε_f	Dissipation rate of turbulent kinetic energy, m^2/s^3
κ_s	Granular conductivity, kg/m.s
ϕ	Specularity coefficient
ρ_f	Fluid density, kg/m^3
ρ_s	Density of particles, kg/m^3
ρ_{sf}	Density ratio
μ_b	Solid bulk viscosity, kg/m·s
μ_S	Solid shear viscosity, kg/m·s
μ_t	Fluid dynamic turbulent viscosity, or eddy viscosity, kg/m s
v_f	Kinematic viscosity, m^2/s
τ_s	Stress tensor of solid phase, N/m^2
τ_{sij}	Solids phase effective stress tensor, N/m^2
τ_{fij}	Fluid phase effective stress tensor, N/m^2
τ_c	Particle–particle collision time, s
τ_f	Flow field timescale, s
τ_p	Particle response timescale, s

References

1. Agrawal, K., Loezos, P. N., Syamlal, M., & Sundaresan, S. (2001). The role of meso-scale structures in rapid gas-solid flows. *Journal of Fluid Mechanics, 445*, 151–185.
2. Alipchenkov, V. M., & Zaichik, L. I. (2000). Modeling of the motion of particles of arbitrary density in a turbulent flow on the basis of a kinetic equation for the probability density function. *Fluid Dynamics, 35*, 883–900.
3. Arastoopour, H., Gidaspow, D., & Abbasi, E. (2017). *Computational transport phenomena of fluid-particle systems*. Springer International Publishing.
4. Boelle, A., Balzer, G., & Simonin, O. (1995). Second-order prediction of the particle-phase stress tensor of inelastic spheres in simple shear dense suspensions. ASME, Gas-Particle Flows, FED -228, 9–18
5. Boxue, P. (2021). *Kinetic theory and numerical simulation of Non-Newtonian fluid particle flow*, Ph. D. Thesis, Harbin Institute of Technology.
6. Brey, J. J., Dufty, J. W., & Santos, A. (1997). Dissipative dynamics for hard spheres. *Journal of Statistical Physics, 87*, 1051–1066.

7. Chapman, S., & Cowling, T. G. (1970). *The mathematical theory of non-uniform gases* (3rd ed.). Cambridge.
8. Derevich, I. V. (2000). Statistical modelling of mass transfer in turbulent two-phase dispersed flows 1: Model development. *International Journal Heat and Mass transfer, 43*, 3709–3723.
9. Derevich, I. V., & Zaichik, L. I. (1988). Particle deposition from a turbulent flow. *Fluid Dynamics, 23*, 722–729.
10. Ding, J., & Gidaspow, D. (1990). A bubbling fluidization model using kinetic theory of granular Flow. *AIChE Journal, 36*, 523–538.
11. Fullmer, W. D., Liu, G., Yin, X., & Hrenya, C. M. (2017). Clustering instabilities in sedimenting fluid-solid systems: Critical assessment of kinetic-theory-based predictions using direct numerical simulation data. *Journal of Fluid Mechanics, 823*, 433–469.
12. Garzó, V., Tenneti, S., Subramaniam, S., & Hrenya, C. M. (2012). Enskog kinetic theory for monodisperse gas-solid flows. *Journal of Fluid Mechanics, 712*, 129–168.
13. Grad, H. (1949). On the kinetic theory of rarefied gases. *Communications on Pure and Applied Mathematics, 2*, 331–407.
14. Gidaspow, D. (1994). *Multiphase flow and fluidization: Continuum and kinetic theory descriptions*. Academic Press.
15. Hadinoto, K., & Curtis, J. S. (2004). Effect of interstitial fluid on particle-particle interactions in kinetic theory approach of dilute turbulent fluid-particle flow. *Industrial and Engineering Chemistry Research, 43*, 3604–3615.
16. Huilin, L., & Gidaspow, D. (2003). Hydrodynamics of binary fluidization in a riser: CFD simulation using two granular temperatures. *Chemical Engineering Science, 58*, 3777–3792.
17. Huilin, L., Gidaspow, D., & Shuyan, W. (2021). *Computational fluid dynamics theory of fluidization: Kinetic theory of granular flow and applications*. Springer International Publishing.
18. Jenkins, J. T., & Richman, M. W. (1985). Grad's 13-moment system for a dense gas of inelastic spheres. *Archive for Rational Mechanics and Analysis, 87*, 355–377.
19. Jenkins, J. T., & Savage, S. B. (1983). A theory for the rapid flow of identical, smooth, nearly elastic spherical particles. *Journal of Fluid Mechanics, 130*, 187–202.
20. Johnson, P. C., & Jackson, R. (1987). Frictional–collisional constitutive relations for granular materials with application to plane shearing. *Journal of Fluid Mechanics, 176*, 67–93.
21. Koch, D. L., & Sangani, A. S. (1999). Particle pressure and marginal stability limits for a homogeneous monodisperse gas-fluidized bed: Kinetic theory and numerical simulations. *Journal of Fluid Mechanics, 400*, 229–263.
22. Kui, L., Chenchen, Z., Yu, W., Zhiyong, P., Libo, L., Hui, J., & Liejin, G. (2024). Experimental investigation on the supercritical water gasification of coal by integrating the hydrogen oxidation process into a supercritical water fluidized bed. *Fuel, 355*, 129503.
23. Lun, C. K. K., Savage, S. B., Effery, D. J. J., & Chepurniy, N. (1984). Kinetic theories for granular flow: Inelastic particles in Couette flow and slightly inelastic particles in a general flow field. *Journal of Fluid Mechanics, 140*, 223–256.
24. Lun, C. K. K., & Savage, S. (2003). Kinetic theory for inertia flows of dilute turbulent gas-solids mixtures. In *Lecture Notes in Physics: Theory of Granular Gases*. Springer.
25. Ma, D., & Ahmadi, G. (1988). A kinetic model for rapid granular flows of nearly elastic particles including interstitial fluid effects. *Powder Technology, 56*, 191–207.
26. Pannala, S., Syamlal, M., & O'Brien, T. J. (2015). *Computational gas-solids flows and reacting systems: Theory, methods and practice*. Engineering Science Reference, Hershey.
27. Reeks, M. W. (1993). On the constitutive relations for dispersed particles in nonuniform flows. I: Dispersion in a simple shear flow. *Physics of Fluids A, 5*, 750–761.
28. Simonin, O., Fevrier, P., & Lavieville, J. (2002). On the spatial distribution of heavy-particle velocities in turbulent flow: From continuous field to particulate chaos. *Journal of Turbulence, 3*, 1–18.
29. Wang, G. Q., & Ni, J. R. (1991). The kinetic theory for dilute solid/liquid two-phase flow. *International Journal of Multiphase Flow, 17*, 273–281.

30. Wang, G. Q., Fu, X. D., Huang, Y. F., & Huan, G. D. (2008). Analysis of suspended sediment transport in open-channel flows: Kinetic-model-based simulation. *Journal of Hydraulic Engineering, 134*, 328–339.
31. Wenjian, C. (2023) *Numerical simulations of low density ratio particle-fluid two-phase fluidization characteristics*. Ph. D. Thesis, Harbin Institute of Technology.
32. Zaichik, L. I. (1999). A statistical model of particle transport and heat transfer in turbulent shear flows. *Physics of Fluids, 11*, 1521–1534.
33. Zaichik, L. I., Pershukov, V. A., Kozelev, M. V., & Vinberg, A. A. (1997). Modeling of dynamics, heat transfer and combustion in two-phase turbulent flows: 1. Isothermal flows. *Experimental Thermal and Fluid Science, 15*, 291–310.
34. Zhang, Y., & Reese, J. M. (2001). Particle-gas turbulence interactions in a kinetic theory approach to granular flows. *International Journal of Multiphase Flow, 27*, 1945–1964.

Chapter 5
Minimum Rate of Energy Dissipation and Interfacial Momentum Closure of Heterogeneous Flow Structures

Abstract This chapter deals with the effects of the presence of particle clusters in fluid-particle flow modeling and numerical simulation. This chapter also introduces key concepts and the fundamental derivations of two approaches for considering the presence of clusters: the subgrid model and the energy minimization model. The minimum rate of energy dissipation (MRED) is proposed as the criterion for the identification of clusters in the grids, called the grid stability condition. Using these approaches, the predicted spatio-temporal evolution of particle clusters is of particular interest in the design and scale-up of gas–solid circulating fluidized beds.

For design and scale-up of commercial fluidized bed reactors and gas–particle flow systems, a computationally feasible approach may be obtained from the averaged continuum equations of motion for both fluid and particles, which is often called Euler-Euler two-fluid model (TFM). The continuum approach generally relies on closures for the solids stresses and interfacial momentum closure. The former constitutive correlations are derived from kinetic theory of granular flow (KTGF) in the kinetic–collisional regime and from soil mechanics in the dense–frictional regime. The latter must develop constitutive relations to model gas–particle interaction forces. The heterogeneous structure of particle clusters has been found to be significant in the gas-particles circulating fluidized bed risers. It is well known that the mass transfer coefficients differ up to seven orders of magnitude due to the formation of particle clusters. Clusters affect particle slip velocity, pressure drop, local solid volume fraction, mixing and heat and mass transfer in risers.

Several heterogeneous drag models have been developed with the aim of accounting for the underlying inhomogeneity in solids concentration. Methods for developing these heterogeneous drag models include averaging fine-grid simulations with homogeneous drag models (e.g., filtered two fluid models), or accounting for non-homogeneous flow structures (clusters and bubbles) as a separate phase and empirically fitting results using homogeneous drag models for both the dilute phase and cluster phase to experimental data (e.g., the energy minimization multi-scale approach and cluster structure-dependent drag model). In the filtered two-fluid models (fTFM), The conservation equations of the fTFM are derived by applying

H. Lu et al., *Computational Transport Phenomena of Multiphase Systems and Fluidization*, Fluid Mechanics and Its Applications 127,
https://doi.org/10.1007/978-981-96-0698-6_5

space filtering operators, assuming the flow of heterogeneous structures dissolved into the spatially resolved parts and unresolved parts [1]. The resolved parts are modeled using kinetic theory-based TFM, while the unresolved parts are solved using filtered sub-grid constitutive models. The constitutive relations of the filtered drag force and solid stress proposed by [29] are solved as a function of filtered solid volume fraction, Froude number and dimensionless filter length based on the terminal settling velocity of particles [30, 42]. In the energy minimization multi-scale (EMMS) method [33] and cluster structure-dependent (CSD) drag model [44], the gas-particle mixture consists of the heterogeneous structures with the particle-rich dense phase as the form of clusters and the gas-rich dilute phase as the form of dispersed particles. The EMMS method assumes that the suspending and transporting energy is minimum, and is known as a stability condition [33]. The drag forces are solved by the set of equations with the stability condition. The drag coefficient is expressed by heterogeneity index which is correlated with solid volume fractions based on empirical correlation [53]. Based on the minimum rate of energy dissipation, the CSD drag model predicts drag coefficient of heterogeneous structures using a set of nonlinear conservation equations [8, 43]. The cluster diameter and velocities are solved from the equation set without additional empirical correlations. These heterogeneous drag models have been developed with the aim of accounting for the underlying inhomogeneity in gas–solid fluidized bed reactors.

5.1 Homogeneous Fluid-Particle Momentum Exchange Coefficient

The homogeneous drag models assume a homogeneous structure of gas and particles suspensions, i.e., negligible solids volume fraction gradients inside the computational grid cell. The available drag models are based either on correlations for the drag coefficient for one particle in a suspension or on the pressure drop per unit length in a suspension. The drag force on a particle in a suspension can be defined either in terms of the relative superficial gas velocity or the relative interstitial velocity. The homogeneous drag models include [16] A and B drag models [15], drag models [26, 48, 53], and they are used in different commercial computational fluid dynamic codes [2, 36]. Homogeneous fluid-particle drag models are physical models describing the momentum exchange via friction for a fluid flowing through a homogeneously distributed particles. In general, homogeneous drag relations must account for the presence of neighboring particles at all solids volume fractions and Reynolds numbers. Hence, the drag force depends on solids volume fraction and Reynolds number.

5.1.1 Gidaspow Drag Coefficient Model

The fluid-particle interaction closures apply in the momentum balances originate from the work on the Newton's second law for particles. For steady state and fully developed flow, a one-dimensional gas momentum balance with a constant cross-sectional area is [16]

$$\frac{d(\rho_g\alpha_g u_g^2)}{dz} = -\alpha_g\frac{dp}{dz} - \rho_g g\alpha_g - \beta_A(u_g - u_s) - \frac{4\tau_{wg}}{D_t} \tag{5.1}$$

where α_g and ρ_g represent the gas volume fraction and density. u_g and u_s denote the velocities of gas and solid phases. β is the gas–solid momentum exchange coefficient. The term on the left-hand side denotes the net rate gas momentum. The first and second terms on the right-hand side represent the rates of momentum due to fluid pressure and gravitational force. The third term shows the rate of momentum dissipation due to relative motion. The last term is the rate of momentum dissipation due to wall shear with the diameter of D_t and wall shear τ_{wg}. With no acceleration, wall friction and gravity, the gas momentum balance of Eq. (5.1) reduces to

$$-\alpha_g\frac{dp}{dz} - \beta_A(u_g - u_s) = 0 \tag{5.2}$$

A general correlation for the pressure drop per unit of length over a fixed bed is introduced by [10], who proposed that:

$$\frac{\Delta p}{\Delta z} = 150\frac{\alpha_s^2}{\alpha_g^3}\frac{\mu_g U_0}{d_p^2} + 1.75\frac{\alpha_s\rho_g U_0^2}{\alpha_g^3 d_p} \tag{5.3}$$

where U_0 is the superficial velocity $U_0 = \alpha_g(u_g - u_s)$, μ_g is the gas dynamic viscosity. The terms on the right-hand side denote the viscous energy loss and the inertial loss (kinetic energy). A comparison of Eqs. (5.2) and (5.3) shows that the exchange coefficient model A is [16]

$$\beta_A = 150\frac{\alpha_s^2\mu_g}{\alpha_g d_p^2} + 1.75\frac{\alpha_s\rho_g|\mathbf{u}_g - \mathbf{u}_s|}{d_p} \tag{5.4}$$

The drag coefficient β_A is inversely proportional to the square of the particle diameter for low flows.

Keep in mind that fluid pressure acts on both the fluid phase and the solid phase in Eq. (5.1). An alternative way to treat fluid pressure is to assume that it acts on fluid phase only. The one-dimensional gas momentum balance is also expressed as follows [16]:

$$\rho_g \alpha_g u_g \frac{du_g}{dz} = -\frac{dp}{dz} - \rho_g g \alpha_g - \beta_B(u_g - u_s) - \frac{4\tau_{wg}}{D_t} \tag{5.5}$$

The relation of drag coefficient can be obtained for the case of zero velocity gradient, zero gravity and zero wall friction. The drag coefficient β_B of model B is [16]

$$\beta_B = \frac{\beta_A}{\alpha_g} \tag{5.6}$$

For dilute flow of particles, Wen and Yu [53] have extended the works of [41] to derive an expression for pressure drop in particulate beds. They studied the voidage function

$$f(\alpha_g) = \frac{F_g - F_s}{F_{dr}} \tag{5.7}$$

where F_g is the gravitational force. F_s is the buoyancy force of fluid acting on the particle and F_{dr} is the drag force of the fluid acting to the particle. The drag force acting on a single particle in a suspension can be written as

$$F_{dr} = \frac{1}{2} C_D \rho_g |u_g - u_s| (u_g - u_s) \frac{\pi d_p^2}{4} \tag{5.8}$$

where C_D is the particle drag coefficient. The investigation was based on experiments done with various gas–solid and liquid–solid systems [53]. The velocity ranges from laminar to turbulent conditions. [53] also studied the minimum fluidization velocity with various gas–solid and liquid–solid systems to develop drag model. Using a correlation to fit data of [41], the momentum exchange coefficient becomes

$$\beta_A = \frac{3}{4} C_D \frac{\alpha_s \alpha_g \rho_g |\boldsymbol{u}_g - \boldsymbol{u}_s|}{d_p} f(\alpha_g) \tag{5.9}$$

$$C_D = \begin{cases} \frac{24}{Re_s}[1 + 0.15(Re_s)^{0.687}] & Re_s < 1000 \\ 0.44 & Re_s \geq 1000 \end{cases} \tag{5.10}$$

$$Re_s = \frac{\alpha_g \rho_g |\boldsymbol{u}_g - \boldsymbol{u}_s| d_p}{\mu_g} \tag{5.11}$$

In Eq. (5.9) the function $f(\alpha_g)$ shows the effect due to the presence of other particles in the fluid and acts as a correction to the Stokes law for free fall of a single particle. Wen and Yu [53] used both fluidized bed and hindered settling experiments to develop function model. Gidaspow and Ettehadieh [17] have used

$$f(\alpha_g) = \alpha_g^{-2.65} \tag{5.12}$$

From Eqs. (5.4) and (5.9), the drag coefficient is given by [16]:

$$\beta_A = \begin{cases} 150\frac{\alpha_s^2 \mu_g}{\alpha_g (\varphi d_p)^2} + 1.75\frac{\alpha_s \rho_g |\boldsymbol{u}_g - \boldsymbol{u}_s|}{(\varphi d_p)} & \alpha_g \le 0.8 \\ \frac{3}{4} C_D \frac{\alpha_s \alpha_g \rho_g |\boldsymbol{u}_g - \boldsymbol{u}_s|}{(\varphi d_p)} \alpha_g^{-2.65} & \alpha_g > 0.8 \end{cases} \tag{5.13}$$

where φ is the sphericity of particles. The Gidaspow drag model is a standard model which covers the whole range of gas volume fractions. The Ergun equation is used where the suspension is dense, whereas the formulation by Wen and Yu is used where the suspension is dilute. It should be also noted that the Gidaspow drag model takes into account the particle sphericity [2, 36].

5.1.2 Syamlal-O'Brien Drag Coefficient Model

A formula for the fluid-solids drag coefficient [47] is derived from a Richardson-Zaki type velocity-voidage correlation [41] according to the [13] equations. The gas–solid momentum exchange coefficient is expressed as a function of drag coefficient C_D [2, 36]

$$\beta_A = \frac{3}{4} \frac{\alpha_s \alpha_g \rho_g}{V_r^2 d_p} C_D |\boldsymbol{u}_g - \boldsymbol{u}_s| \tag{5.14}$$

where V_r is the ratio of the falling velocity of a suspension to the terminal velocity of a single particle. According to Garside and Al-Dibouni formula [13], the multiparticle drag coefficient C_D and its derivative with respect to voidage α_g are continuous, except at the voidage of 0.85 where C_D is continuous, but its derivative is discontinuous. The Syamlal-O'Brien drag model is derived for a single spherical particle in a fluid, and modified with a relative velocity correlation V_r [48]:

$$V_r = 0.5[A - 0.06Re + \sqrt{(0.06Re)^2 + 0.12Re(2B - A) + A^2}] \tag{5.15}$$

$$A = \alpha_g^{4.14} \tag{5.16}$$

$$B = \begin{cases} 0.8\alpha_g^{1.28} & \alpha_g \le 0.85 \\ \alpha_g^{2.65} & \alpha_g > 0.85 \end{cases} \tag{5.17}$$

The single-sphere drag function is

$$C_D = \left(\frac{0.63}{V_r} + \frac{4.8}{\sqrt{V_r Re}}\right)^2 \tag{5.18}$$

The relative velocity correlation V_r is the terminal settling velocity of a particle in a system divided by the terminal settling velocity of a single sphere. The main idea about the drag model is the assumption that the Archimedes number is the same in a single particle and a multiparticle system. The Archimedes number relates the gravitational forces to the viscous forces. This correlation has been used for both fluid-particle multiphase systems, and bubbling and circulating fluidized bed reactors.

As particles are aggregated due to the hydrodynamic interactions and collisional interactions, the momentum exchange between the gas and the particles is altered. The terminal velocity of a cluster exceeds the terminal velocity of a single particle. A correlation for the velocity ratio taking into account the clustering effect is given with a correction of solid mass fluxes.

$$\begin{aligned} V_r &= 0.5\left(A - 0.06Re + \sqrt{(0.06Re)^2 + 0.12Re(2B - A) + A^2}\right) \\ &\quad [1 + C_3 Re\alpha_g \exp(-0.005(Re - 5)^2 - 90(Re - 0.92)^2)] \end{aligned} \tag{5.19}$$

where C_3 is a coefficient, and determined from experimental measurements of circulating fluidized beds.

$$C_3 = \begin{cases} 250 & G_s = 98\ (\mathrm{kg/m^2\ s}) \\ 1500 & G_s = 147\ (\mathrm{kg/m^2\ s}) \end{cases} \tag{5.20}$$

where G_s is the particle recirculation mass flux.

This is a first drag model taking cluster effect into account. Replacing the velocity ratio V_r with the velocity ratio corrected for cluster effects in the model by Syamlal and O'Brien, yields a drag function that takes into account cluster effects. This effect can be seen as a local minimum in the drag function in the gas volume fraction interval 0.7–0.95.

5.1.3 Huilin-Gidaspow Drag Coefficient Model

The Huilin-Gidaspow drag model has extended to avoid switching between the Ergun equation and the Wen-Yu drag relation [26]. The difference between the two models lies in an equation that is used to combine the results of the Ergun and Wen-Yu drag models to blend or smooth the transition when the particle characteristics transition from dilute to dense regimes. The Huilin-Gidaspow drag model is given by [2, 26, 27, 36]

$$\beta_A = (1 - \phi)\beta_E + \phi\beta_{WY} \tag{5.21}$$

where the blending function, ϕ, is

$$\phi = \frac{\arctan[150 \times 1.75(0.2 - \alpha_s)]}{\pi} + 0.5 \tag{5.22}$$

$$\beta_E = 150\frac{\alpha_s^2 \mu_g}{\alpha_g d_p^2} + 1.75\frac{\alpha_s \rho_g \left|\boldsymbol{u}_g - \boldsymbol{u}_s\right|}{d_p} \tag{5.23}$$

$$\beta_{WY} = \frac{3}{4} C_D \frac{\alpha_s \alpha_g \rho_g \left|\boldsymbol{u}_g - \boldsymbol{u}_s\right|}{d_p} \alpha_g^{-2.65} \tag{5.24}$$

The Huilin-Gidaspow drag model shows that the continuum correlation is expressed as the function of Reynolds number and solid volume fractions based on the relative velocity between the fluid phase and solid phase.

5.1.4 Gibilaro Drag Coefficient Model

Gibilaro et al. [15] extended the applicability of Ergun equation to the dilute particle system by relating the energy dissipation in the fluidized bed with the unrecoverable pressure loss to obtain the particle drag force under the fully expanded condition. The pressure drops are correlated with solids volume fractions on the basis of theoretical considerations according to experimental measurements of the fixed bed. It is used to produce drag coefficient correlations for individual particles in the bed which in turn yield a general expression for the drag force on a particle in a fluidized suspension [2, 15, 36].

$$\beta_A = \left(\frac{18}{Re} + 0.33\right) \frac{\alpha_s \rho_g}{d_p} \left|\boldsymbol{u}_g - \boldsymbol{u}_s\right| \alpha_g^{-1.8} \tag{5.25}$$

with the Reynolds number as

$$Re = \frac{\alpha_g \rho_g \left|\boldsymbol{u}_g - \boldsymbol{u}_s\right| d_p}{\mu_g} \tag{5.26}$$

5.2 Particle-Resolved DNS-Based Drag Models

When a fluid flows through the bed with a constant velocity, a particle experiences two forces from the fluid, namely the drag force due to the fluid–solid friction at the surface of the spheres, and a force due to the static pressure gradient, which drives the fluid flow. Both the particle-resolved direct numerical simulation (DNS) and Lattice–Boltzmann method (LBM) have used to calculate the drag force exerted by a fluid flow on a collection of randomly dispersed, fixed particles. Such calculations, repeated for different values of the Reynolds number and solids volume fraction, can be used to derive a drag law. This constitutive law of two-fluid and discrete element models can be determined from first principles. Using DNS and LBM, the calculations were performed over a range of Reynolds numbers and solids volume fractions. The extensive numerical-experimental data constructed drag formulas to different ranges of Reynolds numbers and solids volume fractions. Also care is taken to make the formula go to the known limiting value of single sphere drag.

5.2.1 Hill-Koch-Ladd Drag Coefficient Model

The Hill-Koch-Ladd formulas were used as constitutive relations for gas–solids drag in multiphase flow and fluidization based on LB simulations [4, 22]. It is also referred to as the Koch-Hill drag model in [36]. The gas–solid momentum exchange coefficient β is defined as

$$\beta = 18\mu_g \alpha_s (1-\alpha_s)^2 \frac{F}{d_p^2} \tag{5.27}$$

where μ_g is the gas dynamic viscosity, α_s is the solids volume fraction, d_p is the diameter of particles, and F is the dimensionless drag force per particle. The dimensionless drag force is given as:

$$F = \begin{cases} 1 + (3/8)Re & \alpha_s \le 0.01 \text{ and } Re \le Re_1 \\ F_0 + F_2 Re^2 & \alpha_s > 0.01 \text{ and } Re \le Re_2 \\ F_1 + F_3 Re & (\alpha_s \le 0.01,\ Re > Re_1) \text{ and } (\alpha_s > 0.01,\ Re > Re_2) \end{cases} \tag{5.28}$$

where F_i is the coefficients, and they are defined as follows:

$$F_0 = (1-w)\left[\frac{1 + \sqrt{4.5\alpha_s} + 2.109\alpha_s \ln \alpha_s + 17.14\alpha_s}{1 + 0.681\alpha_s - 8.48\alpha_s^2 + 8.16\alpha_s^3}\right] + \frac{10w\alpha_s}{(1-\alpha_s)^3} \quad 0.01 < \alpha_s < 0.4 \tag{5.29}$$

$$F_1 = (1-w)\left[\frac{1 + \sqrt{4.5\alpha_s} + 2.109\alpha_s \ln \alpha_s + 17.89\alpha_s}{1 + 0.681\alpha_s - 11.03\alpha_s^2 + 15.41\alpha_s^3}\right] + \frac{10w\alpha_s}{(1-\alpha_s)^3} \quad 0.01 < \alpha_s < 0.4 \tag{5.30}$$

$$F_0 = F_1 = \frac{10\alpha_s}{(1-\alpha_s)^3} \quad \alpha_s > 0.4 \tag{5.31}$$

$$F_2 = \begin{cases} 0.025\sqrt{2/\alpha_s} & 0.01 < \alpha_s < 0.4 \\ 0.11 + 0.00051\exp(11.6\alpha_s) & \alpha_s > 0.4 \end{cases} \tag{5.32}$$

$$F_3 = \begin{cases} 0.03667 + 0.9351\alpha_s & 0.01 < \alpha_s < 0.4 \\ 0.0673 + 0.212\alpha_s + 0.0232/(1-\alpha_s)^5 & \alpha_s > 0.4 \end{cases} \tag{5.33}$$

The specific parameters, here they are referred to as Reynolds number Re_1 and Re_2, describe a transition from low Reynolds number to high Reynolds number. They are given as:

$$Re_1 = \frac{F_1 - 1}{(3/8 - F_3)} \tag{5.34}$$

$$Re_2 = \frac{F_3 + \sqrt{F_3^2 - 4F_2(F_0 - F_1)}}{2F_2} \tag{5.35}$$

The weighting factor w is introduced, and correlates to solid volume fraction. The expressions of the weighting factor and Reynolds number are given as:

$$w = e^{-10(0.4-\alpha_s)/\alpha_s} \tag{5.36}$$

$$Re = \frac{\rho_g(1-\alpha_s)\left|\boldsymbol{u}_g - \boldsymbol{u}_s\right|d_p}{2\mu_g} \tag{5.37}$$

From Eq. (5.28), the drag forces are varied with the change of solid volume fraction and Reynolds number. The drag law uses natural connectivity between the different functions at intersection points of Re_1 and Re_2. The weighting factor is used to obtain a smooth transition.

5.2.2 Beestra-van der Hoef-Kuipers Drag Coefficient Model

The Beestra-van der Hoef-Kuipers gas-solids drag model is based on numerical-experimental data using the Lattice-Boltzmann (LB) method [3]. It is also referred to as the BVK model [36]. The flow of gas around the particles is resolved at the surface of the particles. 35 different combinations of the Reynolds number up to 1000 and packing solid volume fractions were studied. The expression of dimensionless drag force, which takes into account the dependence on both the solid volume fraction and the Reynolds number, is given as:

$$F = \frac{10\alpha_s}{(1-\alpha_s)^2} + \frac{(1+1.5\sqrt{\alpha_s})}{(1-\alpha_s)^{-2}} + \frac{0.413Re}{24(1-\alpha_s)^2}\left[\frac{(1-\alpha_s)^{-1} + 3\alpha_s(1-\alpha_s) + 8.4Re^{-0.343}}{1+10^{3\alpha_s}Re^{-(0.5+2\alpha_s)}}\right] \tag{5.38}$$

The drag force clearly exhibits a more complex functional form than the linear scaling with Reynolds number. The first term on the right-hand side increases by increasing the solid volume fraction. The second term increases at first, and then decreases with increasing solid volume fractions. The latter part is larger than the former part at the solid volume fraction less than 0.1. However, it can be neglected at high solid volume fractions. These terms represent drag force contribution in the limit of Stokes flow at low Reynolds numbers. The last term depends on both solid volume fraction and Reynolds number at high Reynolds numbers.

$$Re = \frac{\rho_g(1-\alpha_s)\left|\boldsymbol{u}_g - \boldsymbol{u}_s\right|d_p}{\mu_g} \tag{5.39}$$

The last term describes the inertial contribution on drag force, and it has a clear maximum around $Re = 100$ at high solid volume fractions.

5.2.3 Holloway-Yin-Sundaresan Drag Coefficient Model

The Holloway-Yin-Sundaresan dimensionless average drag force has been performed using Lattice-Boltzmann simulations of low-Reynolds-number fluid flow [24]. It is also referred to as the HYS model [36]. The lubrication effect was added in Lattice-Boltzmann simulations because lubrication interactions between particles are very important [58]. When the separation between particles becomes smaller than the lattice spacing, the lubrication interaction between the particles can be resolved using a lubrication cut-off distance. For perfectly spherical, approaching particle pairs, the lubrication force would diverge near contact, in reality, such a divergence is removed by particle surface roughness. The lubrication force in the normal direction is the primary contribution to the hydrodynamic particle–particle interactions.

To account for the effects of moderate fluid inertia on the dimensionless fluid-particle drag force per particle, Holloway et al. [24] proposed the following expression

$$F = \left[\frac{10\alpha_s}{(1-\alpha_s)^2} + \frac{(1+1.5\sqrt{\alpha_s})}{(1-\alpha_s)^{-2}}\right]\left\{1 + \frac{0.413Re_m[(1-\alpha_s)^{-1} + 3\alpha_s(1-\alpha_s) + 8.4Re_m^{-0.343}]}{[240\alpha_s + 24(1-\alpha_s)^4(1+1.5\sqrt{\alpha_s})][1+10^{3\alpha_s}Re_m^{-(0.5+2\alpha_s)}]}\right\} \tag{5.40}$$

Here, Re_m is the Reynolds number, and defined as follows

$$Re_m = \frac{\rho_g(1-\alpha_s)|\Delta U|d_p}{\mu_g} \tag{5.41}$$

where ΔU is the difference between the local average velocities of the particle and fluid phases.

It is interesting to note that fluid inertia plays a non-negligible role, the fluid-particle drag force is not independent of fluctuating particle motions. Furthermore, the formula for the dimensionless drag force per particle of type i was proposed for polydisperse particles suspension [24]. The fluid-particle drag relation requires the specification of the lubrication cut-off distance, 1×10^{-6} m by default in [36].

5.2.4 Qi-Kuang-Qiu-Yu Drag Coefficient Model

The Qi-Kuang-Qiu-Yu fluid-particles momentum exchange coefficient is referred to as the QKQY model [36]. The drag model is based on about five hundred Lattice Boltzmann simulations [40]. The fluid rheology involves in power-law model to consider the shear-thinning, shear thickening and Newtonian behaviors of fluids. The fluid power-law viscosity model is adopted to describe the fluid rheological behaviors, and the power index n includes shear-thinning ($n < 1$), Newtonian ($n = 1$) and shear-thickening ($n > 1$) fluids. The mathematical form of drag force for fluid-particle systems was formulated taking into account the effect of volume fraction, fluid rheology and Reynolds number.

$$\beta = \frac{3}{4}C_D\frac{\alpha_s\rho_g}{d_p}\left|\boldsymbol{u}_g - \boldsymbol{u}_s\right|(1-\alpha_s)^{-\chi} \tag{5.42}$$

where C_D is the drag coefficient of a single particle in a power-law fluid flow, and is given by

$$C_D = \left(\frac{2.654}{n^{-1.9} + 3.213} + \frac{4.8}{n^{0.08}\sqrt{Re}}\right)^2 \tag{5.43}$$

When $n = 1$, Eq. (5.43) can be reduced to the correlation for Newtonian systems. χ represents the exponent factor used to estimate the increase of drag force for a particle in a multiparticle system. The equation for the exponent factor, considering the effects of Reynolds number, solid volume fraction and index, is given by

$$\chi = 2.65n(2-\alpha_s)^n - [5.3 - 3.5(1-\alpha_s)^n](1-\alpha_s)^2 \exp\left[-\frac{(\log Re - 1.5)^2}{2}\right] + \frac{2.65(2-\alpha_s)^{2-n} - 2.65n(2-\alpha_s)^n}{1+\exp[-2(\log Re - 1.5)]} \tag{5.44}$$

Note that several important factors have been taken account into Eq. (5.44). First, the exponent factor χ should be positive because the drag force on a particle should be larger than that in infinite space due to the presence of surrounded particles. Secondly, the exponent factor should be higher at a lower solid volume fraction for given power index and Reynolds number. Thirdly, when the power index increases, the factor should increase at low Reynolds numbers and decrease at high Reynolds numbers for a given solid volume fraction due to the dependence of viscosity of non-Newtonian fluid on flow condition. Finally, the value of the exponent factor largely remains uniform for low Reynolds number and high Reynolds number flows.

5.2.5 Tenneti-Garg-Subramaniam Drag Model

Using the Lattice-Boltzmann method, a spherical particle is represented by a stair-step lattice approximation. Thus, the exact value of the particle diameter is not known. The drag values obtained from LB simulations are assumed to correspond to an effective hydrodynamic diameter. In order to specify a closure for the inter-phase momentum transfer term, the particle-resolved direct numerical simulation (DNS) can be applied to study interphase momentum transfer in flow past fixed random assemblies of monodisperse spheres using a Navier–Stokes equation [49]. The method is based on the particle-resolved uncontaminated-fluid reconcilable immersed boundary model. The principal advantage of the method is that the fluid stress at the particle surface is calculated directly from the flow solution (velocity and pressure fields) when integrated over the surfaces of all particles yields the average fluid-particle force.

Using DNS, the local profiles of pressure and viscous forces were predicted along the surface of the sphere. Based on the normalized force values obtained from DNS simulations, [49] proposed a correlation for the average fluid–particle force in random arrays of monodisperse spheres. It is also referred to as the TGS model [36]. The dimensionless drag force is given as:

$$F = \frac{1 + 0.15[(1 - \alpha_s)\mathrm{Re}_p]^{0.687}}{(1 - \alpha_s)^3} + \left[\frac{5.81\alpha_s}{(1 - \alpha_s)^3} + \frac{0.48\alpha_s^{1/3}}{(1 - \alpha_s)^4}\right] + \alpha_s^3 \mathrm{Re}_p \left(0.95 + \frac{0.61\alpha_s^3}{(1 - \alpha_s)^2}\right) \tag{5.45}$$

The first term on the right-hand side represents the drag force acting on an isolated sphere moving in an unbounded fluid. The second term denotes the solid volume fraction effect on drag force. The last term relates to solid volume fraction and Reynolds number.

$$\mathrm{Re}_p = \frac{|u_g - u_s|\rho_g(1 - \alpha_s)d_p}{\mu_g} \tag{5.46}$$

The Reynolds number is expressed as a function of the magnitude of the mean slip velocity for flow past homogeneous particle assemblies.

Note that, following [22], Eq. (5.37) for Reynolds number is based on particle radius, rather than particle diameter in Eqs. (5.39) and (5.46) used in dimensionless drag force. A detailed explanation of the DNS-based and LBM-based dimensionless drag force and momentum exchange coefficient can be found in papers by [3, 22, 24, 36, 40, 49, 58].

5.3 Filtered or Subgrid Exchange Coefficient Model

The filtered two-fluid models (fTFM) utilizes multiscale modelling to capture the effects of dynamic particle structures of clusters and bubbles on the transport phenomena inside the fluidized beds. This approach is similar to large-eddy simulations of single-phase turbulent flows. Starting from a kinetic theory based two-fluid model for gas-particle flows, the fTFM was proposed to construct the heterogeneous momentum exchange correlations [29]. The fTFM equations are coarse-grained through a filtering operation that considers to spatial averaging over filter length scale. In these filtered equations, the consequences of the flow structures occurring on a scale smaller than a chosen filter size appear through residual correlations. If the assumptions innate to the filtering methodology hold true, the filtered equations should produce a solution with the same macroscopic features as the resolved kinetic theory model solution.

The filtered approach aims to find an appropriate constitutive relation for the unresolved terms appearing due to filtering of the balance equations for flows of clusters and streamers. According to the method proposed by Igci and Sundaresan [29], the filtered particle volume fraction can be defined as $\alpha_s(\boldsymbol{x}, t)$ denotes the particle volume fraction at location z and time t.

$$\overline{\alpha}_s(\boldsymbol{x}, t) = \int_V G(\boldsymbol{x} - z)\alpha_s(z, t)dz \tag{5.47}$$

where $G(\boldsymbol{x}–z)$ is a weight function that depends on $\boldsymbol{x}$ and z, and V denotes the region over which the gas–particle flow occurs. The weight function satisfies

$$\int G(\boldsymbol{x} - z)dz = 1.0 \tag{5.48}$$

By choosing how rapidly $G(\boldsymbol{x}–z)$ decays with distance measured from $\boldsymbol{x}$, one can change the filter size. The box filter kernel is defined as

$$G(x, z) = \begin{cases} \Delta_f^{-3} & |x - z| \le \Delta_f/2 \\ 0 & \text{otherwise} \end{cases} \tag{5.49}$$

where Δ_f is the filter size. Various markers, such as volume fraction and slip velocity, are employed to classify the sub-filter scale state and averaged to obtain statistics of the filtered quantities. Igci and Sundaresan [30] defined the fluctuation in particle volume fraction as

$$\alpha_s' = \alpha_s - \overline{\alpha}_s \tag{5.50}$$

The filtered phase velocities are defined according to

$$\overline{u}_s(\boldsymbol{x}, t) = \frac{1}{\overline{\alpha}_s(\boldsymbol{x}, t)} \int_V G(\boldsymbol{x} - z)\alpha_s(z, t)u_s(z, t)dz \tag{5.51}$$

$$\overline{u}_g(\boldsymbol{x}, t) = \frac{1}{\overline{\alpha}_g(\boldsymbol{x}, t)} \int_V G(\boldsymbol{x} - z)\alpha_g(z, t)u_g(z, t)dz \tag{5.52}$$

where u_g and u_s denote the local gas phase and solid phase velocities appearing in the microscopic TFM. Applying the filtering operation on the microscopic mass and momentum conservation equations yields the filtered conservation equations. Thus, the filtered two fluid model (fTFM) equations have been deduced by filtering the TFM equations using a filter. The filtered conservation equations are similar in form to their unfiltered counterparts, but contain several residual correlations requiring closure relations, including the solid phase filtered pressure and viscosity and the filtered drag force between the gas phase and the solid phase. The fTFM equations are more suitable for numerical integration at a coarser mesh resolution; however, because of lack of the constitutive models to be employed in conjunction with fTFM it can account for the unresolved subfilter-scale structures through scale-dependent subfilter models. Solving the fTFM numerically, it is commonly assumed that the filter size is the same as the grid size. Based on the Wen and Yu drag coefficient correlation, the algebraic form for the filtered drag coefficient is expressed as below.

$$\beta_{fil} = (1 - c)\frac{3}{4}C_D \frac{\rho_g(1 - \alpha_s)\alpha_s\left|\boldsymbol{u}_s - \boldsymbol{u}_g\right|}{d_p}(1 - \alpha_s)^{-2.65} \tag{5.53}$$

where c is the factor used in the filtered drag coefficient. The quantity c represents the contributions of captured heterogeneous structures on drag forces. It is known as the Igci drag model [36]. Thus, the factor c is most significant in the filtered model, and depends on dimensionless Froude number.

$$c = \frac{h_f Fr_f^{-1.6}}{Fr_f^{-1.6} + 0.4} gg \tag{5.54}$$

where the Froude number Fr_f is based on the filter size, and is defined as

$$Fr_f = \frac{u_t^2}{g\Delta_f} \tag{5.55}$$

where u_t is the terminal settling velocity of particles. The terminal velocity of particles is calculated by:

$$u_t = \frac{gd_p^2(\rho_s - \rho_g)}{18\mu_g} \tag{5.56}$$

For the particle diameter and density of 75 μm and 1500 kg/m^3 with the coefficient of restitution of 0.9, the correlations of the coefficients h_f and gg are expressed as the function of solid volume fractions.

$$gg = \begin{cases} \alpha_s^{0.24}(1.48 + \exp(-18\alpha_s)) & \alpha_s < 0.18 \\ 1 \quad \alpha_s \geq 0.18 & \end{cases} \tag{5.57}$$

$$h_f = \begin{cases} 2.7\alpha_s^{0.234} & \alpha_s < 0.0012 \\ -0.019\alpha_s^{-0.455} + 0.963 & 0.0012 \leq \alpha_s < 0.14 \\ 0.868\exp(-0.38\alpha_s) - 0.176\exp(-119.2\alpha_s) & 0.0014 \leq \alpha_s < 0.25 \\ -4.59 \times 10^{-5}\exp(19.75\alpha_s) + 0.852\exp(-0.268\alpha_s) & 0.25 \leq \alpha_s < 0.455 \\ (\alpha_s - 0.59)(-1501\alpha_s^3 + 2203\alpha_s^2 - 1504\alpha_s + 162) & 0.455 \leq \alpha_s \leq 0.59 \end{cases} \tag{5.58}$$

These correlations indicate that the dimensionless parameter c depends on Froude number and solid volume fraction. Obviously, $c = 0$ corresponds to a homogeneous flow. The larger the value of c, the more inhomogeneous the subfilter-scale structure in gas–solid risers.

Three markers, dimensionless Froude number, solid volume fraction and dimensionless slip velocity, are applied to describe cluster effect on momentum exchange between the gas phase and solid phase in gas–solid circulating fluidized beds. The filter size Δ_f is a key parameter in the coarse-grained closures. Additionally, the filtered particle volume fraction α_s emerges as an important marker. Furthermore, the formation and breaking of particles clusters associate with local gas and solid velocities, indicating the filtered drag coefficient depends on filtered slip velocity. The dimensionless slip velocity is written as

$$u_{gs} = \frac{u_g - u_s}{u_t} \tag{5.59}$$

The filtered drag model proposed by [42] takes the effect of slip velocity on the drag coefficient into account. It is also referred to as SARKAR drag model [36]. The correlation of the factor c is expressed according to three markers of the solid volume fraction, filter size and dimensionless slip velocity.

$$c = \min\left|\left(0.9506 + \frac{0.1708}{|u_{gs}|}\right)\alpha_s^b,\ 0.97\right| \tag{5.60}$$

$$b = 0.049\left(\frac{1}{Fr_f} - 1\right) + \frac{0.3358}{|u_{gs}|} \tag{5.61}$$

It is readily apparent that the factor c varies over a wide range. The factor increases at low solid volume fractions, and reduces at high solid volume fractions. Obviously, the values of the factor c increase with dimensionless slip velocity; i.e., the filtered drag coefficient is smaller at large slip velocities. The expression for the factor c was derived using a sufficiently large filter length, and are applicable to the large-filter-size limit [42]. Therefore, the grid size used in coarse-grid simulations should be sufficiently large to maintain the validity of these large-filter-size-limit expressions.

For Geldart A particles according to the classification [14], the interparticle attractive forces often appear. Because of cohesion, Geldart A particles would form clusters, which in turn reduces the drag force acting on a single particle. The transition from bubbling fluidization to turbulent fluidization was found by plotting of the pressure fluctuation versus the superficial gas velocity. Turbulent fluidization is characterized by two coexisting regions: a bottom dense bubbling region and a dilute top dispersed flow region. Thus, both bubbles and clusters exist in this regime. To better simulate drag force of flow of gas-Geldart A particles suspensions, the Wen-Yu drag model is selected as the base case. The heterogeneity index (normalized drag force) was defined [12], and a filtered gas–solid drag model was constructed using highly-resolved two-fluid model simulations [42]. It is also referred to as GAO drag model [36]:

$$H = \frac{\beta}{\beta_{WY}} = \begin{cases} 1 - 0.95\left[1 - e^{-m(u_{gs}-u_{gs0})^n}\right] & u_{gs} > u_{gs0} \\ 1 & u_{gs} \le u_{gs0} \end{cases} \tag{5.62}$$

where m and n are coefficients, and they relate to solid volume fractions and critical dimensionless slip velocity.

$$m = \frac{(0.755978 + 2.739315\alpha_s - 5.6019645\alpha_s^2 - 1.658538\alpha_s^3 + 16.702992\alpha_s^4)}{1 + e^{100(\alpha_s - 0.55)}} \times \left(1 - \frac{0.44145335}{\Delta_{filter}} + \frac{0.18195034}{(\Delta_{filter})^2}\right)\left(1 - \frac{0.01827347}{u_{gs}^2}\right)\left(1 - e^{-300\alpha_s}\right) \tag{5.63}$$

$$n = (0.47750002 + 0.06279418\alpha_s + 5.13011673\alpha_s^2) \times \left(1 + \frac{0.67680355}{\Delta_{filter}} - \frac{0.54535726}{(\Delta_{filter})^2}\right) \tag{5.64}$$

The critical dimensionless velocity u_{gs0} represents a standard to distinguish between homogeneous flow and heterogeneous flow in the grids. These mean that the gas-particles flow structure in the grid maybe is homogeneous or heterogeneous, depending on the standard. The critical dimensionless velocity is given by:

$$u_{gs0} = \frac{0.28441799 - 1.94357377\alpha_s}{0.01 + \alpha_s^{0.22177961}}\left(1 + \frac{0.3117589}{\Delta_{filter}} - \frac{0.1597196}{(\Delta_{filter})^2}\right) \tag{5.65}$$

where Δ_{filter} is a dimensionless filter size.

$$\Delta_{filter} = \max(Fr_f^{-1}, 0.5) \tag{5.66}$$

The heterogeneity index approaches value of 1.0 at both the maximum and minimum gas volume fraction. It is also found that the heterogeneity index approach 1.0 as the dimensionless slip velocity is less than the critical dimensionless velocity. For these extreme conditions, the gas–solid suspension becomes homogeneous and no sub-grid correlations are needed.

From Eq. (5.62) for gas–solid circulating fluidized beds, we may propose the following threshold to distinguish from the heterogeneous flow and the homogenous flow in the grid.

$$S_c = \frac{u_{gs} - u_{gs0}}{u_{gs0}} = \begin{cases} +0 & \text{Heterogeneous} \\ -0 & \text{Homogeneous} \end{cases} \tag{5.67}$$

This maybe constitute a criterion for stability condition used in the grid of the fTFM. Homogeneous flow of gas-particles suspension occurs when the stability function S_c is negative. However, heterogeneous flow with clusters exists at $S_c > 0$ in the grid, suggesting that the determination of heterogeneity index depends on the grid stability condition. It is anticipated that this grid stability condition provides reliable predictions of flow structures using fTFM in gas-particle circulating fluidized beds.

The remarkable advantages of the fTFM approach are to construct constitutive correlations of the filtered solid phase pressure, filtered solid phase viscosity and filtered reaction rate. The filtered solid phase pressure includes the contributions from the streaming and collisional part captured by the kinetic theory (i.e., filtered kinetic theory pressure $p_{s,k}$) and the subfilter-scale Reynolds-stress-like velocity fluctuations (i.e. subfilter-scale solid pressure $p_{s,fil}$). The filtered solid phase pressure is modelled as follows [30]:

$$p_s = \begin{cases} p_{s,k} + F_{ps}[(\alpha_s - 0.59)(-1.69\alpha_s - 4.61\alpha_s^2 + 11\alpha_s^3)] & \alpha_s \le 0.59 \\ 13.8 - 50.3\alpha_s + 49.5\alpha_s^2 - 6.62\alpha_s^3 & \alpha_s > 0.59 \end{cases} \tag{5.68}$$

$$F_{ps} = 0.48\left(\frac{\rho_s u_t^2}{Fr_f^{0.86}}\right)\left[1 - \exp\left(-\frac{Fr_f^{-1}}{1.4}\right)\right] \tag{5.69}$$

The second term on the right-hand side of Eq. (5.68) represents the subfilter-scale solid pressure $p_{s,fil}$ at α_s less than 0.59. The filtered solid phase pressure resulting from subgrid motions on the scale of the particles is modelled from the filtered kinetic solid pressure [30].

$$\frac{p_{s,k}}{\rho_s u_t^2} = \begin{cases} 0.31\alpha_s - 10.4\alpha_s^2 & \alpha_s \le 0.0131 \\ 0.00232 - 0.000183\alpha_s + 0.066\alpha_s^2 - 0.185\alpha_s^3 & 0.0131 < \alpha_s \le 0.29 \\ 0.00615 - 0.00978\alpha_s & 0.029 < \alpha_s \le 0.595 \end{cases} \tag{5.70}$$

It can be seen that the subfilter-scale solid pressure $p_{s,fil}$ is much larger than the filtered kinetic theory pressure $p_{s,k}$, indicating that one can ignore the kinetic theory contributions to the solid phase pressure. Thus, it approaches the filtered kinetic theory pressure as Fr_f trends to infinity. Within the filtered approach, the filtered kinetic theory pressure is not very important. The filtered solid phase pressure can become three orders of magnitude greater than the filtered kinetic theory pressure.

The solid phase filtered shear viscosity μ_s consists of the filtered kinetic theory viscosity $\mu_{s,k}$ due to the streaming and collision and the contribution of subfilter-scale solid viscosity $\mu_{s,fil}$ due to the subfilter-scale Reynolds-stress-like velocity fluctuations [30].

$$\mu_s = \begin{cases} \mu_{s,k} + \mu_{s,fil} & \alpha_s \le 0.59 \\ \mu_{s,k} & \alpha_s > 0.59 \end{cases} \tag{5.71}$$

$$\mu_{s,fil} = \frac{0.37\left[(\alpha_s - 0.59)\left(-1.22\alpha_s - 0.7\alpha_s^2 - 2\alpha_s^3\right)\right]}{\left(1 + 0.28Fr_f^{-0.43}\right)}\left(\frac{\rho_s u_t^3}{Fr_f^{1.22} g}\right) \tag{5.72}$$

It can be seen that the contributions resulting from the subfilter-scale velocity fluctuations dominate at low solid volume fractions, and the subfilter-scale solid viscosity increases appreciably with filter length. At higher solid volume fractions, the subfilter-scale velocity fluctuations disappear and only the filtered kinetic theory viscosity remains; as a result, the contributions arising from the mesoscale fluctuations are essentially negligible at $\alpha_s > 0.59$. The filtered kinetic theory viscosity becomes independent of the filter length [30].

$$\frac{\mu_{s,k}g}{\rho_s u_t^3} = \begin{cases} 0.00254 - 0.207\alpha_s + 9.81\alpha_s^2 - 215\alpha_s^3 + 1720\alpha_s^4 & \alpha_s \le 0.02 \\ 0.00136 - 0.0296\alpha_s + 0.329\alpha_s^2 - 1.55\alpha_s^3 + 2.72\alpha_s^4 & 0.02 < \alpha_s \le 0.2 \\ 0.000335 - 0.0005\alpha_s + 0.0107\alpha_s^2 - 0.0128\alpha_s^3 & 0.2 < \alpha_s \le 0.6095 \end{cases} \tag{5.73}$$

The dimensionless filtered kinetic pressure and shear viscosity depends on filtered solid volume fractions. The filtered kinetic pressure increases at first, reach maximum at the solid volume fraction of 0.25, and then decreases with increasing solid volume fractions. On the contrary, the filtered kinetic shear viscosity decreases sharply at small solid volume fractions. It reaches minimum at the solid volume fraction of 0.1, and then increases with increasing solid volume fractions. The contributions of the filtered kinetic theory pressure and kinetic theory viscosity to these filtered quantities are negligibly small and so, one need not include the filtered granular energy equation in simulations of gas–solid risers.

The alternative filtered solid pressure and shear viscosity proposed by [42] are expressed as:

$$p_s = \left[\Delta_f^2 \frac{0.01797\alpha_s^{1.645}}{\alpha_{s.\max} - \alpha_s} + \left(\frac{g}{u_t^2}\right)^{3/7} \Delta_f^{17/7} \frac{0.0236\alpha_s^{1.115}}{\alpha_{s.\max} - \alpha_s}\right] \rho_s S_s^2 \tag{5.74}$$

$$\mu_s = \left[\left(\frac{g}{u_t^2}\right)^{-6/7} \Delta_f^{8/7} \frac{0.00307\alpha_s^{1.544}}{\alpha_{s.\max} - \alpha_s} + \left(\frac{g}{u_t^2}\right)^{-2/7} \Delta_f^{12/7} \frac{0.02518\alpha_s^{1.123}}{\alpha_{s.\max} - \alpha_s}\right] \rho_s S_{s,i\neq j} \tag{5.75}$$

where S_s is the solid phase filtered rates of strain. The solid phase pressure and shear viscosity are comprised of two components: the filtered microscopic part and the filtered meso-scale part. The filtered meso-scale pressure of the solid phase represents a Reynolds-stress-like contribution, which arises from sub-filter-scale velocity fluctuations. At the packing limit, the filtered microscopic component diverges whereas the filtered mesoscopic component approaches zero and, therefore, separate closures for the two components would be more desirable. It is noteworthy that the filtered solid phase rate-of-strain has been used to constitute the filtered pressure and viscosity, similar to the sub-grid stress models for single-phase turbulent flows in which the fluid phase sub-grid stress is constituted using the scalar filtered rate-of-strain [39].

For reacting gas–particle flows, the presence of particle clusters leads to an effective reaction rate that is substantially smaller than what would be predicted via coarse-grid simulation. These indicate that the filtered models for effective reaction rates are necessary in order to accurately perform continuum model simulations of reacting gas–particle flows on coarse spatial grids. Without these corrections, coarse-grid continuum model simulations will consistently over-estimate the conversion in a gas–solid reaction process. The ratio of the reaction rate in the fine-grid simulation to that in the coarse-grid simulation is known as the cluster-scale effectiveness factor [23].

$$\eta = \frac{R_{fil}}{R_{hom}} \tag{5.76}$$

where R_{fil} and R_{hom} are the filtered reaction rate and homogeneous reaction rate. For a first order, solid-catalyzed, reaction in a gas–particle reacting flow, the filtered cluster-scale effectiveness factor is given according to solid volume fraction α_s.

$$\eta = \begin{cases} 1 - B\left(\frac{1-(1-\alpha_s/\alpha_{sc})^3}{1+a_1(1-\alpha_s/\alpha_{sc})^3}\right) + A & \alpha_s < \alpha_{sc} \\ 1 - B\left(\frac{(1+a_2)(1-(\alpha_s-\alpha_{sc})/(\alpha_c-\alpha_{sc}))^2}{1+a_2(1-(\alpha_s-\alpha_{sc})/(\alpha_c-\alpha_{sc}))^2}\right) + A & \alpha_{sc} \le \alpha_s < \alpha_c \end{cases} \tag{5.77}$$

where α_{sc} and α_c are the piecewise function about filtered solid volume fraction and constant, respectively. They are recommended the use of 0.42 and 0.59. a_1, a_2, A and B are the coefficients, and they are strong function of filter size Δ_f and terminal settling velocity u_t of an isolated particle.

$$a_1 = \frac{2.76}{\exp(17.2\Phi^2 g\Delta_f/(u_t^2 Sc))} \tag{5.78}$$

$$a_2 = 10.4 + 265\Phi^2 g\Delta_f/(u_t^2 Sc) \tag{5.79}$$

$$A = \frac{(\Delta_f^2/12)\nabla\alpha_s \cdot \nabla\chi_{gA}}{\alpha_s \chi_{gA}} \tag{5.80}$$

$$B = \frac{4.4\sqrt{\Phi^2 g\Delta_f/u_t^2}}{1 + 4.4\sqrt{\Phi^2 g\Delta_f/u_t^2}} \tag{5.81}$$

where χ_{gA} is the mass fraction of species A in the gas phase. Sc and Φ represent the meso-scale Schmidt number and meso-scale Thiele modulus.

$$Sc = \frac{\mu_g}{\rho_g D} \tag{5.82}$$

$$\Phi = \sqrt{kd_p^2/D} \tag{5.83}$$

where D and k represent the diffusivity and reaction rate constant of reacting species. The cluster-scale effectiveness factor has been characterized as a function of Thiele modulus, meso-scale Schmidt number, and filter-size over an order of magnitude range. The profile of cluster-scale effectiveness factor is observed to retain an inverted bell shaped dependence on solid volume fractions in the low and high solid volume fraction limits. At intermediate solid volume fractions a decrease in the cluster-scale effectiveness factor from unity is observed with the magnitude of this reduction being a strong function of meso-scale Thiele modulus and dimensionless filter size with only a weak dependence on Schmidt number. This indicates that the cluster-scale effectiveness factor is substantially smaller than 1.0.

The fTFM-filtered drag model represents a type of the data-driven multiscale modelling method, and aims to resolve the coarse flow structures and macroscale performances which are usually of most interest in practical applications. The effects of unresolved subfilter-scale structures appear in the filtered models through residual correlations, which must be constituted in terms of the filtered variables from three-dimensional CFD-KTGF simulations. The unique features and advantages of the fTFM model are that the approach links the mesoscale structure hydrodynamics to the macroscale performance in the circulating fluidized bed reactors. The radial distribution function is not used in the fTFM.

The fTFM models are the first to take into account the effect of clusters in the solid phase stress and drag force. They account for the effects of microscale structures and resolve macroscale structures in simulations with the focus on probing macroscale gas–particle flow features that are of principal interest in large-scale gas–solid multiphase flows. For more detailed explanation for fTFM constitutive relation expressions, see [12, 29, 38, 42, 46], Fluent theory guide [2] and MFIX [36].

5.4 Energy Minimization Multi-scale Drag Model

The energy minimization multi-scale (EMMS) model was proposed by [32] based on the concept of the coexistence of both dense region and dilute region in a riser. Each region contains its own flow parameters, e.g., superficial gas velocity, superficial particle velocity and gas voidage of the dilute and dense regions. The dense region consists of spherical clusters with volume fraction f and diameter d_c. The dilute region represents disperse particles with low solid volume fraction. The compromise between dominant mechanisms leads to the disparity of structures at different scales and spatiotemporal multiscale behaviors. The multiscale gas–solid interactions can be quantified by the drag coefficients corresponding to the three subsystems of the dilute region and dense region and the interface. The three scale interactions include the micro-scale interaction between gas and solid which is correlated with the size of the constituent particles and is present in both the dense and dilute regions, meso-scale interaction between the dense clusters and dilute region with the size of clusters, and macro-scale interaction between the global gas–solid system and its boundaries. Each dominant mechanism follows its own rule. To understand multi-scale structures and the interaction between different mechanisms, it is necessary to distinguish both dominant mechanisms and scale. Eight parameters are needed to describe the state of a system, among them five parameters describe the dense region, and three parameters explain the dilute one. These parameters are found according to the minimization of the mass-specific energy consumption for suspending and transporting of gas-particles mixture. Therefore, the EMMS method assumes that the suspending and transporting energy is minimum, and is known as a stability condition [33, 51, 52, 57].

From EMMS method, the energy consumption in a gas–solid fluidization system can be resolved into a suspension and transport subsystem and an energy dissipation subsystem. That is, the total energy N_{TE} is expressed as power per unit mass of solids, and it consists of two parts; one describes the energy consumption for suspending and transporting particles N_{st}, and the other includes the energy consumption N_d for particle collision, circulation and acceleration. N_{st} can be further divided into dissipative energies for particle suspension N_s and transport N_t [32]

$$N_{\mathrm{TE}} = N_{st} + N_d = (N_s + N_t) + N_d \tag{5.84}$$

The stability condition is formulated as a mutually constrained extremum; that is, minimization of the energy consumed to suspend and transport particles per unit mass as N_s and N_d are neglected.

$$N_{st} = N_t \rightarrow \min \tag{5.85}$$

This is known as the stability condition for heterogeneous regimes in fluid-particles fluidized beds.

Accordingly, the criterion for the stability condition of heterogeneous flows could be physically expressed as [32]:

$$N_{st} = \frac{\rho_s - \rho_g}{\rho_s} g \left\{ U_g + (fU_c - U_g) \frac{(\alpha_d - \alpha_c) f^2}{1 - \alpha_d + (\alpha_d - \alpha_c) f} \right\} \rightarrow \min \tag{5.86}$$

where α_c and α_d are the gas volume fractions of the dense region and dilute region. U_c is the superficial gas velocities in the dense region, f is the volume fraction of the clusters, and U_g is the superficial gas velocity. From Eq. (5.85), the stability criterion expresses the minimization of the mass-specific power for suspending and transporting particles based on superficial gas velocity U_g. Thus, it maybe refers to as the overall stability condition of gas–solid fluidized bed risers.

The EMMS method treats the fluidized bed riser as an integral. Thus, the drag forces are solved by a set of equations with the stability condition based on Eq. (5.86) in terms of inlet superficial gas velocity U_g and solid mass flux G_s of the fluidized bed risers. The drag coefficient is expressed by heterogeneity index which is correlated with solid volume fraction based on Wen and Yu empirical correlation [53]. The heterogeneity index is defined as the ratio of the structure-dependent, mesoscale drag coefficient β to the homogeneous drag coefficient from the Wen and Yu drag coefficient β_{WY}.

$$\omega = \frac{\beta}{\beta_{WY}} \tag{5.87}$$

The drag expression is expressed as

$$\beta = \begin{cases} \frac{3}{4}\frac{(1-\alpha_g)\alpha_g}{d_p}\rho_g\left|\boldsymbol{u}_g - \boldsymbol{u}_s\right|C_D\omega & \alpha_g > \alpha_{gc} \\ 150\frac{\alpha_s^2\mu_g}{\varepsilon_g d_p^2} + 1.75\frac{\alpha_s\rho_g\left|\boldsymbol{u}_g-\boldsymbol{u}_s\right|}{d_p} & \alpha_g \le \alpha_{gc} \end{cases} \tag{5.88}$$

where ω is the heterogeneity factor. α_{gc} is the voidage at which $\omega(\alpha_g)$ equals the correction factor of Gidaspow's drag model. According to EMMS, the gas tends to flow upward with less resistance, and particles tend to be arrayed with minimal potential energy in gas–solid risers. The compromise between them results in a tendency of minimal energy consumption for suspending and transporting of particles. For gas-particle circulating fluidized bed risers, the correlation of heterogeneous index is expressed as a function of gas volume fractions in the range from 0.553 to 1.0 [31, 32].

$$\omega = \begin{cases} \exp[-4.6495 + 3.66287\alpha_g - 0.55601\alpha_g^2] & 0.553 \le \alpha_g < 0.978 \\ \exp[-265.08307 + 457.73731\alpha_g - 192.62209\alpha_g^2] & 0.978 \le \alpha_g < 0.9997 \\ 1.0 & 0.9997 \le \alpha_g \le 1.0 \end{cases} \tag{5.89}$$

These show that the heterogeneous index is less than 1.0 as the gas volume fraction changes from 0.553 to 0.9997, indicating the heterogeneous flow is significant drag reduction compared to the Wen and Yu drag model. The heterogeneity factor follows a natural exponential function as the function of gas volume fraction. The drag correlations from Eq. (5.89) are obtained at given conditions of the superficial gas velocity U_g of 1.52 m/s, solid flux G_s of 14.3 kg/m^{2}s and material properties with particle diameter and density of 54 μm and 930 kg/m^3. This correlation for heterogeneous index can be enable apply from gas-particles turbulence fluidized beds to circulating fluidized bed risers.

According to the resolution of structures and gas–particles interactions in gas-particles circulating fluidized beds, the total drag force is the sum of the drag force on the particles in the dilute phase, that on the particles in the dense phase, and the drag on the cluster surface. Hou et al. [25] proposed the alternative heterogeneous index based on Gidaspow drag coefficient model from the multiscale method of view in circulating fluidized bed risers [34].

$$\omega = \frac{\alpha_g^{3.7}}{(1-\alpha_g)}\frac{(1-2\frac{d_p}{d_c})f(1-\alpha_c)C_{Dc}|U_{sc}|U_{sc} + (1-f)(1-\alpha_d)C_{Dd}|U_{sd}|U_{sd} + \frac{d_p f}{d_c}C_{Di}|U_{si}|U_{si}}{\left|u_g - u_s\right|(u_g - u_s)C_D} \tag{5.90}$$

where C_{Dd}, C_{Dc} and C_{Di} are the drag coefficients in the dilute phase and dense phase, and the drag coefficient between the cluster and gas phase, respectively. U_{sd} and U_{sc} are the superficial slip velocities between gas and particle in the dilute phase and dense phase. U_{si} is the superficial slip velocity between the cluster and gas phase.

d_c is the cluster diameter. The heterogeneous index is determined according to a set of algebraic equations, assuming the dilute phase gas volume fraction α_d is taken as a constant, and the dense phase gas volume fraction α_c is the sum of α_g plus n-times the standard deviation of voidage, where n is a constant and α_g is the gas volume fraction. The feature of the heterogeneous index model using Eq. (5.90) is that the stability condition of Eq. (5.85) is not required for solving the heterogeneous momentum exchange coefficient, reducing computational costs.

In order to consider transient gas–solid flow of dense phase and dilute phase, three inertial terms (a_c, a_f and a_i) are further introduced into EMMS model, where a_c and a_f are the inertial terms of the dense phase and dilute phase, and a_i is the meso-scale interphase inertial term. Assuming a_f is equal to gravitational acceleration [5], the heterogeneity index is regressed as the function of Reynolds number for the same particle properties, superficial gas velocity and solid mass flux [32]. The final form of the heterogeneity index equation is referred to as EMMS drag model [2], and given by in the Reynolds number range from 0.001 to 35.

$$\omega = a(Re_s + b)^c \tag{5.91}$$

$$Re_s = \frac{\alpha_g \rho_g d_p}{\mu_g}\left|\boldsymbol{u}_g - \boldsymbol{u}_s\right| \tag{5.92}$$

where a, b and c are the coefficients, and they relate to gas volume fraction. As the gas volume fraction varies from 0.545 to 0.9997 in gas–solid circulating fluidized bed risers, the formulas are given by:

$$\left.\begin{array}{r} a = (2124.956 - 2142.349\alpha_g)^{-0.4896} \\ b = (0.8223 - 0.1293\alpha_g)^{13.031} \\ c = (1.0013 - \alpha_g) \\ \left[0.06633 + 9.1391(1.0013 - \alpha_g) - 6.9231(\alpha_g - 1.0013)^2\right]^{-1} \end{array}\right\} 0.545 \le \alpha_g < 0.99 \tag{5.93}$$

$$\left.\begin{array}{r} a = 0.4243 + 0.88\left[1 + \exp\frac{0.9942-\alpha_g}{0.00218}\right]^{-1}\left[1 - (1 + \exp\frac{0.9989-\alpha_g}{0.00003})^{-1}\right] \\ b = 0.01661 + 0.2436\exp\left[-0.5\left(\frac{\alpha_g-0.9985}{0.00191}\right)^2\right] \\ c = 0.0825 - 0.0574\exp\left[-0.5\left(\frac{\alpha_g-0.9979}{0.00703}\right)^2\right] \end{array}\right\} \quad 0.99 \le \alpha_g < 0.9997 \tag{5.94}$$

As the gas volume fractions are larger than 0.9997, the heterogeneity index is equal to 1.0, meaning that gas-particle suspension is homogeneous. The heterogeneity index increases, reaches maximum at the gas volume fraction of 0.998, and then decays with increasing gas volume fractions. Higher Reynolds number generally results in higher values of heterogeneity index, in the sense that formation of aggregates is suppressed by higher slip velocity between gas and particles. By comparing to the heterogeneity index from Eq. (5.89), the major differences of heterogeneity index using Eq. (5.91) contribute to the dense phase inertial term a_c and meso-scale interphase inertial term a_i in the determination of structure-dependent drag coefficient. The inertial terms represent momentum transfer of the dense phase and dilute phase due to hydrodynamic interaction and collisional interaction, showing the theoretical treatments of both a_c and a_i are essential to shed light on the complex behavior and flow structure. These suggest that particle clusters have noticeable inertial effects to determine the solution of heterogeneity index.

The EMMS model determines the structural parameters for steady state gas–solid fluidization [50], meaning that the EMMS method is a separate calculation of the correction for the drag force. These indicate that the EMMS model represents an overall model or zero-dimensional model that describes overall hydrodynamics of heterogeneous structure within gas–solid risers, which depends on material properties and operating conditions of constant inlet gas velocity U_g and solid mass flux G_s. In gas–solid circulating fluidized beds, the solid mass flux G_s consists of the feeding solid mass flux from the reactor inlet feeders and the circulating solid mass flux from the cyclone separators through the loop seal. Gas velocity U_g contains primary air from the distributor, secondary air from the air injection above the bed bottom, and air flux from the loop seal, indicating the gas velocity has a step load changing along riser height. The advantages of EMMS model is that the drag model links the mesoscale structures of clusters to the macroscale performance based on the superficial gas velocity U_g and solid mass flux G_s in gas–solid circulating fluidized beds.

The EMMS drag model correctly evaluates the solid flux and axially S-shaped profiles of voidage in circulating fluidized bed. A detailed discussion and derivation of the EMMS model and heterogeneity index can be found in [2, 5, 25, 32, 34, 50–52, 57].

5.5 Minimum Rate of Energy Dissipation-Based Drag Model

5.5.1 Criterion for Identification of Clusters

High-velocity gas-particle flows in circulating fluidized beds do show persistent fluctuations and are accompanied by the presence of clusters and streamers. These non-uniform structures arise as a result of local instabilities and not because of instabilities associated with the entire circulation loop of circulating fluidized beds. Therefore, the local gas and particles suspension flow at least resolves into two flow states in the grid, seeing Fig. 5.1. One is known as the heterogeneous flow with clusters if the solids volume fraction α_s is larger than the threshold α_{cr} in the cell [45]. This

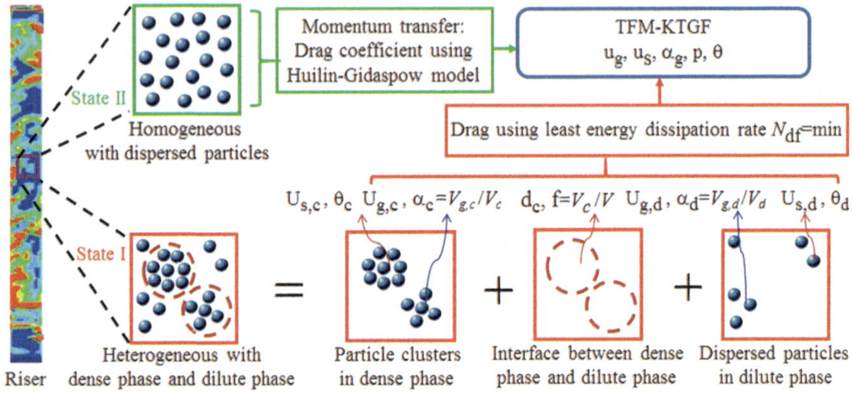

Fig. 5.1 Homogeneous and heterogeneous structures of the grid with structure parameters of the dense and dilute phases and gird parameters in the state I and II

is behaved by the state I. The other is the homogeneous flow without clusters when α_s is smaller than α_{cr} in the cell. This represents at the state II. Ether the state I or the state II describe flow of gas and particles mixture in the control volume based on the criterion. This indicates that the state of gas-particle flow is either a heterogeneous structure with clusters of the state I or a homogeneous without clusters of the state II of the control volumes in each grid.

For the heterogeneous structure of the state I, the grid is dissolved into a dilute phase where the gas volume fraction is larger than α_g, a dense phase in which the gas volume fraction is less than α_g and the interface between the dilute phase and the dense phase in the control volume V. For the state I, the dense phase volume part V_c consists of spherical clusters with a mean cluster diameter d_c, gas superficial velocity $U_{g,c}$, clusters superficial velocity $U_{s,c}$, gas volume $V_{g,c}$ and granular temperature of particles within the clusters θ_c. The dilute phase volume part V_d of dispersed particles is described by the gas superficial velocity $U_{g,d}$, gas volume $V_{g,d}$, dispersed particles superficial velocity $U_{s,d}$ and dispersed particles granular temperature θ_d. The dense phase volume fraction inside of the grid defines $f = V_c/V$. The gas volume fractions of the dense phase and the dilute phase are $\alpha_c = V_{g,c}/V_c$ and $\alpha_d = V_{g,d}/V_c$ in the control volume. Thus, the heterogeneous flow of gas and particles mixture is characterized by ten structure parameters which are related to five gird variables (i.e., u_g, u_s, α_g, p and θ). We see that the criterion must be required to distinguish between the heterogeneous suspension of the state I and the homogeneous suspension of the state II in the computational cell.

When the heterogeneous structure exists in the control volume of the grid, the state I is resolved into the dilute phase and the dense phase with an interphase between them. The different scale interactions include a micro-scale interaction and meso-scale interaction. The micro-scale interactions between gas and particles are correlated with the size of the constituent particles and presents in both the dense phase drag force F_c and dilute phase drag force F_d. These interaction forces are expressed as the magnitude of the forces acting on a single particle in the dense and dilute phases. The meso-scale interaction F_i between the dense clusters and dilute broth surrounding them is correlated with the size of clusters, indicating the identification and characterization of particle clusters must be required.

The kinetic energy equation of solid phase was derived according to the kinetic theory [11, 16]. For flow of heterogeneous structures, the granular energy E_s consists of the energy component of dispersed particles in the dilute phase $E_{s,d}$ and the energy component of clusters in the dense phase $E_{s,c}$ according to the velocities of the dilute phase $u_{s,d}$ and dense phase $u_{s,c}$.

$$E_{s,d} = \frac{1}{2}\boldsymbol{u}_{s,d}\boldsymbol{u}_{s,d} + \frac{3}{2}\theta_d \tag{5.95}$$

$$E_{s,c} = \frac{1}{2}\boldsymbol{u}_{s,c}\boldsymbol{u}_{s,c} + \frac{3}{2}\theta_c \tag{5.96}$$

The heterogeneous flow is described using transient one-dimensional conservation equations of gas-particles suspension. The transient one-dimensional kinetic energy equation of solid phase is given by

$$\begin{aligned}
&\frac{\partial E_s}{\partial t} + \frac{\partial}{\partial z}\left[f^2\rho_s E_{s,c}U_{s,c} + (1-f)^2\rho_s E_{s,d}U_{s,d}\right] \\
&+ \alpha_s u_s \frac{\partial p}{\partial z} + \frac{\partial}{\partial z}(q_{s,d} + q_{s,c}) \\
&+ \frac{\partial}{\partial z}\left[\left(\frac{fU_{s,c}}{1-\alpha_c}\right)p_{s,c} + \left(\frac{(1-f)U_{s,d}}{1-\alpha_d}\right)p_{s,d}\right] \\
&+ [f^2\rho_s U_{s,c} + (1-f)^2\rho_s U_{s,d}]g \\
&= \left[n_d F_d\left(\frac{(1-f)U_{s,d}}{1-\alpha_d}\right) + n_c F_c\frac{fU_{s,c}}{1-\alpha_c} + n_i(1-f)F_iU_i\right] + (\Gamma_{s,d} + \Gamma_{s,c})
\end{aligned} \tag{5.97}$$

$$E_s = f(1-\alpha_c)\rho_s E_{s,c} + (1-f)(1-\alpha_d)\rho_s E_{s,d} \tag{5.98}$$

The first term on the left-hand side denotes the rate of accumulation of kinetic energies of particles in the dense phase and dilute phase within the control volume, the second term is the kinetic energy increase by convection of particles in the dense and dilute phases, and the third and fifth terms denote the rates of work done through fluid and particles pressure forces of the dense and dilute phases. The fourth term represents the net rate of kinetic energy addition due to kinetic energy fluxes of particles in the dense and dilute phases, and the last term represents rate of work by external body force of particles in the dense and dilute phases. The first term on the right-hand side denotes the rate of the work done on the solid phase within the control volume by drag forces, and the last term denotes energy dissipation rate of particles of the dense and dilute phases in the control volume. The rates of the collisional energy dissipations of particles and clusters in the dilute phase and dense phase are given by

$$\Gamma_{s,d} = \frac{3(1-e^2)[(1-f)(1-\alpha_d)]^2 \rho_s \alpha_{s,\max}^{1/3} \theta_d}{\alpha_{s,\max}^{1/3} - [(1-f)(1-\alpha_d)]^{1/3}} \left[\frac{4}{d_p}\sqrt{\frac{\theta_d}{\pi}} - \frac{\partial}{\partial z}\frac{(1-f)U_{s,d}}{1-\alpha_d} \right] \tag{5.99}$$

$$\Gamma_{s,c} = \frac{3(1-e^2)[f(1-\alpha_c)]^2 \rho_s \alpha_{s,\max}^{1/3} \theta_c}{\{\alpha_{s,\max}^{1/3} - [f(1-\alpha_c)]^{1/3}\}} \left[\frac{4}{d_p}\sqrt{\frac{\theta_c}{\pi}} - \frac{\partial}{\partial z}\frac{fU_{s,c}}{1-\alpha_c} \right] \tag{5.100}$$

where e is the restitution coefficient of particles of the dilute and sense phases. $\alpha_{s,max}$ is the solids volume fraction at closest packing.

A similar operation is performed to analyze the gas energy E_g in the dilute phase and dense phase. The conservation energy equation of gas phase is given by

$$\begin{aligned} &\frac{\partial E_g}{\partial t} + \frac{\partial}{\partial z}[f^2 \rho_g E_{g,c} U_{g,c} + (1-f)^2 \rho_g E_{g,d} U_{g,d}] \\ &+ \alpha_g u_g \frac{\partial p}{\partial z} + \frac{\partial}{\partial z}(q_{g,d} + q_{g,c}) + [f\rho_g U_{g,c} + (1-f)\rho_g U_{g,d}]g \\ &= -[n_d F_d \frac{(1-f)U_{g,d}}{\alpha_d} + n_c F_c \frac{fU_{g,c}}{\alpha_c}] + (\Gamma_{g,d} + \Gamma_{g,c}) \end{aligned} \tag{5.101}$$

$$E_g = f\alpha_c \rho_g E_{g,c} + (1-f)\alpha_d \rho_g E_{g,d} \tag{5.102}$$

The first term on the left-hand side denotes the rate of accumulation of gas kinetic energies in the dense phase and dilute phase within the control volume, the second term denotes gas kinetic energy increase by convection in the dense and dilute phases, and the third term represents the rates of work done through fluid pressure forces. The fourth term represents the net rate of kinetic energy addition due to gas kinetic energy fluxes in the dense and dilute phases, and the last term represents rate of work by gas external body force in the dense and dilute phases. The first term on the right-hand side denotes the rate of the work done by gas drag forces in the dense and dilute, and the last term denotes the rate of work done by the gas viscous forces within the control volume. The rates of energy dissipations of gas in the dilute and dense phases are given according to effective gas viscosity μ_g.

$$\Gamma_{g,d} = \frac{4}{3}(1-f)\alpha_d \mu_g \left[\frac{\partial}{\partial z}\left(\frac{(1-f)U_{g,d}}{\alpha_d} \right) \right]^2 \tag{5.103}$$

$$\Gamma_{g,c} = \frac{4}{3} f\alpha_c \mu_g \left[\frac{\partial}{\partial z}\left(\frac{fU_{g,c}}{\alpha_c} \right) \right]^2 \tag{5.104}$$

Finally, the addition of these equations yields the energy balance of gas-particles suspension in the control volume as follows:

$$\begin{aligned}
&\frac{\partial}{\partial t}(E_s + E_g) + \frac{\partial}{\partial z}[f^2(\rho_s U_{s,c} E_{s,c} + \rho_g U_{g,c} E_{g,c}) \\
&(1-f)^2(\rho_s U_{s,d} E_{s,d} + \rho_g U_{g,d} E_{g,d})] \\
&+ (\alpha_g u_g + \alpha_s u_s)\frac{\partial p}{\partial z} + \frac{\partial}{\partial z}[(\frac{f U_{s,c}}{1-\alpha_c})p_{s,c} \\
&+ (\frac{(1-f)U_{s,d}}{1-\alpha_d})p_{s,d}] + \frac{\partial}{\partial z}[(q_{s,d} + q_{s,c}) + (q_{g,d} + q_{g,c})] \\
&+ [f^2 \rho_s U_{s,c} + (1-f)^2 \rho_s U_{s,d}]g \\
&+ [f \rho_g U_{g,c} + (1-f)\rho_g U_{g,d}]g \\
&= [n_d F_d U_d + n_c F_c U_c + n_i(1-f)F_i U_i] \\
&+ (\Gamma_{s,c} + \Gamma_{s,c}) + (\Gamma_{g,d} + \Gamma_{g,c})
\end{aligned} \tag{5.105}$$

From Eq. (5.98), the kinetic energy components of the dilute phase $E_{g,d}$ and the dense phase $E_{g,c}$ are unclosed, and must be found from the transport equations, leading to Eq. (5.105) is more complicated than in the single gas phase energy conservation equation. But, we observe that the terms at the right-hand side represent the rate of energy dissipations of gas-particles suspension through three contributions by hydrodynamic forces, interactions of collisions of particles and gas viscous forces of heterogeneous structures.

$$\begin{aligned}
\Omega = &[n_d F_d U_d + n_c F_c U_c + n_i F_i U_i(1-f)] \\
&+ (\Gamma_{s,d} + \Gamma_{s,c}) + (\Gamma_{g,d} + \Gamma_{g,c})
\end{aligned} \tag{5.106}$$

The energy dissipation rate per unit mass of gas-solids mixture is defined to be

$$\mathrm{N}_{df} = \frac{[n_d F_d U_d + n_c F_c U_c + n_i F_i U_i(1-f)] + (\Gamma_{s,d} + \Gamma_{s,c}) + (\Gamma_{g,d} + \Gamma_{g,c})}{\alpha_g \rho_g + (1-\alpha_g)\rho_s} \tag{5.107}$$

To keep the energy dissipation rate as low as possible, particles tend to aggregate into clusters to reduce the gas flow resistance and the collisional energy dissipations, while gas tends to bypass the clusters instead of penetrating through clusters [16, 28]. The gas and particles have to compromise in such a way gas attempt to minimize resistance, and particles try to minimize collisions. Hence, the criterion of the existence of heterogeneous structures defines that the rate of energy dissipation per unit mass of mixture is minimum:

$$N_{df} = \frac{\Omega}{\alpha_g \rho_g + (1-\alpha_g)\rho_s} = \Omega_{dr} + \Omega_{co} + \Omega_{vi} \rightarrow \min \tag{5.108}$$

This is a criterion that the rate of energy dissipation of heterogeneous flow must be minimization when the particle clusters occur. The threshold is known as the minimum rate of energy dissipation (MRED). The heterogeneous flow would thus be identified. Equation (5.108) is an extreme value equation of the rate of energy dissipation including the energy dissipation component of drag forces Ω_{dr}, collisional energy dissipation component Ω_{co} and gas viscous energy dissipation component Ω_{vi}.

$$\Omega_{dr} = \frac{(1-f)F_d U_{g,d} + F_i(1-f)U_{g,d} + fF_c U_{g,c}}{\alpha_g \rho_g + (1-\alpha_g)\rho_s} \tag{5.109}$$

$$\Omega_{co} = \frac{(\Gamma_{s,d} + \Gamma_{s,c})}{\alpha_g \rho_g + (1-\alpha_g)\rho_s} \tag{5.110}$$

$$\Omega_{vi} = \frac{(\Gamma_{g,d} + \Gamma_{g,c})}{\alpha_g \rho_g + (1-\alpha_g)\rho_s} \tag{5.111}$$

where $U_{gs,d}$ is the slip velocity between particle and gas in the dilute phase, and $U_{gs,c}$ is the slip velocity between particle and gas in the dense phase.

For gas–solid flows, the different mechanisms may be responsible for rates of energy dissipation. These main mechanisms are (I) kinetic energy dissipation due to the drag forces, (II) dissipation rate of energy due to collisions of particles, (III) increase of the gas turbulent viscosity due to the presence of particles and (IV) dissipation of turbulent kinetic energy by particles and (V) enhancement of the velocity gradient between particles and fluid moving with the particles as added fluid mass to particles. These mechanisms may not be independent and, moreover, mechanisms IV and V might be of minor importance in gas–solid flow. These energy dissipations increase with increasing volume fractions of particles. Thus, for homogeneous flow without particle clusters, the energy dissipation rate N_{df} does not trend to minimum.

$$N_{df} \neq \min \tag{5.112}$$

From Eqs. (5.108) and (5.112) for flow of gas-particles mixture in the control volume at the state I and II, the criterion of the discrimination of homogeneous flow and heterogeneous flow is given by

$$N_{df} = \Omega_{dr} + \Omega_{co} + \Omega_{vi} \begin{cases} = \min & \text{heterogeneous} \\ \neq \min & \text{homogeneous} \end{cases} \tag{5.113}$$

The homogeneous flow without particle clusters and heterogeneous flow with clusters can be identified in the grids as the gas-particles suspension satisfies the criterion of minimization of energy dissipation rate from Eq. (5.113). The threshold describes a discrimination between the state I for heterogeneous flow with clusters and the state II for homogeneous flow without clusters.

The MRED indicates that the heterogeneous structures arise because of (a) instability associated with the relative motion between the gas phase and particle phase and (b) dissipation of fluctuating energy of particles by inelastic collisions and the gas viscous damping. From Eq. (5.113), the principle of the MRED reveals a multiscale nature involving three different mechanisms: collisional dominate, hydrodynamic dominate and viscous dominate. The compromise among them leads to the heterogeneous flow structures at different scales and spatio temporal multiscale behaviors in gas–solid fluidized beds.

5.5.2 MRED-Based Drag Coefficient

For heterogeneous flow of gas and particles with clusters, the drag forces acting on a suspension of particles can be formulated in terms of the forces acting on particles in the dense and dilute phases and interaction force between the dense phase and the dilute phase from momentum balance equations of the dense phase and dilute phase [28], the minimum rate of energy dissipation-based multiscale drag coefficient, i.e. dynamic cluster structure-dependent (DCSD) drag coefficient, is given by:

$$\beta_{DCSD} = \frac{n_c F_c + n_d F_d + n_i F_i}{\left|u_g - u_s\right|} \quad N_{df} = \min \tag{5.114}$$

where F_c, F_d and F_i are the drag force components of the dilute phase, dens phase and interface under the minimal energy dissipation rate. n_c, n_d and n_i are the number of particles in the dense phase and dilute phase, and clusters. The number densities of cluster and dispersed particles are given by according to particle diameter d_p.

$$n_c = \frac{f(1-\alpha_c)}{\pi d_p^3/6}, n_d = \frac{(1-f)(1-\alpha_d)}{\pi d_p^3/6} \tag{5.115}$$

The drag forces of gas and particles in the dilute and dense phases are given by

$$F_d = \frac{\pi d_{\mathrm{p}}^2 \rho_{\mathrm{g}}}{8\alpha_d^{2.65}}\left[\frac{24}{Re_d} + \frac{3.6}{Re_d^{0.313}}\right]\left|\frac{U_{\mathrm{g,d}}}{\alpha_{\mathrm{d}}} - \frac{U_{\mathrm{s,d}}}{1-\alpha_{\mathrm{d}}}\right| \left(\frac{U_{\mathrm{g,d}}}{\alpha_{\mathrm{d}}} - \frac{U_{\mathrm{s,d}}}{1-\alpha_{\mathrm{d}}}\right) \tag{5.116}$$

$$F_{\mathrm{c}} = \frac{\pi d_{\mathrm{p}}^2 \rho_{\mathrm{g}}}{8\alpha_c^{2.65}}\left[\frac{24}{Re_{\mathrm{c}}} + \frac{3.6}{Re_c^{0.313}}\right]\left|\frac{U_{\mathrm{g,c}}}{\alpha_{\mathrm{c}}} - \frac{U_{\mathrm{s,c}}}{1-\alpha_{\mathrm{c}}}\right| \left(\frac{U_{\mathrm{g,c}}}{\alpha_{\mathrm{c}}} - \frac{U_{\mathrm{s,c}}}{1-\alpha_{\mathrm{c}}}\right) \tag{5.117}$$

where the local dense phase and dilute phase Reynolds numbers read

$$\mathrm{Re}_c = \frac{\rho_g d_p U_{gs,c}}{\mu_g}, \mathrm{Re}_d = \frac{\rho_g d_p U_{gs,d}}{\mu_g} \tag{5.118}$$

The superficial gas velocity in the dense phase $U_{g,c}$ and the superficial gas velocity in the dilute phase $U_{g,d}$ are given by

$$U_{g,c} = U_{gs,c} + \frac{\alpha_c U_{s,c}}{1-\alpha_c},\ U_{g,d} = U_{gs,d} + \frac{\alpha_d U_{s,d}}{1-\alpha_d} \tag{5.119}$$

The superficial particle velocities in the dense phase and dilute phase are expressed as

$$U_{s,c} = \frac{(1-\alpha_g)u_s - (1-f)U_{s,d}}{f} \tag{5.120}$$

$$U_{s,d} = \frac{(1-\alpha_d)(1-\alpha_c)}{(\alpha_d-\alpha_c)(1-f)} \left[\alpha_g u_g - \frac{\alpha_c}{1-\alpha_c}(1-\alpha_g)u_s - (1-f)U_{gs,d} - fU_{gs,c}\right] \tag{5.121}$$

Similarly, the number density and the drag force of the interface of clusters can be defined according to cluster diameter d_c.

$$n_i = \frac{f}{\pi d_c^3/6} \tag{5.122}$$

$$F_i = \frac{\pi d_c^2 \rho_g \alpha_d^2}{8(1-f)^{2.65}} \left[\frac{24}{Re_i} + \frac{3.6}{Re_i^{0.313}}\right] \left|\left(\frac{U_{g,d}}{\alpha_d} - \frac{U_{s,c}}{1-\alpha_c}\right)(1-f)\right| \left[\left(\frac{U_{g,d}}{\alpha_d} - \frac{U_{s,c}}{1-\alpha_c}\right)(1-f)\right] \tag{5.123}$$

In the same way according to cluster diameter, the interface Reynolds number reads

$$Re_i = \frac{\rho_g d_c}{\mu_g} \left[\left(U_{g,d} - \frac{\alpha_d U_{s,c}}{1-\alpha_c}\right)(1-f)\right] \tag{5.124}$$

For homogeneous flow of gas and particles mixture, the drag coefficient is calculated using Huilin-Gidaspow drag correlation with the switch function ϕ:

$$\beta_{HG} = \phi\left(\frac{150\alpha_s(1-\alpha_g)\mu_g}{\alpha_g d_p^2} + \frac{1.75\alpha_s\rho_g|\boldsymbol{u}_s - \boldsymbol{u}_g|}{d_p}\right) + (1-\phi)\frac{3C_D\alpha_s\alpha_g\rho_g|\boldsymbol{u}_s - \boldsymbol{u}_g|}{4d_p}\alpha_g^{-2.65} \quad N_{df} \neq \min \tag{5.125}$$

From Eqs. (5.114) and (5.125), the drag coefficient is modeled based on the MRED between the DCSD and the Huilin-Gidaspow exchange coefficient models as follows:

$$\beta = \begin{cases} \beta_{DCSD} & N_{df} = \min \\ \beta_{HG} & N_{df} \neq \min \end{cases} \tag{5.126}$$

This indicates that the momentum exchange coefficient depends on the criterion of the discrimination of particle clusters for heterogeneous flow based on the principal of minimum rate of energy dissipation in each grid.

5.5.3 *Transient Equations of Dilute and Dense Phases*

In order to solve the criterion of Eq. (5.108), the forces acting on a single particle of the dense phase F_c and the dilute phase F_d, and the force acting on a cluster through the interface F_i are first evaluated in the control volume. These forces are determined from momentum balance of gas and solids phases in the grid.

For fluid catalytic cracking (FCC) particles, the solid velocity is close to the gas velocity in the direction of the flow [16, 19, 28]. Thus, one-dimensional flow is assumed along the mean flow direction (vertical z direction) in a first approximation. Although one-dimensional equations of heterogeneous structures are easier to obtain from the momentum equations of gas and solid phases, the equations are solved to be useful for further investigations of three-dimensional equations of hydrodynamics of heterogeneous flow structures.

5.5.3.1 Heterogeneous Transient Momentum Equations

The momentum conservation equation of particles includes momentum equations of particles in the dilute phase and dense phase. The one-dimensional transient momentum equation of dispersed particles in the dilute phase is given by [56]

$$\begin{aligned} &\frac{\partial}{\partial t}[(1-f)^2\rho_s U_{s,d}] + \frac{\partial}{\partial z}\left[(1-f)^2\rho_s U_{s,d}\frac{(1-f)U_{s,d}}{(1-\alpha_d)}\right] \\ &= -(1-f)(1-\alpha_d)\frac{\partial p}{\partial z} + n_d F_d - (1-f)(1-\alpha_d)\rho_s g \\ &\quad - \frac{\partial p_{s,d}}{\partial z} - \beta_{dc}\left[\frac{fU_{s,c}}{1-\alpha_c} - \frac{(1-f)U_{s,d}}{1-\alpha_d}\right] \end{aligned} \tag{5.127}$$

The second term on the right-hand side is the drag force in the dilute phase. The forth term is the dispersed particles pressure, and the last term is momentum exchange between the dilute phase and the dense phase. The first and second terms on the left-hand side denote the temporal acceleration and convective acceleration, representing the rate of velocity change with respect to time and the spatial gradients of velocity of dispersed particles in the dilute phase. They are positive or negative, representing the acceleration or deceleration of particles. β_{dc} is the drag coefficient of the solids-solids momentum transfer between the dilute phase and the dense phase. $p_{s,d}$ is the solid pressure part in the dilute phase. Both drag coefficient of particles between the dilute phase and the dense phase and solid pressure in the dilute phase are given by

$$\beta_{dc} = \frac{3(1+e_s)\rho_s f(1-\alpha_d)(1-f)(1-\alpha_c)}{4d_p} \left[\frac{1}{\alpha_g} + \frac{3(2-\alpha_d-\alpha_c)}{2\alpha_g^2}\right]\left|\frac{fU_{s,c}}{1-\alpha_c} - \frac{(1-f)U_{s,d}}{1-\alpha_d}\right| \tag{5.128}$$

$$p_{s,d} = \rho_s(1-f)\alpha_{s,d}\theta_d[1 + 2g_{0,d}(1-f)\alpha_{s,d}(1+e_s)] \tag{5.129}$$

$$g_{0,d} = \frac{\alpha_{s,max}^{1/3}}{\{\alpha_{s,max}^{1/3} - [(1-f)(1-\alpha_d)]^{1/3}\}} \tag{5.130}$$

We assume that the off-diagonal elements of the gradient of the solid stress tensor of clusters are zero in a first approximation. The one-dimensional transient momentum equation of clusters is given by:

$$\begin{aligned} &\frac{\partial}{\partial t}(f^2\rho_s U_{s,c}) + \frac{\partial}{\partial z}\left[(f^2\rho_s U_{s,c})\frac{fU_{s,c}}{(1-\alpha_c)}\right] \\ &= -f(1-\alpha_c)\frac{\partial p}{\partial z} + n_c F_c + n_i F_i - \frac{\partial p_{s,c}}{\partial z} \\ &\quad - f(1-\alpha_c)\rho_s g + \beta_{dc}\left[\frac{fU_{s,d}}{1-\alpha_d} - \frac{(1-f)U_{s,c}}{1-\alpha_c}\right] \end{aligned} \tag{5.131}$$

The second and third terms on the right-hand side represent the drag forces in the dense phase and the interface between the dilute phase and the dense phase. The first and second terms on the left-hand side denote cluster temporal acceleration and convective acceleration of the dense phase, representing the rate of cluster velocity change with respect to time and the spatial gradients of velocity of clusters due to the change of position. Both accelerations are positive or negative. The solid pressure in the dense phase is given by

$$p_{s,c} = \rho_s(1-f)\alpha_{s,c}\theta_c[1 + 2g_{0,c}(1-f)\alpha_{s,c}(1+e_s)] \tag{5.132}$$

$$g_{0,c} = \frac{\alpha_{s,max}^{1/3}}{\{\alpha_{s,max}^{1/3} - [f(1-\alpha_c)]^{1/3}\}} \tag{5.133}$$

Similar to the assumptions used in the momentum equations of dispersed particles and clusters, the one-dimensional transient momentum equations of gas phase in the dilute phase and the dense phase along the direction of the flow are given by:

$$\frac{\partial}{\partial t}[(1-f)^2\rho_g U_{g,d}] + \frac{\partial}{\partial z}\left[(1-f)^2\rho_g U_{g,d}\frac{(1-f)U_{g,d}}{\alpha_d}\right]$$
$$= -(1-f)\alpha_d\frac{\partial p}{\partial z} - n_d F_d - n_i F_i - (1-f)\alpha_d\rho_g g \tag{5.134}$$

$$\frac{\partial}{\partial t}(f^2\rho_g U_{g,c}) + \frac{\partial}{\partial z}\left[(f^2\rho_g U_{g,c})\frac{fU_{g,c}}{\alpha_c}\right]$$
$$= -f\alpha_c\frac{\partial p}{\partial z} - n_c F_c - f\alpha_c\rho_g g \tag{5.135}$$

5.5.3.2 Mass Equations of Heterogeneous Structures

Mass conservation of gas and particles includes mass balance in the dilute and dense phases. The gas and particles mass balance equations in the dense phase and dilute phase are expressed in the grid.

$$f\rho_g U_{g,c} + (1-f)\rho_g U_{g,d} = \alpha_g\rho_g u_g \tag{5.136}$$

$$f\rho_s U_{s,c} + (1-f)\rho_s U_{s,d} = (1-\alpha_g)\rho_s u_s \tag{5.137}$$

where u_g and u_s are the gas phase and solid phase velocities. α_g is the gas volume fraction. They represent grid parameters, and are obtained from mass and momentum conservation equations.

The gas volume fraction requires that the sum of volume fraction of gas and particles be unity in the computational cell, that is,

$$\alpha_g = f\alpha_c + (1-f)\alpha_d \tag{5.138}$$

5.5.3.3 Granular Temperatures of Heterogeneous Structures

The fluctuating energy equations describe granular temperature equations for dispersed particles and clusters. The algebraic expression for granular temperature assumes that the granular energy is dissipated locally; neglecting the convection and diffusion contributions, and retaining the generation and dissipation terms. In present study, the algebraic expression model is extended to predict granular temperatures of dispersed particles θ_d and granular temperature of particles within the clusters θ_c

Table. 5.1 Expression of coefficients for dilute and dense phases

Dense phase	Dilute phase	
$K_{1,c} = 2(1+e)\rho_s g_{0,c}$	$K_{1,d} = 2(1+e)\rho_s g_{0,d}$	(T1)
$K_{2,c} = \frac{d_p\rho_s B_c}{(3\sqrt{\pi})} - \frac{d_p\rho_s}{18}[\frac{\sqrt{\pi}A_c}{(3-e)} + \frac{3B_c}{5\sqrt{\pi}}]$	$K_{2,d} = \frac{d_p\rho_s B_d}{(3\sqrt{\pi})} - \frac{d_p\rho_s}{18}[\frac{\sqrt{\pi}A_d}{(3-e)} + \frac{3B_d}{5\sqrt{\pi}}]$	(T2)
$K_{3,c} = \frac{d_p\rho_s}{4}\left\{\frac{\sqrt{\pi}A_c}{3(3-e)} + \frac{B_c}{5\sqrt{\pi}}\right\}$	$K_{3,d} = \frac{d_p\rho_s}{4}\left\{\frac{\sqrt{\pi}A_d}{3(3-e)} + \frac{B_d}{5\sqrt{\pi}}\right\}$	(T3)
$K_{4,c} = 12(1-e^2)\rho_s g_{0,c}(\sqrt{\pi}d_p)^{-1}$	$K_{4,d} = 12(1-e^2)\rho_s g_{0,d}(\sqrt{\pi}d_p)^{-1}$	(T4)
$K_{5,c} = \frac{12(1-e^2)f(1-\alpha_c)\rho_s g_{0,c}}{\sqrt{\pi}d_p}$	$K_{5,d} = \frac{12(1-e^2)(1-f)(1-\alpha_d)\rho_s g_{0,d}}{\sqrt{\pi}d_p}$	(T5)
$K_{6,c} = \frac{[96(1-e^2)f(1-\alpha_c)\rho_s g_{0,c}]^2}{(\sqrt{\pi}d_p)^2}$	$K_{6,d} = \frac{[96(1-e^2)(1-f)(1-\alpha_d)\rho_s g_{0,d}]^2}{(\sqrt{\pi}d_p)^2}$	(T6)
$A_c = 1+0.4(1+e)(3e-1)f(1-\alpha_c)g_{0,c}$	$A_d =$ $1+0.4(1+e)(3e-1)(1-f)(1-\alpha_d)g_{0,d}$	(T7)
$B_c = 8f(1-\alpha_c)g_{0,c}(1+e)$	$B_d = 8(1-f)(1-\alpha_d)g_{0,c}(1+e)$	(T8)

in a first approximation [55].

$$\theta_d = \left\{\frac{-K_{1,d}}{2K_{4,d}}\frac{\partial}{\partial z}\left[\frac{(1-f)U_{s,d}}{(1-\alpha_d)}\right] + \sqrt{\left(\frac{K_{1,d}^2}{4K_{4,d}^2} + \frac{K_{2,d}}{K_{5,d}}\right)\frac{\partial^2}{\partial z^2}\left[\frac{(1-f)U_{s,d}}{(1-\alpha_d)}\right] + \frac{2K_{3,d}}{K_{6,d}}\frac{\partial}{\partial z}\left[\frac{(1-f)U_{s,d}}{(1-\alpha_d)}\right]^2}\right\}^2 \tag{5.139}$$

$$\theta_c = \left\{\frac{-K_{1,c}}{2K_{4,c}}\frac{\partial}{\partial z}\left[\frac{fU_{s,c}}{(1-\alpha_c)}\right] + \sqrt{\left(\frac{K_{1,c}^2}{4K_{4,c}^2} + \frac{K_{2,c}}{K_{5,c}}\right)\frac{\partial^2}{\partial z^2}\left[\frac{fU_{s,c}}{(1-\alpha_c)}\right] + \frac{2K_{3,c}}{K_{6,c}}\frac{\partial}{\partial z}\left[\frac{fU_{s,c}}{(1-\alpha_c)}\right]^2}\right\}^2 \tag{5.140}$$

The coefficients $K_{i,d}$ and $K_{i,c}$ of the dilute phase and the dense phase ($i = 1-6$) relate to properties of gas and particles, and they are listed in Table 5.1, where e is the coefficient of restitution of particles.

5.5.4 Solution of Minimum Energy Dissipation Rates

Using two-fluid model (TFM) with kinetic theory of granular flow (KTGF), the phase properties in terms of velocities u_g and u_s, fluid volume fraction α_g, fluid pressure p, and granular temperature θ, are obtained when modeling the conservation of mass, momentum and granular temperature under the threshold from Eq. (5.108) in any arbitrary grids. Hence, the minimization of energy dissipation rate of Eq. (5.113) is used to close TFM simulations.

The importance of the criterion of Eq. (5.108) is to predict distributions of cluster parameters in fluidized bed reactors. As indicated in Fig. 5.2 using N_{df} solver and β_{DCSD} solver, ten parameters (α_c, α_d, f, $U_{s,c}$, $U_{g,c}$, $U_{s,d}$, $U_{g,d}$, d_c, θ_d and θ_c) involve to describe such a heterogeneous structure. Here, $U_{s,c}$, $U_{g,c}$, α_c, f, θ_c and d_c are for the dense phase, and $U_{g,d}$, $U_{s,d}$, θ_d and α_d are for the dilute phase. However, analysis of continuity and momentum conservation in such a multiscale structure yield only eight equations in the equation set, which are insufficient to solve the ten parameters. This implies that there are multiple solutions satisfying the mass and momentum conservation equations. To define the state of heterogeneous flow from these multiple solutions, the threshold of minimal energy dissipation rate had to be satisfied.

Recall that Eq. (5.108) represents a constrained optimization problem of the objective function N_{df} with equality constraints. Using N_{df} solver, the first step in establishing a procedure for the solution of optimization problem is the definition of an objective function. The objective function is the mathematical representation of an aspect under evaluation, which must be minimized (or maximized). For the identification of clusters, the objective function N_{df} is minimum as a function of dense

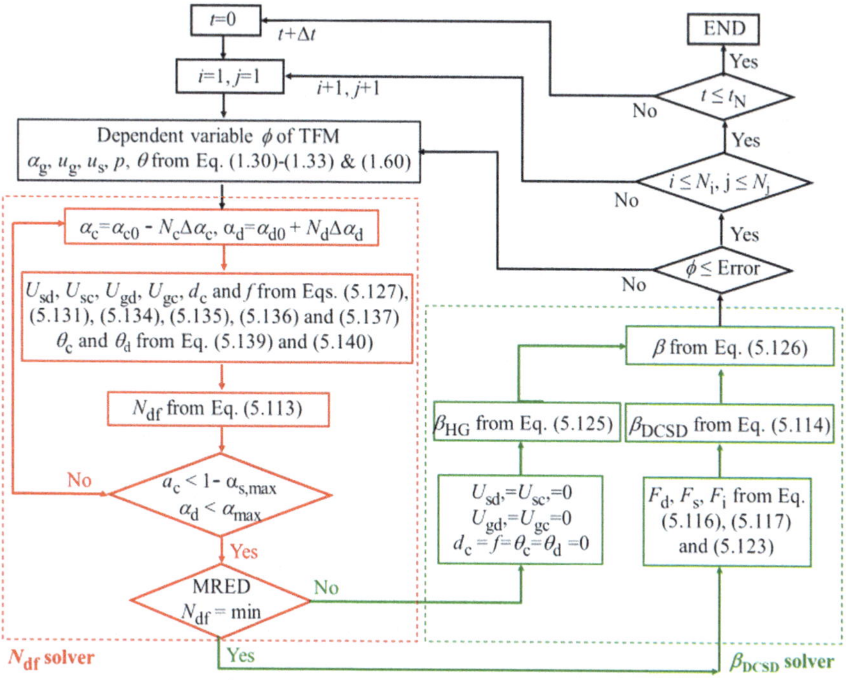

Fig. 5.2 Flow chart outlining the process of computational threshold of minimum energy dissipation rate and DCSD drag coefficient

and dilute phase gas volume fractions of α_c and α_d. This problem is equivalent to the following constrained optimization problem with constraints

$$\begin{aligned} &\min \ N_{df} \\ &\text{subject to } \alpha_g < \alpha_c \le \alpha_{s,mf}, \quad \alpha_g < \alpha_d \le \alpha_{max} \end{aligned} \tag{5.141}$$

where $\alpha_{s,mf}$ is the solid volume fraction at the minimum fluidization. α_{max} is the gas volume faction for particle aggregating, and its value is taken as 0.9997. The schematic showing a computational threshold of minimum energy dissipation rate and drag coefficient is shown in Fig. 5.2, where N_c and N_d are the iteration numbers of gas volume fractions of the dense phase and dilute phase. Δt is a time step. ϕ represents the grid variables of gas and solid phases. $\Delta\alpha_c$ and $\Delta\alpha_d$ denote the steps of dense and dilute gas volume fractions. N_i and N_j are the grid numbers along axial and lateral directions in two-dimensional simulations. The initial volume fractions of dense and dilute phases are calculated in terms of gas volume fraction in the grid.

$$\alpha_{d0} = \alpha_g + R_0\sigma_v \tag{5.142}$$

$$\alpha_{c0} = \alpha_g - R_0\sigma_v \tag{5.143}$$

where R_0 is a constant, and ranges from 0.0001 to 0.01. σ_v represents the standard deviation of volume fraction fluctuation. From the approximate statistical mechanical theory developed by [7] for dense gases composed of identical hard spheres, the following correlation for the concentration standard deviation was first proposed by [6] in fluidized beds.

$$\begin{aligned} \sigma_v &= \sqrt{\alpha_s^2[1 + 2\alpha_s\frac{4-\alpha_s}{(1-\alpha_s)^4}]^{-1}} \\ &= \sqrt{\frac{(1-\alpha_g)^2\alpha_g^4}{[1 + 4(1-\alpha_g) + 4(1-\alpha_g)^2 - 4(1-\alpha_g)^3 + (1-\alpha_g)^4]}} \end{aligned} \tag{5.144}$$

The computational program contains two modules with the TFM-KTGF and the criterion of minimum rate of energy dissipation. The former solves for volume fraction, velocity, pressure and granular temperature from the fluid and solid phases mass, momentum and granular fluctuating kinetic energy equations in kth grid with a built-in the minimum energy dissipation rate subroutine (i.e., N_{df} solver). The latter solves for the minimum energy dissipation rate in order to obtain gas–solid momentum transfer coefficient in kth grid (i.e., β_{DCSD} solver). These indicate the fluid phase and solid phase conservation equations coupling with the criterion of heterogeneous flow provide not only distributions of volume fractions, velocities, pressure and granular temperature distributions, but also cluster velocity, volume fraction and diameter distributions within fluidized bed risers. Hence, TFM-KTGF with MRED-based drag approach links the gas–solid CFD to the criterion of minimum rate of energy dissipation in computational simulations of gas–solid fluidized bed reactors.

5.5.5 Minimum Rate of Energy Dissipation of Clusters

From Eq. (5.113), the solutions provide to both values of gas volume fractions of the dilute phase α_d and dense phase α_c at the minimum energy dissipation rate $N_{df,min}$ (i.e., min N_{df}). The solutions of the transient momentum equation of dispersed particles in the dilute phase of Eq. (5.127) and clusters in the dense phase of Eq. (5.131), transient momentum equations of gas phase in the dilute phase of Eq. (5.134) and the dense phase of Eq. (5.135), mass balance equations of Eqs. (5.136) and (5.137), and granular temperature equations of Eq. (5.139) for particles in the dilute phase and Eq. (5.140) for clusters in the dense phase provide to parameters of U_{sc}, U_{gc}, d_c, f and θ_c for clusters of the dense phase and U_{sd}, U_{gd} and θ_d of the dilute phase. Hence, all parameters for both dense phase and dilute phase are determined using an equation set and the criterion in each grid.

The instantaneous contours of solids volume fractions and rate of energy dissipation $N_{df,min}$ at the minimum value point (min N_{df}) are shown in Fig. 5.3 at the inlet gas velocity and inlet particles mass flux of 4.0 m/s and 75.0 kg/m^{2}s, respectively. The riser height and diameter are 6.58 m and 75 mm. The particle diameter and density are 75 μm and 1654 kg/m^3, respectively. Simulations reveal that the low local volume fractions give way to high volume fractions accompanying with the occurrence of clusters. The variations from small to large solids volume fractions represent that an initially homogeneous flow (HF) of gas and particles suspension trends to a heterogeneous flow with clusters, resulting in large fluctuations in volume fractions. The occurrence of clusters is based on the criterion for the threshold of the minimization of energy dissipation rate. As the energy dissipation rate of gas-particles mixture occurs at the minimum value point $N_{df,min}$, i.e. minimum rate of energy dissipation, it can be said with certainty that the particles cluster exists in the grids. Otherwise, the HF of gas and particles suspension representing white color exists in the grids. The variation of $N_{df,min}$ reveals the formation and breaking-down of particle clusters.

The computed rates of energy dissipation of N_{df} is displayed in Fig. 5.4 as the function of dense and dilute gas volume fractions of α_c and α_d. Both cases show that the gas and particles velocities are positive, representing gas and particles co-current upward flow. For the case (a), the energy dissipation rates decrease with increasing dilute gas volume fraction α_d and decreasing dense gas volume fractions α_c, indicating that there is no extreme point of the function of energy dissipation rate, i.e. $N_{df} \neq$ min in the grid. Definitely, this represents a gas-particles mixture homogeneous flow (HF) in the grid. The drag coefficient β is calculated using Huilin-Gidaspow drag model, and the value is 7152.048 kg/m^{3}s. For the case (b), three extreme points are found with changing dilute and dense gas volume fractions. The values of the rate of energy dissipation are 0.872 (W/kg) at the α_c and α_d of 0.9195 and 0.9989 at the

Fig. 5.3 Instantaneous solids volume fractions and minimum energy dissipation rate (The different color of $N_{df,min}$ represents flow of clusters, and the white color means homogeneous flow without clusters at $N_{df,min} \neq 0.0$ in the grids.)

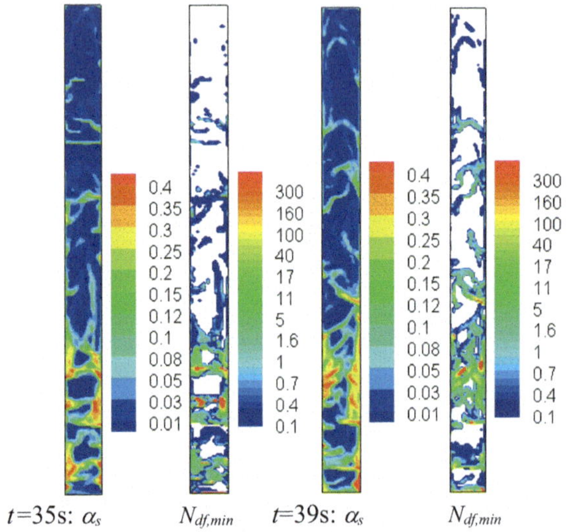

first extreme point, 1.075 (W/kg) at the α_c and α_d of 0.8834 and 0.9981 at the second extreme point, and 0.7527 (W/kg) at the α_c and α_d of 0.8075 and 0.9973 at the third extreme point. The minimum value point $N_{df,min}$ for the function of energy dissipation rate is determined from three extreme points, indicating that the heterogeneous flow exists in the grid.

For countercurrent flows of particles downflow and gas upflow, the computed energy dissipation rates N_{df} are displayed in Fig. 5.5 with a positive gas velocity and negative particle velocity in the grid. For the case (a), the computed rate of energy dissipations increases at first, and tends to a maximum. Finally, it decreases with increasing dilute gas volume fraction and decreasing dense gas volume fraction. There is an extreme point that it has maximum energy dissipation rate. The value of energy dissipation rate is 93.258 (W/kg) at the dense and dilute gas volume fraction

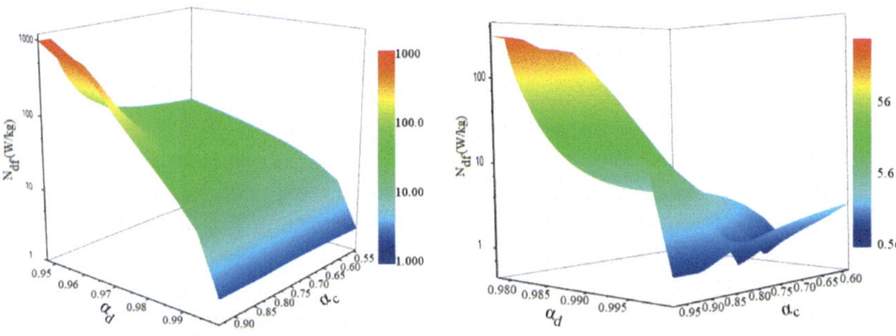

(*a*)u_g=4.336m/s, u_s=0.348m/s (*b*)u_g=5.817m/s, u_s=2.732m/s

Fig. 5.4 Energy dissipation rates of co-current upflow of gas and particles in two grids

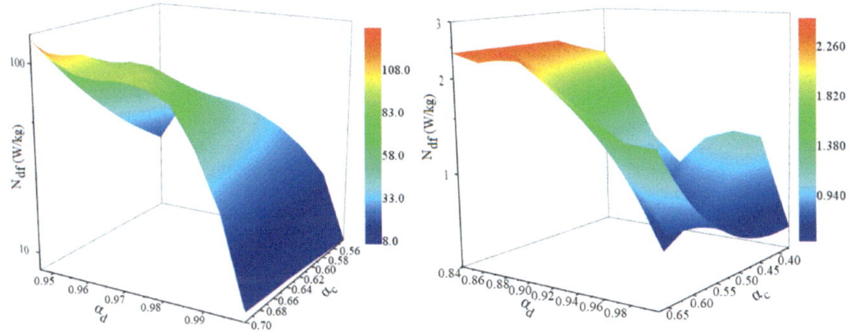

(a)u_g=1.785m/s, u_s=-0.537m/s (b)u_g=0.617m/s, u_s=-2.11m/s

Fig. 5.5 Energy dissipation rates of countercurrent flow of gas and particles in two grids

of 0.6761 and 0.9696 at the maximum extreme point. This extreme point is non equilibrium point of gas-particles mixture because the function of energy dissipation rate N_{df} is monotonically decreasing from the maximum energy dissipation rate with changing dilute and dense gas volume fractions, indicating that the gas-particles suspension tends to keep homogeneous flow (HF). For the case (b), the function of energy dissipation rate drops at first, and then trends to a minimum. Finally, it increases with increasing dilute volume fraction α_d and decreasing dense volume fraction α_c. There is an extreme point which represents a minimum value point of energy dissipation rate. The value of the minimum energy dissipation rate is 0.768 (W/kg) at the dense and dilute gas volume fractions of 0.748 and 0.998. The drag coefficients β are 14,301.1 kg/m^{3}s for homogeneous flow and 3038.4 kg/m^{3}s for heterogeneous flow.

5.6 Compromise in Competition of Multiple Mechanisms

The formation of particle clusters originates from the energy dissipation of the dynamic interactions of particles through gas and other particles. The particles inelastic collisions result in kinetic energy dissipation Ω_{co} and the gas-particles hydrodynamic interactions dissipate energy dissipation Ω_{dr}. The viscous energy dissipation Ω_{vi} represents the rate at which the dilute and dense gas stresses do work. The minimization of energy dissipation rate is dissolved into minimum hydrodynamic, collisional and viscous energy dissipation rate components of $\Omega_{dr,min}$, $\Omega_{co,min}$ and $\Omega_{vi,min}$ at the minimum value point.

$$N_{df,\min} = \Omega_{co,\min} + \Omega_{dr,\min} + \Omega_{vi,\min} \tag{5.145}$$

The instantaneous minimum energy dissipation rate components are displayed in Fig. 5.6. We can see that the minimum viscous energy dissipation rate component

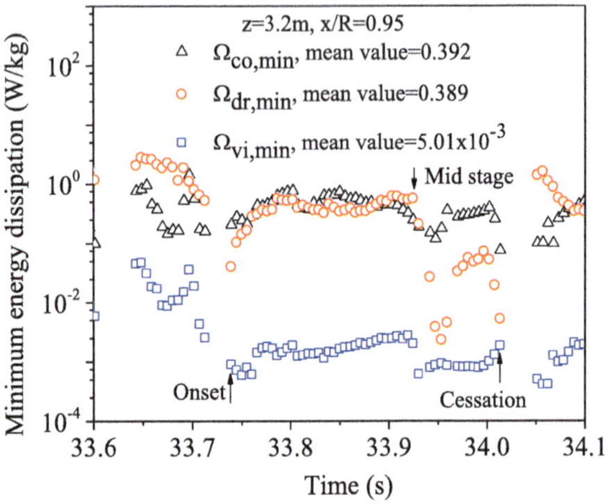

Fig. 5.6 Instantaneous minimum rate of energy dissipation components at two different grids

$\Omega_{vi,min}$ is much smaller than the minimum collisional and drag energy dissipation rate components because of low gas Reynolds number. The mean minimum hydrodynamic, collisional and viscous energy dissipation components are 0.389, 0.392 and 5.01×10^{-3} W/kg, respectively. This indicates that the dissipative collision is dominant in comparison to the contribution of gas viscous energy dissipation rate component. At the onset and cessation, the collisional energy dissipation rate is largest. These indicate that the forming and breaking-down of cluster is caused by collisional interaction of particles. On the other hand, when the cluster duration is at the middle period, the hydrodynamic energy dissipation rate component is higher than the collisional energy dissipation rate, indicating that the sustaining maintenance of particle cluster flow is induced by hydrodynamic interactions.

The instantaneous minimum hydrodynamic energy dissipation component $\Omega_{dr,min}$ and collisional energy dissipation rate component $\Omega_{co,min}$ are displayed in Fig. 5.7 where the white color presents homogeneous flow without particle clusters in the grids. The instantaneous snapshots of minimum energy dissipations rate $N_{df,\ min}$ give temporal and spatial cluster flow because of the compromise in competition through the dissipative drag mechanism and the dissipative collision mechanism. Simulated minimum energy dissipation rate components show that the minimum hydrodynamic energy dissipation components $\Omega_{dr,min}$ are large in the center region because of the high gas velocities. It is further observed that the minimum collisional energy dissipation rate components $\Omega_{co,min}$ are large near the wall region because of large solids volume fractions.

The threshold of the clusters formation means that the rate of energy dissipation of gas and particles suspension tends to minimum. Because of low viscous energy dissipation rates, Eq. (5.108) is simplified as follows:

$$N_{df} \approx \Omega_{co} + \Omega_{dr} = \min \quad (5.146)$$

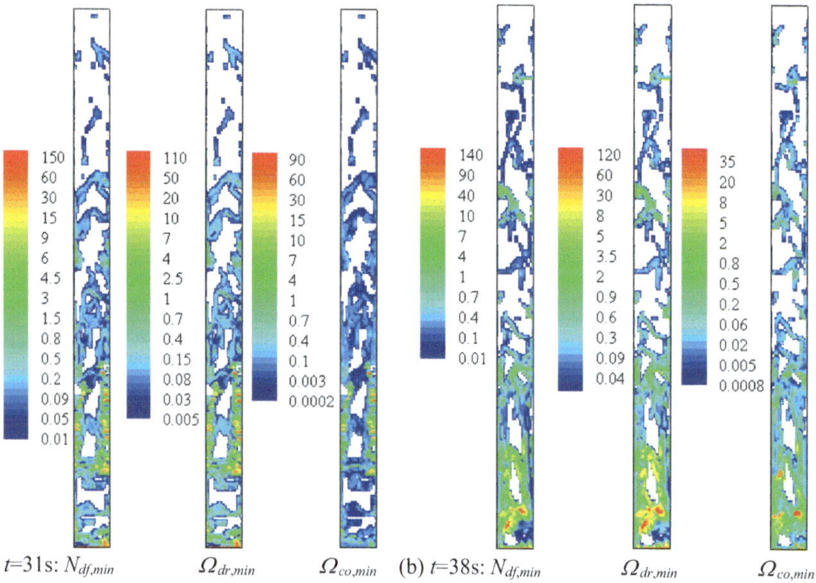

Fig. 5.7 Instantaneous counters of energy dissipation rate $N_{\mathrm{df},min}$, energy dissipations by collisions $\Omega_{co,\mathrm{min}}$ and drag forces $\Omega_{dr,min}$. $N_{df,min} = \Omega_{dr,min} + \Omega_{co,min}$. The white color represents homogeneous flow without clusters

Assuming that the rate of energy dissipation of collisions Ω_{co} is lower than the rate of energy dissipation of hydrodynamic interactions Ω_{dr}, the dissipative drags play a dominant role in comparison to collision dissipation, representing a dissipative drag mechanism. The gas-particles minimum energy dissipation rate is

$$N_{df} \approx \Omega_{dr} = \min \tag{5.147}$$

The minimization of gas and particles energy dissipation rate N_{df} is approximate to be the minimization of hydrodynamic energy dissipation component Ω_{dr}. It is named as a hydrodynamic dominant (HD) mechanism causing the formation of particle clusters. It appears that $\Omega_{dr} > \Omega_{co}$ is satisfied to be a valid approximation. The formation of clusters arises from the hydrodynamic forces of heterogeneous structures. Wylie and Koch [54] found that the formation of clusters was driven by the energy dissipation due to interactions surrounding gas and particles. The dilute clusters with low solids volume fraction increase slip velocity between the gas and particles, thereby, tend to be destroyed due to the increasing gas flow within the cluster [18, 37]. Interactions between gas and particles promote energy dissipation component by drag forces due to the velocity difference between the gas phase and the solid phase [9].

Assuming that the minimization of gas-particles energy dissipation rate N_{df} is the same as the minimization of energy dissipation rate of collisions Ω_{co} as the hydrodynamic energy dissipation component is negligible:

$$N_{df} \approx \Omega_{co} = \min \tag{5.148}$$

The dissipative collision mechanism plays a dominant role in comparison to its drag counterpart. It is named as a collisional dominant (CD) mechanism causing the formation of particle clusters. The minimization of energy dissipation rate is equivalent to the minimization of energy dissipation component of collisional interactions of heterogeneous structures. It appears that $\Omega_{dr} < \Omega_{co}$ must be satisfied to be a valid approximation. This indicates the interactions of collisions are dominant. In the granular homogeneous cooling system (HCS) with high solids volume fractions, the particles are agglomerated because the granular temperature decreases due to particle–particle collisions [20, 35, 59]. The formation of heterogeneous structure could arise when the collisions between the particles are sufficiently inelastic.

From Eqs. (5.147) and (5.148), the dominant mechanism for the occurrence of the cluster is

$$\frac{\Omega_{co,\min}}{\Omega_{dr,\min}} = \begin{cases} > 1.0 & \text{CD cluster} \\ < 1.0 & \text{HD cluster} \end{cases} \tag{5.149}$$

Figure 5.8 shows the probability of the ratio of the minimum collisional energy dissipation to the minimum hydrodynamic energy dissipation. The vertical ordinate denotes the probability that the ratio falls within the indicated ratio range. The value is calculated by dividing the number of clusters flow in the ratio range by the total number of particle clusters. It is found that a large proportion of the occurrence of particle clusters is in the minimum energy dissipation ratio range from 0.2 to 0.8, reflecting that the value for the minimum hydrodynamic energy dissipation is an order of magnitude larger than that for the minimum collisional energy dissipation. The particle clusters are formed by the dissipative drags mechanism. More important is that the second peak of the probability of minimum energy dissipation rate ratio occurs at the range of the minimum energy dissipation ratio from 1.2 to 1.8, indicating that the dominant mechanism for flow of clusters is dissipative collisions. It further indicates that the ratios differ up to four orders of magnitude. The formation of particle clusters appears at low energy dissipation rate ratio, indicating hydrodynamic energy dissipation is dominant in comparison with dissipative collisions.

The dynamic cluster structure-dependent drag model indicates the drag coefficient depends upon the solution of energy dissipation N_{df}. If there is a solution of the extreme value of a function N_{df}, the drag coefficient is calculated at minimal value using β_{DCSD}. Otherwise, the drag coefficient is predicted by means of β_{HG}. The instantaneous drag coefficients are shown in Fig. 5.9. Both drag coefficients oscillate with times. Roughly, the instantaneous drag coefficient β_{DCSD} is smaller than β_{HG}, that is the drag coefficient with temporal and spatial clusters is reduced in comparison with uniformly dispersed particles.

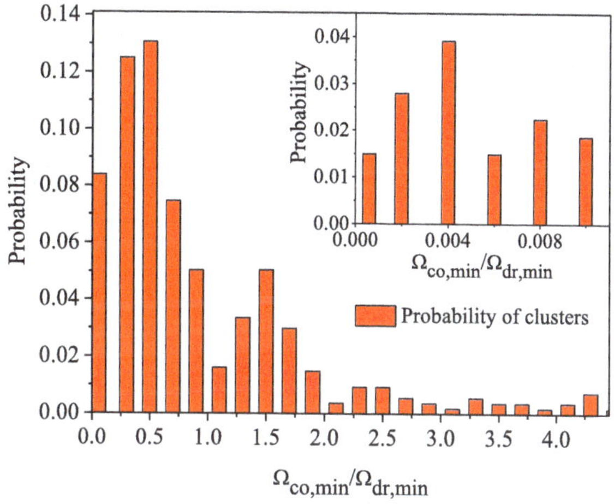

Fig. 5.8 Probability of clusters occurrence as a function of minimum ratio of energy dissipations

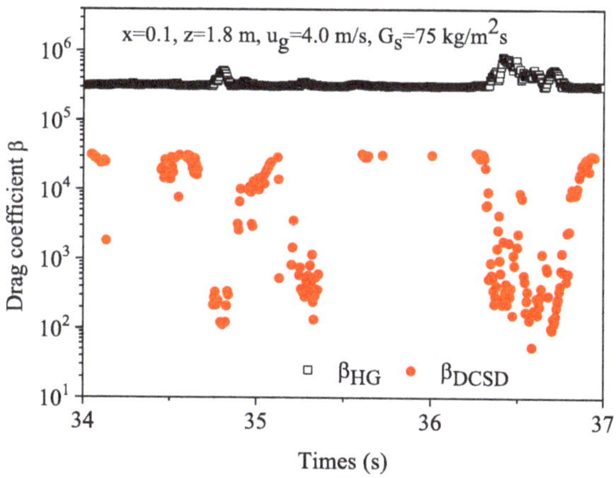

Fig. 5.9 Instantaneous drag coefficients in a cell

Figure 5.10 shows the distribution of predicted volume fraction of dense phase f and solids volume fraction of clusters $\alpha_{s,c}$ as a function of solids volume fraction. Generally speaking, both predicted volume fractions of dense phase and solids volume fraction of clusters increase with the increase of solids volume fraction. The high solids volume fraction of clusters represents more particles within the cluster, indicating the clusters are dense with the increase of solids volume fraction. The correlation proposed by Harris et al. [21] showed that the solids volume fraction of clusters increases rapidly at low solid volume fraction, and then increase slightly at high solid volume fraction. Both the correlation and the simulations give the solids volume fractions of clusters increase with the increase of solids volume fraction.

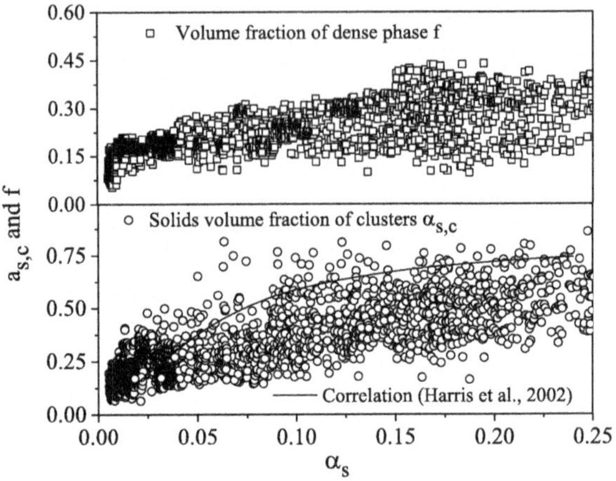

Fig. 5.10 Distribution of volume fractions of clusters and dense phase

5.7 Summary

The macroscale flow structures, termed coherent flow structures, can have a large effect on the overall flow hydrodynamics in gas-particles fluidized beds. At the same time, the mesoscale structures of clusters are also important as they affect the emergence of the macroscale structures. As a result, accurate simulations of TFM equations require spatial resolution from mesoscale of clusters to microscale of particles. In order to identify clusters during computational simulations, the criterion of the stability condition is proposed to identify heterogeneous flow with clusters and homogeneous flow without clusters in the grids. The minimum rate of energy dissipation (MRED) is proposed as a criterion [55]. The threshold defines that the rate of energy dissipations is minimum distinguishing between the homogeneous flow and the heterogeneous flow. Both drag coefficient and cluster parameters can be computed using TFM-KTGF combined with the criterion. Hence, the criterion of the minimum rate of energy dissipation acts as a bridge between TFM-KTGF approach and predictions of heterogeneous flow of clusters in gas–solid circulating fluidized beds.

A remarkable feature of the criterion of the minimum rate of energy dissipation (MRED) links to TFM-KTGF approach. This approach offers both the heterogeneous drag force and momentum exchange coefficient β_{DCSD} using dynamic cluster structure-dependent drag coefficients model and the spatial and temporal distribution of heterogeneous structure properties, including cluster diameter d_c, cluster velocity U_c and volume fraction f of clusters. These clusters distributions play an important role in the determination of mass and heat transfer coefficients and chemical reaction rates in computational simulations of fluid-particles circulating fluidized bed reactors.

Nomenclature

C_D	Drag coefficient
d	Particle diameter, m
d_c	Cluster diameter, m
e	Restitution coefficient
E	Kinetic energy, m^2/s^2
$E_{s,d}$	Dispersed particles kinetic energy, m^2/s^2
$E_{s,c}$	Clusters kinetic energy, m^2/s^2
f	Volume fraction of dense phase
$\boldsymbol{F}$	Force acting on each particle or cluster, N
Fr	Froude number
g	Gravity, m/s^2
g_0	Radial distribution function at contact
G	Weight function
G_s	Solids mass flux, kg/m^2 s
k_s	Conductivity of fluctuating energy, kg/m s
n_{den}	Number of particles in the dense phase per unit volume
n_{dil}	Number of particles in the dilute phase per unit volume
N_{df}	Rate of energy dissipation per unit mass, W/kg
p	Gas pressure, Pa
p_s	Particle pressure, Pa
Re	Reynolds number
$\boldsymbol{u}_g$	Gas velocity, m/s
$\boldsymbol{u}_s$	Particle velocity, m/s
U	Superficial gas velocity, m/s
$U_{g,c}$	Gas superficial velocity in dense phase, m/s
$U_{g,d}$	Gas superficial velocity in dilute phase, m/s
U_{int}	Superficial velocity of interface, m/s
$U_{s,c}$	Clusters superficial velocity, m/s
$U_{s,d}$	Dispersed particles superficial velocity in dilute phase, m/s
V_{den}	Volume of dense phase in the control volume, m^3
V_{dil}	Volume of dilute phase in the control volume, m^3
x	Transverse distance from axis, m
z	Height along vertical direction, m

Greek letters

β	Drag coefficient, kg/m^3 s
γ	Collisional energy dissipation, kg/m s^3
α_c	Porosity in the dense phase
α_d	Porosity in the dilute phase

α_g	Porosity
α_{max}	Maximum porosity for particle aggregating
α_s	Solids volume fraction
θ	Granular temperature, m^2/s^2
μ_g	Gas shear viscosity, kg/m s
μ_s	Solids shear viscosity. kg/m s
ρ_g	Gas density, kg/m^3
ρ_s	Density of solid phase, kg/m^3
τ_g	Gas stress tensor, Pa
τ_s	Particle stress tensor, Pa
Ω	Energy dissipation rate, W/m^3
Ω_{co}	Collisional energy dissipation rate, W/kg
Ω_{dr}	Rate of energy dissipation of drag force, W/kg
Ω_{vi}	Viscous energy dissipation, W/kg

Subscripts

c	Dense phase with clusters
d	Dilute phase
g	Gas phase
s	Solid phase

Abbreviations

CD	Collisional dominant
CFD	Computational fluid dynamics
DCSD	Dynamic cluster structure-dependent
HCS	Homogeneous cooling system
HD	Hydrodynamic dominant
HF	Homogeneous flow
KTGF	Kinetic theory of granular flow
MRED	Minimum rate of energy dissipation
TFM	Two-fluid model

References

1. Agrawal, K., Loezos, P. N., Syamlal, M., & Sundaresan, S. (2001). The role of meso-scale structures in rapid gas-solid flows. *Journal of Fluid Mechanics, 445*, 151–185.
2. Ansys. (2024). *Ansys fluent theory guide* (p. 15317). ANSYS Inc.
3. Beetstra, R., van der Hoef, M. A., & Kuipers, J. A. (2007). Drag force of intermediate Reynolds number flow past mono- and bidisperse arrays of spheres. *AIChE Journal, 53*, 489–501.
4. Benyahia, S., Syamlal, M., & O'Brien, T. J. (2012) *Summary of MFIX equations 2012-1*. Report U.S. Department of Energy, Office of Fossil Energy, Morgantown Energy Technology Center.
5. Bona, L., Wei, W., & Jinghai, L. (2009). Searching for a mesh-independent sub-grid model for CFD simulation of gas–solid riser flows. *Chemical Engineering Science, 64*, 3437–3447.
6. Buyevich, Y. A., & Kapbasov, S. K. (1994). Random fluctuations in a fluidized bed. *Chemical Engineering Science, 49*, 1229–1243.
7. Carnahan, N. F., & Starling, K. E. (1969). Equation of state for non-interacting rigid spheres. *Journal of Chemical Physics, 51*, 635–637.
8. Dan, L., Shuyan, W., Guodong, L., Huilin, L., Xiaoxue, J., Ming, T., & Zhenjie, L. (2018). A dynamic cluster structure-dependent drag coefficient model applied to gas-solid risers. *Powder Technology., 325*, 381–395.
9. Eaton, J. K., & Fessler, J. R. (1994). Preferential concentration of particles by turbulence. *International Journal of Multiphase Flow, 20*, 169–209.
10. Ergun, S. (1952). Fluid flow through packed columns. *Chemical Engineering Progress, 48*, 89–94.
11. Fox, R. O. (2014). On multiphase turbulence models for collisional fluid-particle flows. *Journal of Fluid Mechanics, 742*, 368–424.
12. Gao, X., Li, T., Sarkar, A., Lu, L., & Rogers, W. A. (2018). Development and validation of an enhanced filtered drag model for simulating gas-solid fluidization of Geldart A particles in all flow regimes. *Chemical Engineering Science, 184*, 33–51.
13. Garside, J., & Al-Dibouni, M. R. (1977). Velocity-voidage relationships for fluidization and sedimentation in solid-liquid systems. *Industrial Engineering Chemistry Process Design and Development, 16*, 206–214.
14. Geldart, D. (1973). Types of gas fluidization. *Powder Technology, 7*, 285–292.
15. Gibilaro, G. I., Felice, D. R., Waldram, P. S., & Foscolo, U. P. (1985). Generalized friction factor and drag coefficient correlations for fluid-particle interactions. *Chemical Engineering Science, 40*, 1817–1823.
16. Gidaspow, D. (1994). *Multiphase flow and fluidization: Continuum and kinetic theory descriptions*. Academic Press.
17. Gidaspow, D., & Ettehadieh, B. (1983). Fluidization in two-dimensional beds with a jet, Part II: Hydrodynamics modeling. *Industrial and Engineering Chemistry Fundamentals, 22*, 193–201.
18. Good, G. H., Gerashchenko, S., & Warhaft, Z. (2012). Intermittency and inertial particle entrainment at a turbulent interface: The effect of the large-scale eddies. *Journal of Fluid Mechanics, 694*, 371–398.
19. Grace, J. R., Avidan, A. A., & Knowlton, T. M. (1997). *Circulating fluidized beds*. Springer.
20. Haff, P. K. (1983). Grain flow as a fluid mechanical phenomenon. *Journal of Fluid Mechanics, 134*, 401–948.
21. Harris, A. T., Davidson, J. F., & Thorpe, R. B. (2002). The prediction of particle cluster properties in the near wall region of a vertical riser. *Powder Technology, 127*, 128–143.
22. Hill, R., Koch, D., & Ladd, A. (2001). Moderate-Reynolds-number flows in ordered and random arrays of spheres. *Journal of Fluid Mechanics, 448*, 243–278.
23. Holloway, W., & Sundaresan, S. (2012). Filtered models for reacting gas–particle flows. *Chemical Engineering Science, 82*, 132–143.
24. Holloway, W., Yin, X., & Sundaresan, S. (2010). Fluid-particle drag in inertial polydisperse gas–solid suspensions. *AIChE Journal, 56*, 1995–2004.
25. Hou, B., Wang, X., Zhang, T., & Li, H. (2016). Eulerian simulation of a circulating fluidized bed with a new flow structure-based drag model. *Chemical Engineering Journal, 284*, 1224–1232.

26. Huilin, L., & Gidaspow, D. (2003). Hydrodynamics of binary fluidization in a riser: CFD simulation using two granular temperatures. *Chemical Engineering Science, 58*, 3777–3792.
27. Huilin, L., Gidaspow, D., Bouillard, J., & Wentie, L. (2003). Hydrodynamic simulation of gas-solid flow in a riser using kinetic theory of granular flow. *Chemical Engineering Journal, 95*, 1–13.
28. Huilin, L., Gidaspow, D., & Shuyan, W. (2021). *Computational fluid dynamics theory of fluidization: Kinetic theory of granular flow and applications*. Springer International Publishing.
29. Igci, Y., Andrews, A. T., Sundaresan, S., Pannala, S., & O'Brien, T. (2008). Filtered two-fluid models for fluidized gas-particle suspensions. *AIChE Journal, 54*, 1431–1448.
30. Igci, Y., & Sundaresan, S. (2011). Constitutive models for filtered two-fluid models of fluidized gas- particle flows. *Industrial and Engineering Chemistry Research, 50*, 13190–14132.
31. Kun, H., Shi, Z., Wang, W., & Li, J. (2013). A structure-dependent multi-fluid model (SFM) for heterogeneous gas–solid flow. *Chemical Engineering Science, 99*, 191–202.
32. Li, J., & Kwauk, M. (1994). *Particle-fluid two-phase flow: The energy-minimization multi-scale method*. Metallurgical Industry Press.
33. Li, J., & Kwauk, M. (2003). Exploring complex systems in chemical engineering-the multi-scale methodology. *Chemical Engineering Science, 58*, 521–535.
34. Liu, W., Li, H., Zhu, Q., & Hu, C. (2019). Sensitivity analysis of mesoscale structural parameters for simulation of fluidized beds. *Particuology, 47*, 77–93.
35. McMillan, J., Shaffer, F., Gopalan, B., Chew, J. W., Hrenya, C. M., Hays, R., Karri, S. B. R., & Cocco, R. (2013). Particle cluster dynamics during fluidization. *Chemical Engineering Science, 100*, 39–51.
36. MFIX. (2020). *MFiX-Exa documentation*. MFS Development Group, National Energy Technology Laboratory.
37. Moran, J. C., & Glicksman, L. R. (2003). Experimental and numerical studies on the gas flow surrounding a single cluster applied to a circulating fluidized bed. *Chemical Engineering Science, 58*, 1879–1886.
38. Ozarkar, S. S., Yan, X., Wang, S., Milioli, C. C., Milioli, F. E., & Sundaresan, S. (2015). Validation of filtered two-fluid models for gas-particle flows against experimental data from bubbling fluidized bed. *Powder Technology, 284*, 159–169.
39. Pope, S. B. (2000). *Turbulent flows*. Cambridge University Press.
40. Qi, Z., Kuang, S. B., Qiu, T. S., & Yu, A. B. (2020). Lattice Boltzmann investigation on fluid flows through packed beds: Interaction between fluid rheology and bed properties. *Powder Technology, 369*, 248–260.
41. Richardson, J. F., & Zaki, W. N. (1954). Sedimentation and fluidization: Part I. *Transactions of the Institution of Chemical Engineers, 32*, 35–53.
42. Sarkar, A., Miloioli, F. E., Ozarkar, S., Li, T., Sun, X., & Sundaresan, S. (2016). Filtered sub-grid constitutive models for fluidized gas-particle flows constructed from 3-D simulations. *Chemical Engineering Science, 152*, 443–456.
43. Shuai, W., Guangbo, Z., Guodong, L., Huilin, L., Feixiang, Z., & Tianyu, Z. (2014). Hydrodynamics of gas-solid risers using cluster structure-dependent drag model. *Powder Technology, 254*, 214–227.
44. Shuai, W., Huilin, L., Guodong, L., Zhiheng, S., Pengfei, X., & Gidaspow, D. (2011). Modeling of cluster structure-dependent drag with Eulerian approach for circulating fluidized beds. *Powder Technology, 208*, 98–110.
45. Soong, C. H., Tuzla, K., & Chen, J. C. (1994) Identification of particle clusters in circulating fluidized bed. *Circulating Fluidized Bed Technology, IV*, 615–620
46. Sundaresan, S., Ozel, A., & Kolehmainen, J. (2018). Toward constitutive models for momentum, species, energy transport in gas–particle flows. *Review of Chemical and Biomolecular Engineering, 9*, 61–81.
47. Syamlal, M., & O'Brien, T. J. (1987) *A generalized drag correlation for multiparticle systems*. Report U.S. Department of Energy, Office of Fossil Energy, Morgantown Energy Technology Center.

48. Syamlal, M., Rogers, W., & O'Brien, T. J. (1993). *MFIX documentation theory guide*. DOE/METC-94/1004 NTIS/DE94000087 Department of Energy, Office of Fossil Energy, Morgantown Energy Technology Center Morgantown, WV, Springfield.
49. Tenneti, S., Garg, R., & Subramaniam, S. (2011). Drag law for monodisperse gas–solid systems using particle-resolved direct numerical simulation of flow past fixed assemblies of spheres. *International Journal of Multiphase Flow, 37*, 1072–1092.
50. Tian, Y., Geng, J., & Wang, W. (2021). On the choice of mesoscale drag markers. *AIChE Journal, 68*, 1–13.
51. Wang, J., Ge, W., & Li, J. (2008). Eulerian simulation of heterogeneous gas-solid flows in CFB risers: EMMS-based sub-grid scale model with a revised cluster description. *Chemical Engineering Science, 63*, 1553–1571.
52. Wei, G., Wei, W., Ning, Y., Jinghai, L., Mooson, K., Feiguo, C., Jianhua, C., Xiaojian, F., Li, G., Xianfeng, H., Xinhua, L., Yaning, L., Bona, L., Jian, W., Junwu, W., Limin, W., Xiaowei, W., Qingang, X., Ming, X., … Guangzheng, Z. (2011). Meso-scale oriented simulation towards virtual process engineering (VPE)—The EMMS paradigm. *Chemical Engineering Science, 66*, 4426–4458.
53. Wen, C. Y., & Yu, Y. H. (1966). Mechanics of fluidization. *American Institute of Chemical Engineers Symposium Series, 62*, 100–111.
54. Wylie, J. J., & Koch, D. L. (2000). Particle clustering due to hydrodynamic interactions. *Physics of Fluids, 12*, 964–970.
55. Xiaoxue, J. (2022). *Minimum energy dissipation rate drag coefficient model and numerical simulation on the active and passive regulation of cluster in riser*. Ph. D. Thesis, Harbin: Harbin Institute of Technology.
56. Xiaoxue, J., Dan, L., Shuyan, W., Hassan, M., Wenjian, C., Weiqi, C., & Huilin, L. (2020). Comparative analysis of heterogeneous gas-solid flow using dynamic cluster structure-dependent drag model in risers. *International Journal of Multiphase Flow, 122*, 103126.
57. Yang, N., Wang, W., Ge, W., & Li, J. (2003). CFD simulation of concurrent-up gas-solid flow in circulating fluidized beds with structure-dependent drag coefficient. *Chemical Engineering Journal, 96*, 71–80.
58. Yin, X., & Sundaresan, S. (2009). Fluid-particle drag in low-Reynolds-number polydisperse gas–solid suspensions. *AIChE Journal, 55*, 1352–1368.
59. Yin, X., Zenk, J. R., Mitran, P. P., & Hrenya, C. M. (2013). Impact of collisional versus viscous dissipation on flow instabilities in gas-solid systems. *Journal of Fluid Mechanics, 727*, 1–12.

Chapter 6
Applications of Computational Fluid Dynamics to Multiphase Flows

Abstract This chapter demonstrates the unique application of the multiphase approaches to the predictions of several chemical and energy related processes. These topics include cluster regulations using the active and passive methods in risers, CFD-KTGF-DEM simulations in fluidized beds, low-stokes-number KTGF simulations in supercritical fluid and molten salt fluid-particles fluidized beds, hydraulic transportation of Non-Newtonian fluid and cuttings in well drilling, and particles brownout phenomenon induced by hovering helicopter. They demonstrate cohesively and constructively how the innovative application of multiphase transport equations produces an approach for design and scale-up of these processes.

Transport phenomena of fluid-particle systems and fluidization emerge as a methodology that quantifies the interplay between transport phenomena and kinetics for various reactor performance. The ability to establish reactor models linking between reactor performance and input and operational variables is essential in optimizing the operating conditions for proper reactor design and scale-up of fluidized bed reactors. CFD modeling and simulation can be applied to many multiphase flows, including coal and biomass gasification and combustion, fluid catalytic cracking reactor, supercritical water and carbon dioxide fluidized beds. This can be included in initial stages of process and product design. Industrial processes also can be successfully analyzed using CFD technique. This can largely decrease the development costs, increase the productivity, and shorten the marketing time. With knowledgeable users and powerful computers, modeling and simulation of more complex geometries will become believable in the variety of fluid-particles systems and fluidized beds.

H. Lu et al., *Computational Transport Phenomena of Multiphase Systems and Fluidization*, Fluid Mechanics and Its Applications 127,
https://doi.org/10.1007/978-981-96-0698-6_6

6.1 Regulations of Particle Clusters in Circulating Fluidized Beds

Gas–solid fluidized bed reactors are dissipative processes, and develop complex spatiotemporal patterns of particles clusters, representing a dynamically structured clusters. Particles clusters organize into a scalable structure that is highly predictable and responsive to changes in operation parameters and reactor structural design. A dynamically structured clusters offer opportunities for process intensification through the regulation of the formation and distribution of clusters. This informs new intensified designs using the criterion of the minimum rate of energy dissipation (MRED) for fluidized bed processes.

Circulating fluidized beds (CFBs) are applied in biomass and coal gasification and combustion, fluid catalytic cracking and materials drying. The advantages of CFBs include a high heat transfer and chemical reactions rate as well as a controllable residence time distribution of particles. In the process of the transformation of the crude oil mixture into useful products, the crude oil feed into the fluid catalytic cracking (FCC) unit where they are distilled into products such as liquefied petroleum gas, gasoline, kerosene, diesel and some fuel oils. In the riser reactor, hot porous catalyst particles from the regenerator are brought into contact with the gas oil feed and the lifting steam. Upon contact, the gas oil is vaporized and cracked into lower molecular weight compounds. Additionally, coke is also produced as a by-product of the cracking reactions, which lowers the activity of the catalyst as it blocks access of the vaporized gas oil to the active site on the catalyst. The deactivated catalyst is sent to the regenerator where coke on the catalyst is burned off. In spite of the successful industrial applications of CFBs, it is difficult to correlate CFB performance with reactions and heat and mass transfer of particles, which is significantly affected by the occurrence of particle clusters. Particle clusters with high local solid volume fraction cause non-uniformity in the distributions of gas-particles suspension along radial and axial directions, with significant effects on productions of CFB reactors.

The performance of fluidized bed reactors depends on the contact pattern between gaseous and solid species. Mixing and residence time vary drastically between different designs, raw materials and scales. The solid motion is affected by the formation of clusters or bubbles in the fluidized beds. A rising clusters or bubble conveys the solids up and towards the center, after which solids recirculate downward near the walls. This affects on gas–solid contact, because the solids mix inside the clusters or bubble wakes, but, while large clusters or bubbles create mixing, they also allow part of the gas to bypass the solid phase, thus reducing the time available for reaction or interfacial transport. The techniques for regulating the characteristics of gas bubbles and particle clusters in fluidized beds may be classified into two categories: active method and passive method. Figure 6.1 shows a general technical approach of both methods according to how they apply the additional sources through collision and hydrodynamic interaction modes. The active technique requires external power to change the hydrodynamics of gas-particles distributions with particle clusters and gas bubbles. The additional sources of energy include magnetic field [35, 55], sound field [10] and electric field [25, 36]. On the contrary, the passive techniques do not require external power sources to adjust gas bubbles and particle clusters distributions, but apply the modification of the wall surface structures and configurations. The passive methods include ring internals, two-stage arrangement, tapered shape riser and secondary air jets [30, 32, 39]. The main aim of both methods is to move the particle clusters from the wall region to the center region of the riser, resulting in a uniform cluster distribution and improving uniformly solid volume fractions along axial and radial directions. Aside the external magnetic, electric and sound fields, the mechanical vibration and pulsation also belong to active methods. This dynamic energy source can be applied through a gas-particle drag force and interactions of collisions. Using these active and passive methods equip with a new set of tools or degrees of freedom to manipulate the operation of the bed and reaction conversion or selectivity, not only improving performance but also enabling better control and process integration.

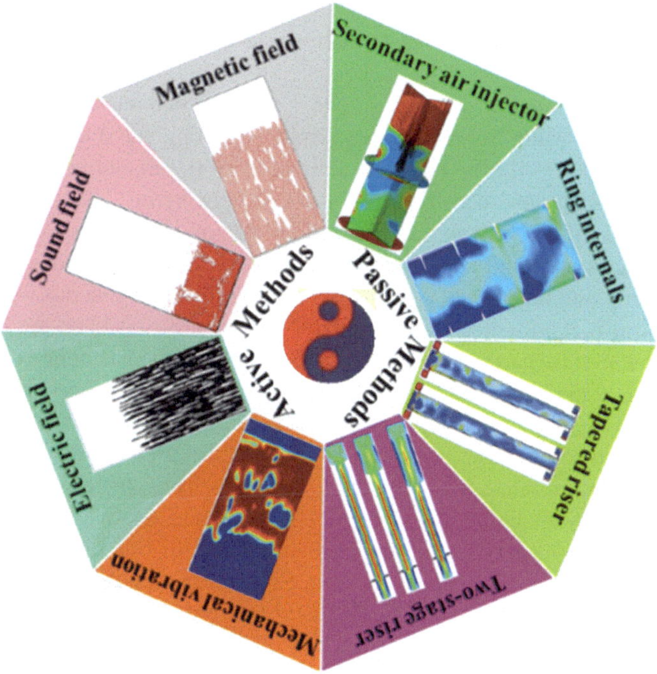

Fig. 6.1 Summary of active and passive methods of clusters and bubbles regulation in fluidized beds

6.1.1 Active-Method Cluster Regulation of Risers

Pulsation, supplying through the inlet gas oscillation, is an example of the active methods to modify both gas-particles interphase and particles interactions. The gas pulsation is achieved by means of different approaches before the plenum chamber, to create an oscillation of inlet gas flow rate. The unique hydrodynamic behavior of pulsed fluidized beds is observed when the inlet gas flux rate which relates with inlet gas frequency and amplitude is compared with a constant inlet gas mass flux in conventional fluidized beds. The rational pulsing frequency and amplitude improve the performance of fluidized beds reactors. It can also be speculated that the inlet gas pulsations play very important roles on cluster distributions in gas-particles fluidized bed risers.

The inlet gas velocity profile for pulsation is expressed as follows [54]:

$$u_g = u_m + u_a \sin(2\pi f_{in} t) \tag{6.1}$$

where u_m is the mean gas velocity. f_{in} and u_a are the frequency and amplitude of inlet gas velocity.

The instantaneous solid volume fractions α_s and minimal rates of energy dissipation (i.e. min N_{df}) using Eq. (5.108) are displayed in Fig. 6.2 at different inlet gas velocity amplitudes [54]. The riser has a diameter of 0.2 m and height of 14.2 m. The particle diameter and density are 76 μm and 1712 kg/m^3. Using the computational threshold (min N_{df}) from Eq. (5.108), the rate of energy dissipations N_{df} are obtained at the minimum value point when the particle cluster is identified in the grid. Contrarily, the function of N_{df} has no solution for homogeneous flow without clusters, and it is represented by a white color. Generally speaking, the high solid volume fraction cause particle clusters formation in the grids. The number of particle clusters is less at the top than that at the bottom along height. The energy dissipation caused by hydrodynamic and collisional interactions originates from relative velocity between gas and solid phases and granular temperature. Both relative velocity and granular temperature relate to solid volume fractions. Thus, the energy dissipation rate is varied with changing solid volume fractions and relative velocities at the bottom of the riser, leading to the minimum energy dissipation rates to be larger at the bottom than at the upper part of the riser. Comparing with the constant inlet gas velocity, the number of clusters is significantly reduced and the energy dissipation rate is increased at pulsating inlet gas velocity in the riser.

The distributions of energy dissipation rates are displayed in Fig. 6.3 for two grids at the pulsing amplitude of 3.0 m/s. For the grid at $x/R = 0.7$, one extreme minimum value is occurred from solutions of function N_{df} with changing dense and dilute gas volume fractions. This extreme value denotes only a minimum rate of energy dissipation in the range of dense phase gas volume fraction α_c from α_g to 0.6 and dilute phase gas volume fraction α_d from α_g to 0.999. The dense and dilute gas volume fraction at the extreme value are $\alpha_c = 0.86$ and $\alpha_d = 0.996$, representing heterogeneous flow with clusters. On the contrary, the rate energy dissipation rate

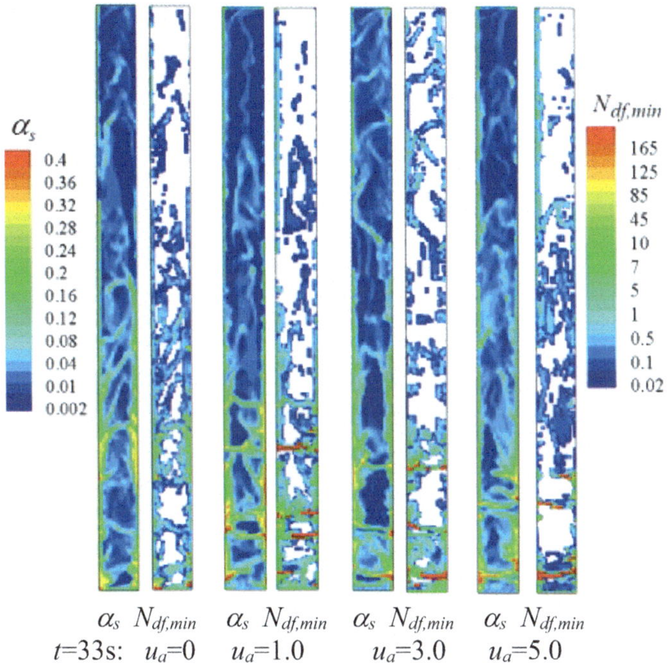

Fig. 6.2 Instantaneous solid volume fractions and minimum energy dissipation rates at different amplitudes ($f_{in} = 8.0$ Hz. The color of $N_{df,min}$ represents energy dissipation rate at the minimum value point, i.e. min N_{df} for heterogeneous flow. The white color means homogeneous flow at N_{df} $\neq$ min.)

decreases continuously with increasing dense and dilute gas volume fractions. There are no extreme minimum values at $x/R = 0.0$, representing a homogeneous flow without clusters.

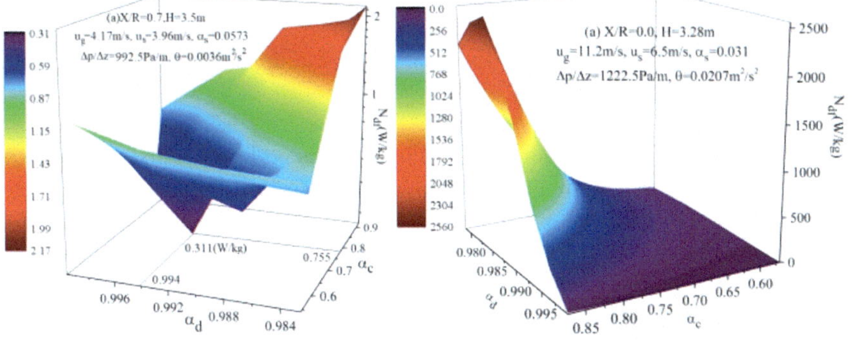

Fig. 6.3 Energy dissipation rates at the amplitude of 3.0 m/s in two grids ($t = 40$ s, $f = 8.0$ Hz)

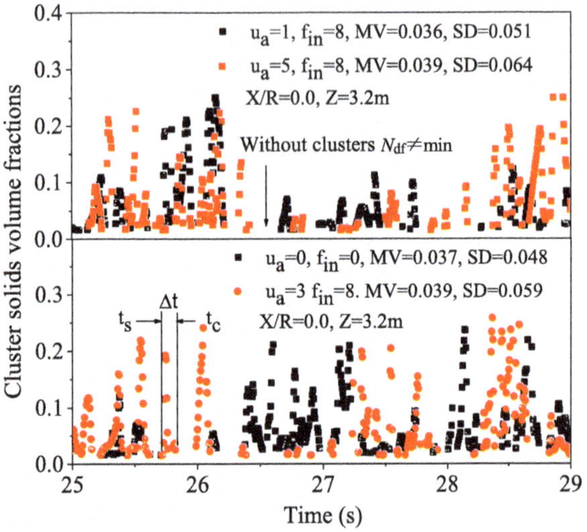

Fig. 6.4 Instantaneous cluster solid volume fractions at four different amplitudes

The cluster fraction f and dense gas volume fraction α_c are obtained from the solution of the equation set at N_{df} = min. The cluster solid volume fraction is $f(1-\alpha_c)$ in the grids. Figure 6.4 shows the instantaneous cluster solid volume fractions at four inlet gas velocity amplitudes. Both dense clusters with large solid volume fractions and the looser clusters with low solid volume fractions are observed in the riser, and the cluster solids volume fractions change with changing inlet gas velocity amplitudes in comparison to a constant inlet gas velocity. The cluster existence time fraction is calculated according to the cluster occurrence start time t_s, and cessation time t_c. The values are 0.037, 0.036, 0.029 and 0.039 at the center of $x/R = 0.0$ as the inlet gas velocity amplitude ranges from 0.0 m/s to 5.0 m/s, indicating that the time interval of the clusters varies with changing inlet gas velocity amplitudes.

The cluster is identified based on the computational criterion at N_{df} = min. The occurrence of heterogeneous flow with clusters is observed from the start time t_s to the end time t_e from Fig. 6.4. The cluster duration is

$$T_c = \frac{1}{N}\sum_{i=1}^{N}\Delta t_i = \frac{1}{N}\sum_{i=1}^{N}(t_c - t_s)_i \tag{6.2}$$

where N represents the total number of clusters observed in a statistical analysis period. Δt_i is the time interval of particle clusters in the grid with the area A_i.

The cluster existence time fraction defines as follows.

$$F_c = \frac{1}{T_0}\sum_{i=1}^{N}\Delta t_i \tag{6.3}$$

where T_0 is the computing time within which the clusters are detected.

The cluster frequency is computed according to the cluster number observed per second over the total computing time.

$$f_c = \frac{N}{T_0} \tag{6.4}$$

The heterogeneous flow with clusters is discriminated according to the computational criterion ($N_{df} = \min$). The total grid number with heterogeneous flow N_t is detected at the time t in a statistical analysis period. The number fraction of the heterogeneous flow with clusters is calculated at the time t according to the total grid number N_0 of the riser.

$$\varphi_c = \frac{\sum_{i=1}^{N} A_i}{\sum_{i=1}^{N_0} A_i} \tag{6.5}$$

where A_i is the grid area ($A_i = \Delta x_i \Delta z_i$ along lateral and axial directions in the two-dimensional riser). For the homogeneous flow of the riser, the number fraction is zero, while it is 1.0 for the complete heterogeneous flow in the riser. For a real gas-particles flow in the riser, the value of the number fraction is $0.0 < \varphi_c < 1.0$.

The distributions of particle cluster existence time fractions F_c are displayed in Fig. 6.5 at different inlet gas velocity amplitudes. All cases illustrate that the existence time fractions of particle clusters are small within the center region and large near the wall region, indicating that the ascending particle clusters have low existence time fractions at the center region while the descending particle clusters have high existence time fractions near the wall region. The cluster duration is low at the center, and increases toward the walls. It reaches maximum at the walls. Computational simulations indicate that the cluster existence time fractions are larger at the bottom zone than at the upper part of the riser. The existence of particle clusters is dominant in the bottom along riser height. For four different inlet gas velocity amplitudes, the difference among the existence time fractions is obvious at the riser height of 2.8 m. However, the effect of inlet gas velocity amplitudes on the cluster existence time fractions can be neglected at the height of 9.0 m, indicating that the inlet gas pulsation is effective for short-heighted risers.

The cluster diameters are shown in Fig. 6.6 at different pulsating amplitudes. The diameters of clusters are largest at the constant inlet gas velocity ($u_a = 0.0$), and they are smallest at the inlet gas pulsating amplitude of 1.0 m/s. The cluster diameters lie in between, at the inlet gas pulsating amplitudes of 3.0 and 5.0 m/s. These results indicate that the diameters of clusters can be manipulated using the regulation and control of pulsating amplitudes of inlet gas velocity from the inlet at the bottom.

The instantaneous solid volume fractions and rates of energy dissipation $N_{df,\min}$ at the minimum value points are displayed in Fig. 6.7 at the inlet gas velocity of 5.2 m/s and solids mass flux of 81.5 kg/m^{2}s. All cases show that the high solid

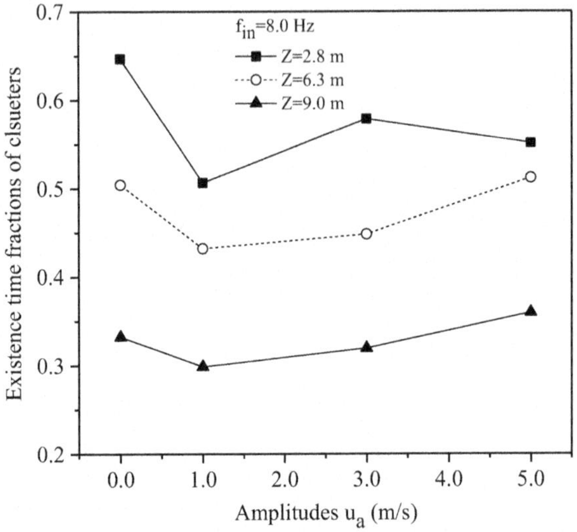

Fig. 6.5 Distribution of cluster existence time fractions at different amplitudes

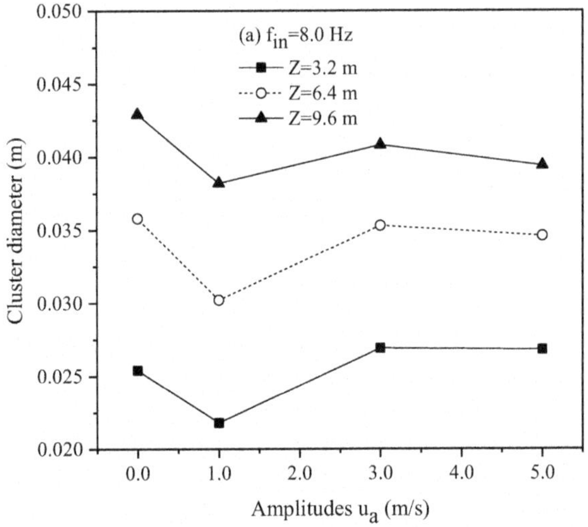

Fig. 6.6 Profile of averaged diameters of particle clusters

volume fractions appear at the bottom zone, while the low solid volume fractions exist at the upper part of the riser. The values of energy dissipation rates at the minimum value points (i.e., min N_{df}) are expressed by different contour plots, the white color represents homogeneous flow without particle clusters in the grid. More particle clusters are found near the bottom of the riser. Furthermore, it is found that more clusters exist near the walls at the upper part of the riser. The values of energy dissipation rates at the minimum values points for three different frequencies increase in comparison with the minimum energy dissipation rates at the constant inlet gas

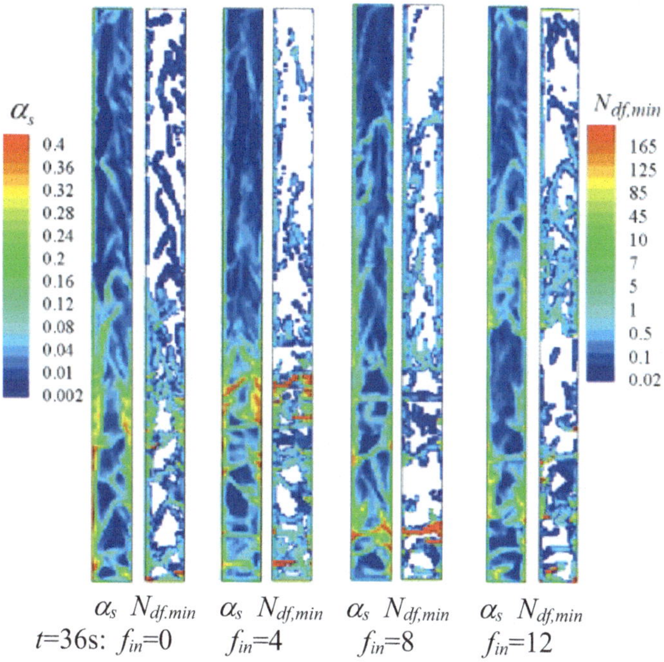

Fig. 6.7 Instantaneous solids volume fractions and minimum energy dissipation rates at different frequencies (The color of $N_{df,\min}$ represents the values of energy dissipation rates at min N_{df} for heterogeneous flow, and the white color means homogeneous flow in the grids.)

velocity, indicating that the kinetic energies are supplied by inlet gas velocities with different pulsating frequencies are dissipated mainly near the bottom of the riser.

The distributions of existence time fractions of particle clusters are illustrated in Fig. 6.8 at different pulsating frequencies. All cases show that the cluster existence time fractions are lower at the center region than near the wall region. The walls are expected to be covered by particle clusters with largest existence time fractions. It is found that the trends are similar at the two inlet gas velocity frequencies along lateral direction. Because the inlet gas kinetic energy from the bottom cannot propagate far enough without damping through collisions of particles, the degree of sensitivity to pulsating frequencies is reduced at the height of 9.0 m compared to the existence time fraction of particle clusters at the height of 2.8 m, reflecting that the effect of inlet gas velocity frequency on cluster existence time fractions can be neglected in the upper part of the tall riser. Furthermore, the cluster exist time fractions are observed to be lowest at the inlet gas pulsating frequency of 4.0 Hz, and largest at the constant inlet gas velocity. The cluster existence time fractions at the frequencies of 8.0 and 12.0 Hz fall in between. The cluster existence time fraction reduces by about 0.189–0.213, indicating that the inlet gas velocity pulsation contributes to the reduction of cluster duration times in the riser.

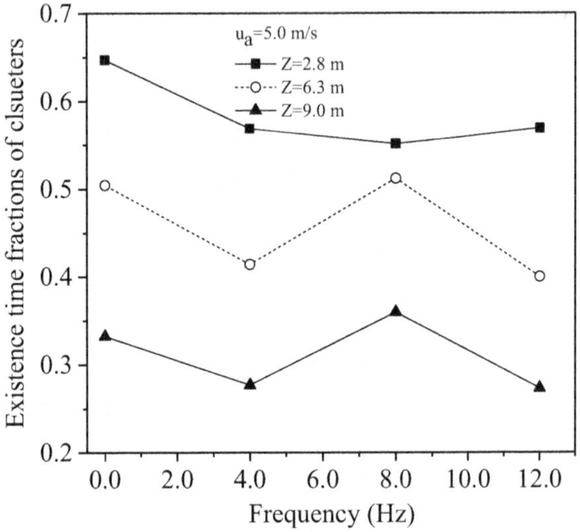

Fig. 6.8 Distribution of cluster existence time fractions

The distributions of cluster diameters are illustrated in Fig. 6.9 at three different riser heights. The cluster diameters are lowest at the inlet gas pulsating frequency of 4.0 Hz, and largest at the inlet gas pulsating frequency of 12.0 Hz, indicating that the cluster diameter will become large at high pulsating frequencies. It is also found that the cluster diameter decreases along flow direction in the riser. These indicate that the inlet gas velocity frequency contributes to the reduction of particle cluster diameter and increment of gas volume fractions of clusters in pulsating fluidized bed risers.

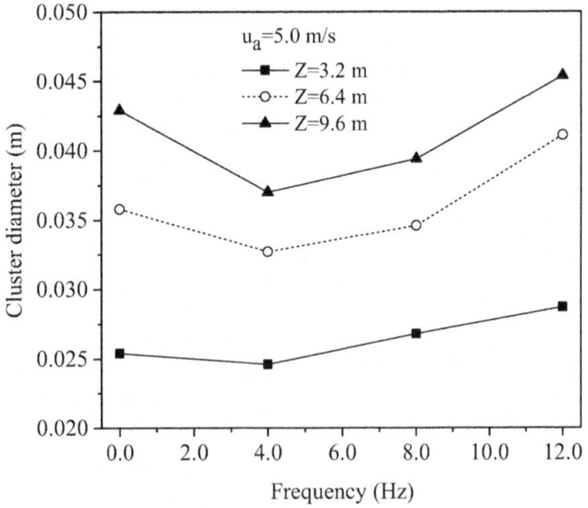

Fig. 6.9 Profile of sectional area-averaged diameter of clusters

The cluster diameter and cluster solid volume fraction, and existence time fractions are varied with changing inlet gas pulsating amplitude and frequency, indicating that the hydrodynamics of particle clusters can be improved in pulsating fluidized bed risers. This approach represents a new avenue for the intensified design of gas–solid fluidized bed operations–not only the optimization of existing processes, but also the creation of new designs and operations. This is based on the ability to externally fine-tune a dynamic flow structure of clusters using the periodic gas flow oscillations as a self-organizing principle in nature.

6.1.2 Passive-Method Cluster Regulation of Two-Stage Riser

The distributions of instantaneous volume fractions of solid phase α_s and particle clusters $\alpha_{s,c}$ are displayed in Fig. 6.10 in three different size ratios of D_1/D_2, where D_1 and D_2 are the sizes of the first and second chambers [53]. In the contracting two-stage riser, the first chamber has a large diameter in comparison to the diameter of the second chamber. For a constant-diameter riser, the size is kept constant along height. For the expanding two-stage riser, the diameter is smaller in the first chamber than it is in the second chamber. The heights of the first and second chambers are H_1 with the diameter D_1 and H_2 with the diameter D_2. The transition section with the height H_{12} connects between the first and the second chambers. All risers have the same size of inlet at the bottom of the risers. Gas and particles are introduced through the inlet at the bottom, and exited through two symmetric outlets at the top of the riser [54].

For contracting-diameter two-stage riser which is shown in Fig. 6.10a, the diameter of the first chamber is larger than that of the second chamber. The diameter of the first chamber D_1 is 76 mm with the diameter of second chamber D_2 of 66 mm and 56 mm, respectively. The heights of first and second chamber are 4.94 m and 4.94 m, respectively. Particle diameter and density are 76 μm and 1780 kg/m^3. The inlet gas velocity and sold mass flux are 5.0 m/s and 100 kg/m^{2}s.

When compared to distribution of solid volume fractions in the constant-diameter riser, the solid volume fractions are large in the first chamber, and low in the second chamber, indicating that the solid phase becomes denser in the first chamber and looser in the second chamber in the contracting-diameter two-stage risers. The volume fraction of particle clusters $\alpha_{s,c}$ expresses the fractional volume of the dense phase occupied by clusters. Using Eq. (5.108), the solution of the equation set gives a value of volume fraction $\alpha_{s,c}$ at the minimum value point ($N_{df} = \min$). On the other hand, no solution is found using Eq. (5.112), as represented by the white color at $N_{df} \neq \min$ in Fig. 6.10 (b) along riser height. It is further observed that more particle clusters formed near the bottom. The formation of particle clusters become much lesser in the second chamber than in the first chamber. Furthermore, it is found that the particle clusters are much lower in the contracting-diameter two-stage risers compared to the cluster formation in the constant-diameter riser. Particle clusters occur due to the change in flow area, hence due to hydrodynamic interactions between the gas and

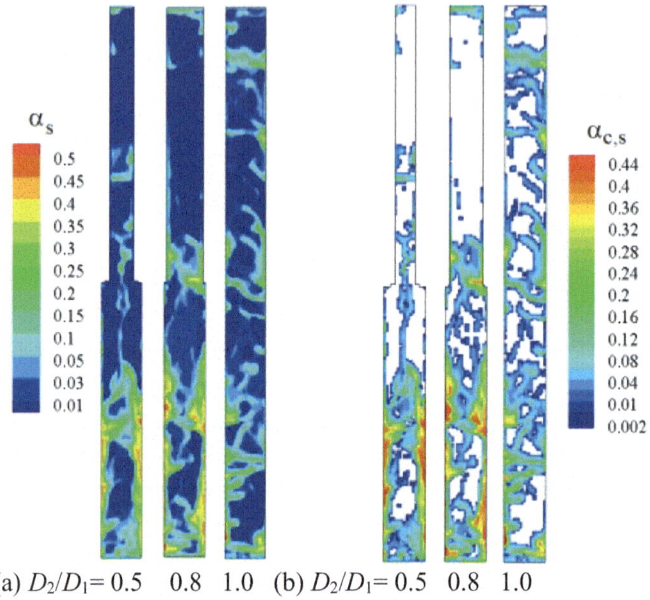

Fig. 6.10 Instantaneous volume fractions of solid phase α_s and clusters $\alpha_{s,c}$ at t = 38 s in contracting-diameter riser (The colors of $\alpha_{s,c}$ represent flow of clusters at N_{df} = min, and the white color means homogeneous flow at $N_{df} \neq$ min)

solid phases as well as interactions of collisions of particles. Thus, a reduction in flow area will reduce particle clusters in the second chamber compared to the upper part of the constant-diameter riser.

A comparison of cluster volume fractions and velocities is displayed in Fig. 6.11 at three different size ratios D_2/D_1 in two-stage risers. The mean cluster solid volume fractions are similar in the first chamber, and reduce from 0.066 in the constant-diameter riser to 0.058 at the second chamber in the contracting-diameter two-stage riser with D_2/D_1 of 0.5, indicating that the particle clusters are diluted in the second chamber. For the contracting-diameter two-stage riser of D_2/D_1 equals 0.8, the mean solid volume fractions fall in between them. From the start time t_s and the end time t_e, the existence time fractions of clusters are computed. The cluster existence time fractions are similar in the first chamber. Whereas it reduces from 0.349 in the constant-diameter riser to 0.132 in the contracting-diameter two-stage riser with D_2/D_1 of 0.5, illustrating that the existence time interval of clusters is short at the second chamber.

The axial distributions of cluster existence time fractions are displayed in Fig. 6.12 at three different risers. The mean existence time fraction is largest at the constant-diameter riser, and smallest at the contracting-diameter two-stage riser with D_2/D_1 = 0.5. For the contracting-diameter two-stage riser with D_2/D_1 = 0.8, the cluster existence time fraction falls in between them at the second chamber. Furthermore,

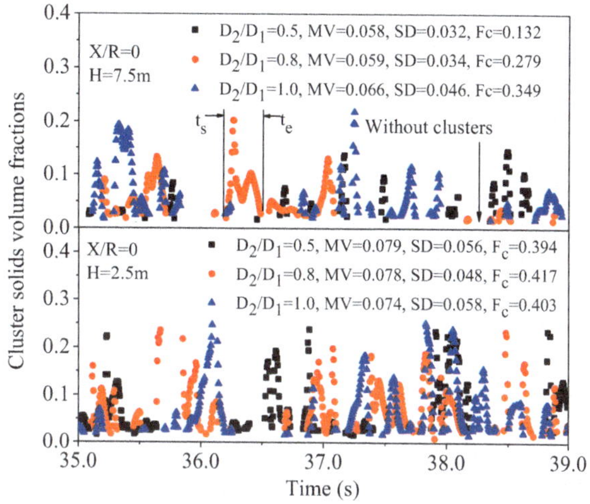

Fig. 6.11 Cluster volume fractions and velocities at three different size ratios in two-stage risers

it is found that the profile of existence time can be divided into five regimes. They include inlet and outlet regimes of I and V, developed flow regimes of II and IV in the first and second chambers and transition regime of III between the first chamber and the second chamber. The cluster existence time fractions reduce at the transition regime III from the first chamber to the second chamber. It is also observed that the cluster existence time fractions drop in the second chamber compared to the first chamber, indicating that the contracting-diameter two-stage riser provides a low existence time fractions of particle clusters.

For expanding-diameter two-stage riser, the diameter of the first chamber is smaller than that of the second chamber. The instantaneous volume fractions of

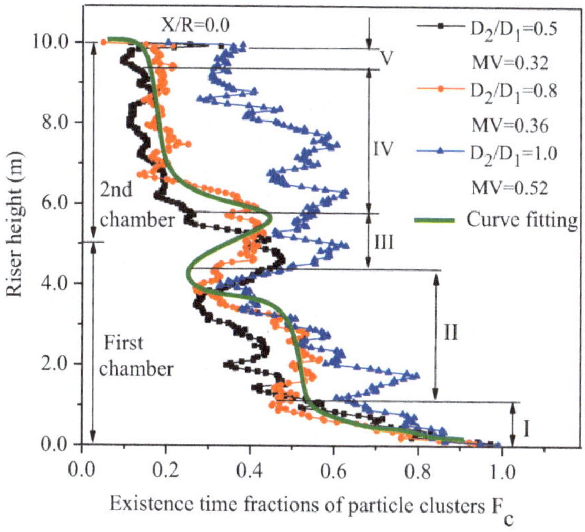

Fig. 6.12 Profiles of cluster existence time fractions at three contracting-diameter risers

solid phase and particle clusters are shown in Fig. 6.13 at three different risers. The diameter of first chamber D_1 is 76 mm. The diameter of second chamber D_2 is 84 mm and 94 mm, respectively. The heights of first and second chambers are 4.94 and 4.94 m. the diameter and density of particles are 76 μm and 1780 kg/m^3. Inlet gas velocity and sold mass flux are 5.0 m/s and 100 kg/m^{2}s. All cases illustrate that the solid volume fractions are similar in the first chamber. For both expanding-diameter two-stage risers, the instantaneous solid volume fractions are large in the second chamber compared to the contour plots in the constant-diameter riser. It is found that the dilute flow with low solid volume fractions and the dense flow with high solid volume fractions exits along lateral direction of the second chamber. The volume fractions of particle clusters are also large in both expanding-diameter two-stage risers, reflecting that larger diameter contributes to the formation and growth of particle clusters in the second chamber.

The instantaneous solid volume fractions and velocities of particle clusters are displayed in Fig. 6.14 at two different two-stage risers. The time interval of particle clusters is discriminated from the start time t_s to the end time t_e. The cluster frequencies are calculated, and they are 5.48 and 5.79 (1/s) at $D_2/D_1 = 1.25$ and 1.5 in the first chamber. For the second chamber, the cluster frequencies are 7.96 and 7.81 (1/s) at $D_2/D_1 = 1.25$ and 1.5, indicating that the number of particle clusters increased, compared to the rate of formation of clusters in the first chamber. Furthermore, it is found that the mean values of solid volume fractions and velocities of clusters are larger in the second chamber compared to values obtained from cluster flow in the

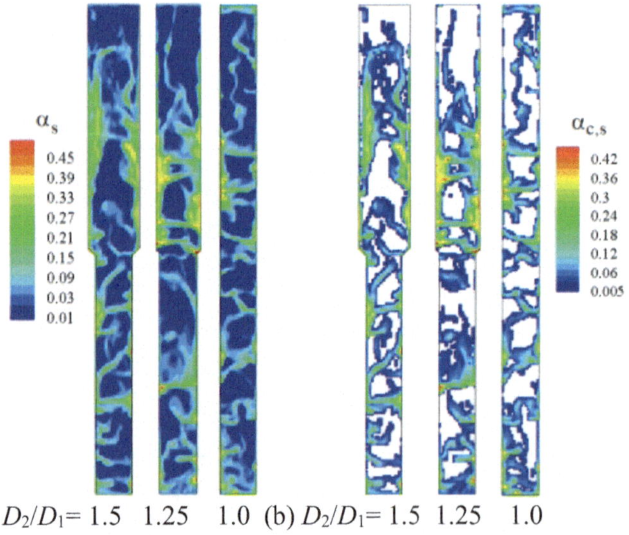

Fig. 6.13 Instantaneous volume fractions of solid phase α_s and clusters $\alpha_{s,c}$ at $t = 35$ s in expanding-diameter risers (The colors of $\alpha_{s,c}$ represent heterogeneous flow with clusters at $N_{df} = \min$, and the white color represents homogeneous flow without clusters at $N_{df} \neq \min$)

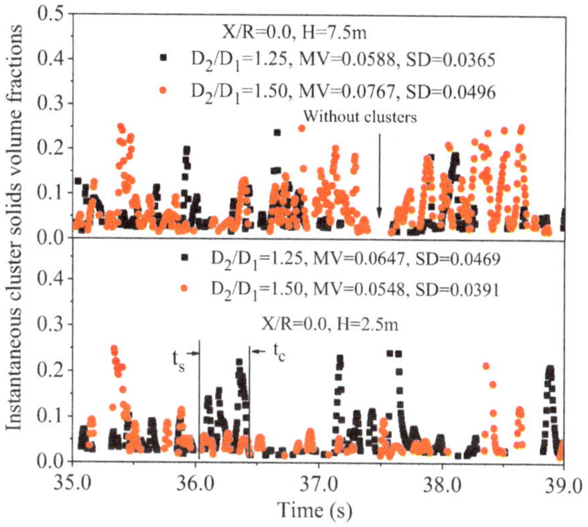

Fig. 6.14 Instantaneous cluster solid volume fractions in expanding-diameter risers

first chamber. These results illustrate that increasing the diameter ratio D_2/D_1 leads to reduction of cluster velocities in the second chamber.

Figure 6.15a compares the distributions of existence time fractions of particle clusters in two different expanding-diameter two-stage risers. The axial distribution of existence time fractions of particle cluster can be divided into five different regimes. The inlet and outlet regimes of I and V are at the bottom and exit of the riser. The developed flow regime of II is located in the middle of the first chamber. The transition regime III exists between the first chamber and the second chamber. The descent regime IV is formed at the middle of the second chamber, where the existence time fraction falls gradually. The cluster existence time fractions rise at the transition regime III from the first chamber to the second chamber, indicating that large diameter promotes reduction of particle clusters due to the effect of the core-annulus flow of particles in the second chamber.

The number fraction φ_c represents the ratio of the computational grid number occupied by heterogeneous flow with clusters to the total computational grid number of the riser. Figure 6.15b shows a comparison of the number fractions in two different two-stage risers. The larger the number fraction is, the higher the grid number of heterogeneous flow with clusters is. It is found that cluster number fractions φ_c are similar at two different ratios D_2/D_1 in the first chamber. However, the time-averaged number fraction φ_c is larger in the second chamber, compared to the grid number with heterogeneous flow in the first chamber, indicating that increasing the diameter of the second chamber is beneficial to formation of particle clusters. Furthermore, the mean value φ_c is lower at $D_2/D_1 = 1.25$ than that at $D_2/D_1 = 1.50$, indicating that larger diameter of the second chamber promotes the formation of particle clusters.

In summary, compared to distributions of solid volume fractions along height of a constant-diameter riser, a transition region exists between the first chamber and the second chamber.

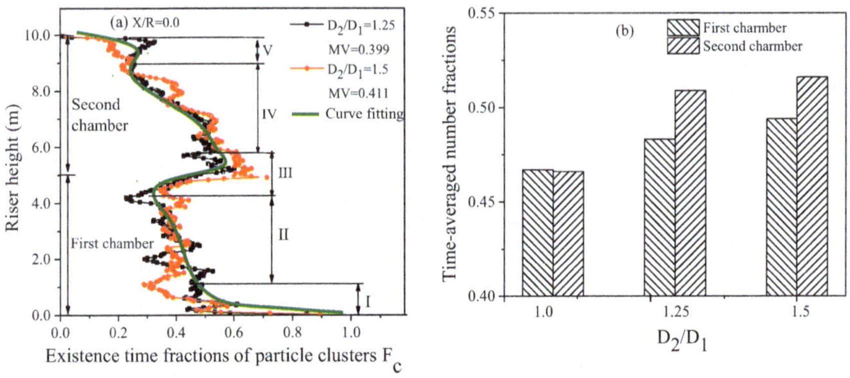

Fig. 6.15 Profiles of existence time fractions of particle clusters in expanding-diameter risers

Compared to distributions of particles clusters in a constant-diameter riser, the cluster solid volume fractions and diameters are decreased in the contracting-diameter two-stage risers. The cluster velocities increase and the cluster existence time fractions reduce in the second chamber.

For expanding-diameter two-stage riser, the solid volume fraction and diameter of clusters increase comparing to a constant-diameter riser. The cluster velocities are reduced and the cluster existence time fractions are increased in the second chamber.

6.1.3 Cluster Regulation of Gas Nozzle Stripper

The strippers are widely used in fluid catalytic cracking (FCC) units. In fluidized bed risers, strippers are employed to increase gas–solid contacting, to prevent short-circuiting of solids, to reduce gas bypassing at the centre of the column and to help distribute the gas radially, reducing axial dispersion and thereby increasing stripping efficiency. Hence, the stripper performance is greatly influenced through proper design practices. It is also very important that the stripper performance is greatly influenced by the catalyst properties and operating conditions.

Gas jet strippers are essential by an upflowing gas removes product entrained in a downflowing stream of the dense phase of clusters. The role of the stripper unit is to facilitate the transfer of the gas from the dense phase of the stripping gas. The main functions of the strippers are to (i) break up and slow down dense phase of clusters in order to enhance the interchange of gas between the dense phase and the dilute phase; (b) divide a fluidized bed into several axial sections (i.e., staging), so as to make it operate more in a plug flow mode to reduce gas and solids back-mixing; (c) add or remove heat in fluidized bed reactors and combustors; and (d) strip gaseous products entrained in the dense phase of clusters. The counter-current contacting between the down-flowing dense phase of clusters and the up-flowing dilute phase of gases is enhanced.

For the stripper with horizontal gas jets, the stripping efficiency depends on the arrangements of horizontal gas jets. The distributions of cluster properties are simulated in a gas nozzles-assisted fluidized bed riser with a height and diameter of 8.1 and 0.186 m. The inlet of gas and particles locates at the bottom of the riser, and the exit is placed at the right-hand side wall at the top of the riser. The injection nozzles are mounted at the height of 2.64 m from the inlet. The diameter of the nozzle is 0.027 m. Three different configurations of the injection nozzles are simulated. One is a co-current flow of jet gas and fluidizing gas. The upward direction of gas jet stream from the nozzles inclines at the angle of 30 degree. The second case is a counter-current flow of jet gas and fluidizing superficial gas. The downward direction of gas jet stream is ejected into fluidizing gas and particles mixture. The last case is for a horizontal gas jets. All nozzles-assisted fluidized bed risers are divided into two zone. One is the upstream zone from the inlet to the injection nozzles. The other is the downstream zone from the injection nozzles to the outlet [53].

From the criterion of Eq. (5.108), the value of instantaneous rate of energy dissipation $N_{df,\min}$ (i.e. $N_{df} = \min$) is computed, indicating that the heterogeneous flow of clusters exists in the cell. On the other hand, the homogeneous flow is identified according to Eq. (5.112) in the cell. Figure 6.16 shows the instantaneous solid volume fractions and energy dissipation rate $N_{df,\min}$ (i.e. a solution of $N_{df} = \min$) in the cells of the riser. Particle diameter and density are 65 μm and 1310 kg/m^3, respectively. The inlet gas velocity and solid mass flux are 3.28 m/s and 113 kg/m^{2}s, respectively. The white color represents homogeneous flow. The clusters occur as the solid volume fraction is high comparing to instantaneous solid volume fraction, indicating that more clusters are formed with increasing solid volume fractions. Furthermore, much more clusters are formed in the inlet and dense regions comparing to the transition and dilute regions, indicating that the formation of clusters depend on jet direction along riser height.

According to the criterion from Eq. (5.141), the cluster diameter is determined from the gas volume fractions of dense phase α_c and dilute phase α_d at the solution of $N_{df} = \min$. For a homogeneous flow without clusters, there is no solution for dense and dilute gas volume fractions in the cell. The axial distributions of cluster diameters are displayed in Fig. 6.17 at three different jet directions. All cases show that the cluster diameters increase from the inlet to the maximum in the inlet region. After that, the cluster diameter decreases gradually in the dense region. In the jet region, the cluster diameters are rapidly dropped toward the minimum, and then rise sharply due to the additional energy added by gas jets. The cluster diameter decreases gradually in the transition region. The cluster diameters tend to constants in the mixing region. Furthermore, the cluster diameters are lower for the downward gas jets than that for the upward and horizontal gas jets. All cases show that the cluster diameters are larger in the dense region than that in the transition and mixing regions, indicating gas injection reduce cluster diameters in the transition and mixing regions. The averaged cluster diameters are calculated, and they are 0.0166, 0.0187

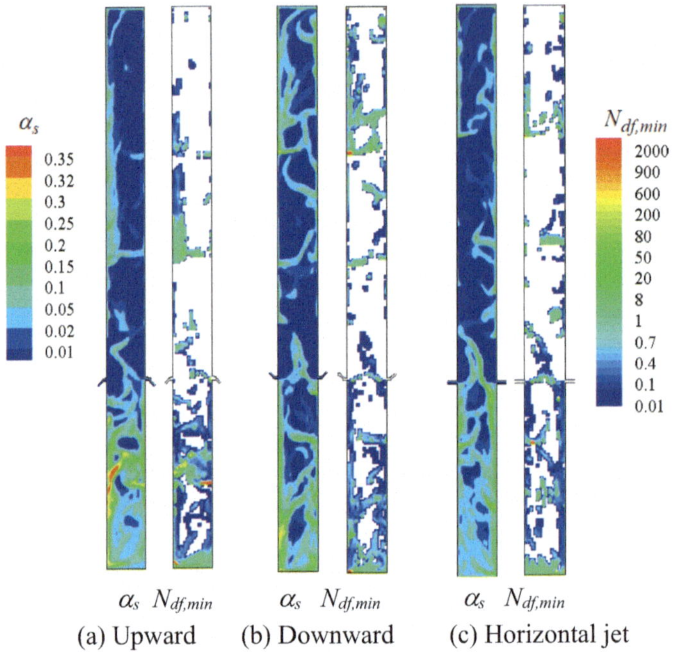

Fig. 6.16 Instantaneous solid volume fractions and minimum energy dissipation rates at $t = 23$ s (The color represents N_{df} = min for heterogeneous flow. The white denotes $N_{df} \neq$ min for homogeneous flow)

and 0.0199 m for the downward, horizontal and upward jets, reflecting that the gas jet direction affects cluster diameters in the fluidized bed risers.

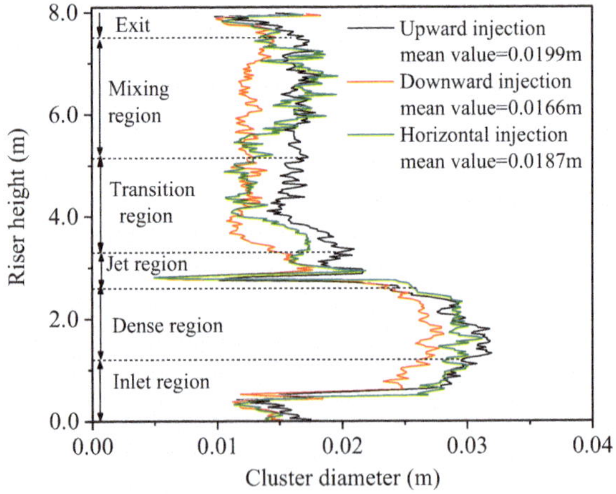

Fig. 6.17 Profiles of cluster diameters along axial directions

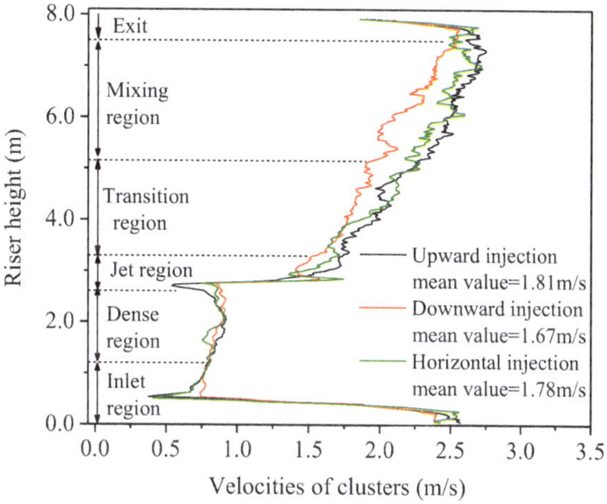

Fig. 6.18 Cluster velocities at three gas jet directions

The distributions of cluster velocities are illustrated in Fig. 6.18 at three different gas jet directions. The cluster velocities decrease from the bottom inlet to minimum, and then increase in the inlet region. In the dense region, the cluster velocities are gradually increased at first, and then decrease. From the jet region to mixing region, the cluster velocities increase at three different gas jets directions. The calculated mean values of cluster velocities are 1.67, 1.78 and 1.81 m/s for the downward, horizontal and upward gas jets, indicating that the downward gas jets contribute to reduce cluster velocity in the riser.

From simulations at injection point, a small fraction of the gas flow leads to localized micromixing. This creates smaller clusters. Gas jet creates a degree of mixing related to the local gas flow rate. Judicious placement of the gas jets provides a new degree of freedom to control mixing, decoupling it from the chaotic evolution when all clusters emerge within the riser. For scalability, it is important that the local conditions around each gas jets are similar or independent of each other. Delivering the gas flux through a set of gas jets ensures a homogeneous distribution. This structure is able to self-regulate, so that it can be scaled up, maintaining the same local environment at every gas jets.

In summary, the different regions are observed for three different gas jets directions along riser height. They include an inlet, a dense, jet, transition, and mixing and exit regions from the bottom to the top exit. The solid volume fractions are large at the downstream and low at the upstream of the gas jets.

In the transition and mixing regions at the downstream of the gas jet, the cluster diameters and velocities are smallest for the downward gas jets, and largest for the upward gas jets. The cluster solid volume fractions are largest for the upward gas jets, and lowest for the horizontal gas jests in the mixing region.

Computational simulations show that the gas jets direction play a major role in the regulation and control of clusters in the fluidized bed risers. The jet gas velocities are also lead to variations of formation of clusters in scale-up processes that has not been studied. Thus, numerical analyses for different gas jets directions and velocities are an essential step in future scale-up investigations of fluidized bed risers.

6.2 CFD-KTGF-DEM Simulations of Gas–Solid Fluidized Beds

The statistical descriptions for fluid-particles two–phase flows can be classified into two broad categories: Euler–Euler (E-E) and Euler–Lagrangian (E-L) representations. In the E-E two fluid method (TFM) approach, the kinetic theory for granular flows (KTGF) is applied [21, 41] for describing solid particulates. From a numerical viewpoint, simulations that solve the continuum conservation equations for both phases have traditionally been called as TFM-KTGF simulations. In the E-L statistical description, the dispersed phase is treated as being composed of discrete particles using discrete element method (DEM), and is known as CFD-DEM. However, some difficulties, associated with reliance of these methods on an empirical constant of restitution coefficient in TFM-KTGF and turbulence effect in CFD-DEM, are still not adequately solved in computational simulations.

CFD-KTGF-DEM approach is a coupling method to combine TFM-KTGF model and CFD-DEM model. The approach consists of linking numerical models of different scales together in a single combined model in which the meso-scale model of control volume uses information obtained by the more detailed micro-scale model of discrete particles. The micro-scale model communicates directly and instantly with the meso-scale model through some coupling procedure. Hence, the CFD-KTGF-DEM approach tries to tackle these challenges and to provide a comprehensive methodology allowing for correct modeling of multiphase systems and fluid–solid fluidized bed reactors. Application of the developed CFD-KTGF-DEM approach as an integration technique to enrich the computational results would be an avenue worth exploring.

6.2.1 Central Gas Jet Fluidized Bed

The flow behavior of bed materials is measured using a particle image velocimetry and positron emission particle tracking in a central gas jet fluidized bed [9]. The pseudo-3D central gas jet fluidized bed has a width, depth and height of 145, 20 and 1000 mm with a spout jet at the center of the distributor, as shown in Fig. 6.19a. The diameter and density of particles are 3.0 mm and 2505 kg/m^3 with the minimum fluidization velocity of 1.9 m/s. Gas is supplied through the spout jet with spouting velocity u_{sp} and the distributor with background velocity u_{bg} [44].

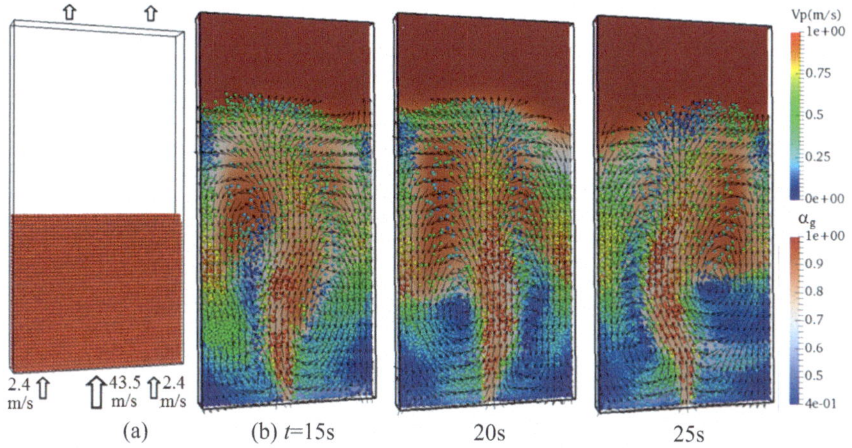

Fig. 6.19 Instantaneous volume fraction contour of Euler gas phase, velocity vector of Euler granular phase and velocity color of Lagrangian discrete particles

Using CFD-KTGF-DEM approach, the hydrodynamics of gas phase, granular phase and discrete particles are predicted in gas–solid fluidized beds [44]. The instantaneous volume fraction contour of Euler gas phase, velocity vector of Euler granular phase and velocity color of Lagrangian discrete particles are shown in Fig. 6.19b at the inlet spouting velocity u_{sp} and background velocity u_{bg} of 43.5 and 2.4 m/s. The colors of particles represent the velocity magnitude of Lagrangian discrete particles. The numerical simulations show that the dense annulus region near the walls and the dilute spout region at the center exist in the central gas jet fluidized bed. The inlet background velocity is around the minimum fluidization velocity and the inlet spout velocity is high to enable the spout channel to form. This leads to the dilute spout channel with the upward flow of Euler granular phase and Lagrangian discrete particles in the center. However, the down-flow of Euler granular phase and Lagrangian discrete particles and the up-flow of Euler gas phase exist in the annular region. The circulation of Euler granular phase and Lagrangian discrete particles is formed in the bed. It is also found that the central jet region is unstable, and moves the spout channel from side to side due to the competition between the collisional interactions and the hydrodynamic interaction of the central gas jet.

The instantaneous velocity vector of Euler gas phase, granular temperature contour of Euler granular phase and rotational velocity color of Lagrangian discrete particles are shown in Fig. 6.20 in a central gas jet fluidized bed [44]. The colors of particles represent the rotational velocity magnitude of Lagrangian discrete particles. Gas ejects from the center jet of the distributor. The high rotational velocity of Lagrangian discrete particles is formed in the spout because the discrete particles flow from the annular region to the central gas jet region. Numerical simulations further indicate that the granular temperatures of the Euler granular phase are high at the gas jet region and low at the annular region.

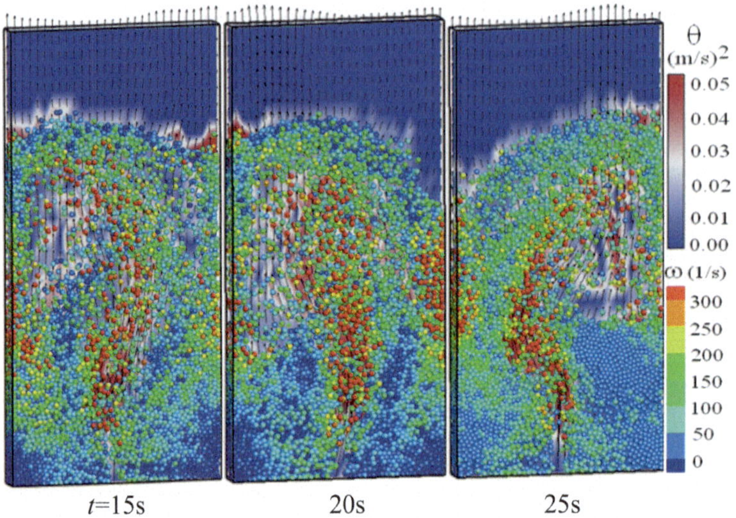

Fig. 6.20 Instantaneous granular temperature contour of Euler granular phase, velocity vector of Euler gas phase and rotational velocity color of Lagrangian discrete particles

The simulated and measured axial velocities of bed materials are shown in Fig. 6.21 along bed width. The simulated axial velocities of Euler granular phase and Lagrangian discrete particles match with experimental measurements. Both simulations and measurements show the axial velocities are positive at the gas jet region and negative at the annular region. However, the difference of axial velocities exists between simulations and experiments at two heights. The axial velocities of Lagrangian discrete particles are smaller than that of Euler granular phase at the dilute spout region. The simulated axial velocities are larger than that of measurements at the interface between the central gas jet region and the annular region, indicating that the numerical simulations over-predict the gas jet region size in comparison with measurements in the gas jet fluidized bed.

The simulated instantaneous velocity components of Euler granular phase and Lagrangian discrete particles are shown in Fig. 6.22. The large variations along axial direction exist because of the lateral movement of gas-particles jet. Both Euler granular phase and Lagrangian discrete particles show same trends within the bed.

6.2.2 *Internally Circulating Fluidized Bed*

The internally circulating fluidized bed (ICFB) is a bubbling fluidized bed with two chambers as illustrated in Fig. 6.23a. The bed is divided into two chambers termed as the high velocity (HV) bed or the reaction chamber and the low velocity (LV) or the heat exchange chamber by placing a plate called baffle at some distance above the

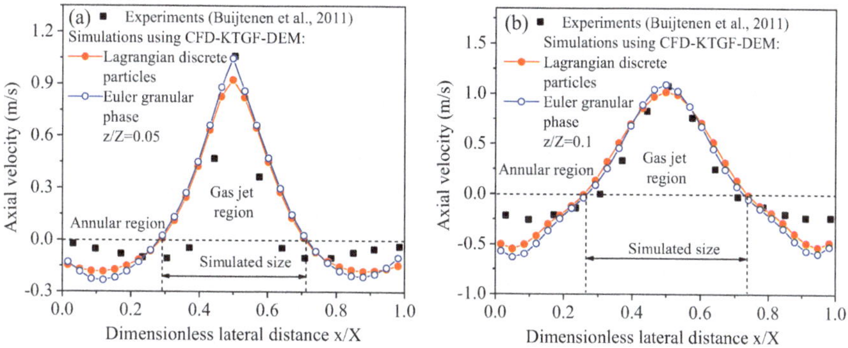

Fig. 6.21 Comparison of simulated and measured axial velocities at two heights

Fig. 6.22 Time series of velocity components for Euler granular phase and discrete particles

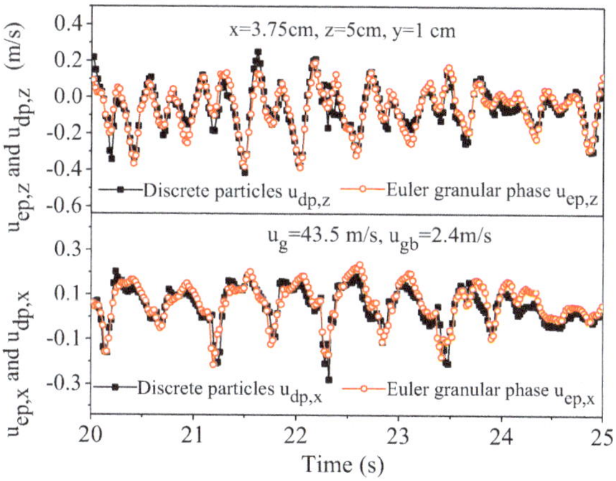

gas distributor forming a slot for the solid circulation between the chambers. Thus the solid circulation process in the reactor is developed by applying unequal gas velocities to these chambers. Such a process is beneficial to several applications in the ICFB. Due to its special layout and features, ICFBs possess several advantages over conventional fluidized beds such as reduced height, compact size and comparatively less heat loss from the reactor. These advantages enable ICFBs to be implemented in coal/biomass combustion and gasification, solid waste disposal, production of highly pure polysilicon particles for the photovoltaic industry and flue gas desulfurization.

For simulations, the HV chamber has a 24 mm × 6 mm in cross-section, and the LV chamber has a 24 mm × 6 mm in cross-section [45]. The height of ICFB is 250 mm. The two chambers are separated by a vertical baffle with a slot in the lower section. The height of the slot is 15 mm from the distributor, and the baffle height is 75 mm. The diameter and density of particles are 1.0 mm and 1050 kg/m^3. The number of discrete particles is 20000 within the packing height of 70 mm of the

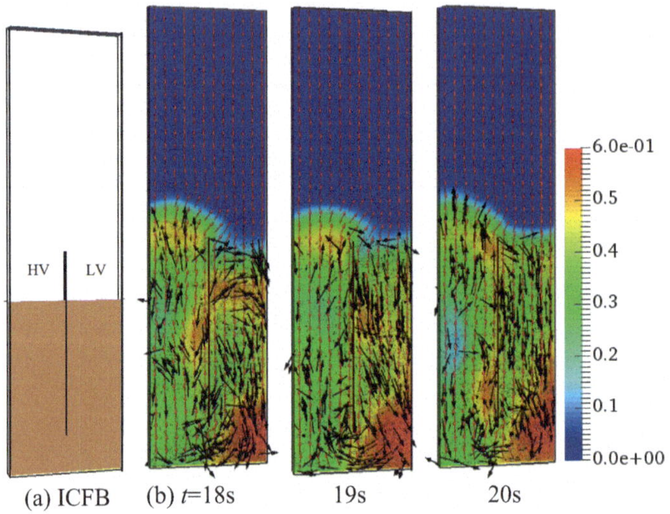

Fig. 6.23 Simulated instantaneous velocities of discrete particles and Euler gas phase and volume fraction of Euler granular phase (black vector for discrete particles, red color vector for Euler gas phase, contour for Euler granular phase)

HV and LV chambers. The calculated packing solid volume fraction based on the particle number is 0.519 which is used as the initial solid volume fraction.

Figure 6.23b shows snapshots of simulated velocity vectors of Euler gas phase and Lagrangian discrete particles and the volume fractions of Euler granular phase in the ICFB at the inlet velocity u_{gHV} and u_{gLV} of 1.3 and 0.5 m/s, respectively. Initially, the simulated Euler granular phase and Lagrangian discrete particles sink by the side of the bed. When gas enters into the chambers, the formation of bubbles is observed from the motion of discrete particles and the variations of volume fractions of Euler granular phase. A passing bubble lifts the discrete particles and Euler granular phase preventing it to reach the bottom of the fluidized bed. At a certain instant, the discrete particles detach from the rising bubble starting a new sinking path. Finally, the discrete particles are raised again by the action of passing bubble. From the simulated instantaneous volume fraction of Euler granular phase, the bubbles are initially formed at the distributor, and then gathered into large bubbles due to bubble coalescence. Afterwards, these bubbles break at the bed surface. The flow profile of discrete particles around a bubble includes the upward motion at the wake and nose of a bubble and the downward motion at the left and right sides of the bubble. Such processes are repeated continuously during the simulations. Nevertheless, simulations are able to reproduce the rising and sinking behavior of Euler granular phase and Lagrangian discrete particles as well as the motion of bubble of Euler gas phase in the chambers.

The instantaneous Euler gas volume fractions, Euler granular phase velocity and Lagrangian discrete particle velocity are shown in Fig. 6.24 in an ICFB. The velocity vectors of both Euler granular phase and Lagrangian discrete particles show that they flow from the HV chamber to the LV chamber, forming a circulation in the ICFB. The velocities of Euler granular phase and Lagrangian discrete particles v_p are higher in the HV chamber than that in the LV chamber. Both Euler granular phase and Lagrangian discrete particles fluctuate strongly in the HV chamber. In the LV chamber, the velocity fluctuation of Euler granular phase and Lagrangian discrete particles are gentle. These lead that Euler gas phase volume fractions α_g are larger in the HV chamber than that in the LV chamber. Both Euler granular phase and Larrangian discrete particles move downflow in the LV chamber, flow into the HV chamber through the slot, and rise fast in the upflow HV chamber. Finally crosses over the partition to fall into the downflow LV chamber. These indicate that the hydrodynamics of bed materials operate different fluidization states in the HV and LV chambers.

The instantaneous velocities of Euler granular phase u_s and Lagrangian discrete particles u_p are shown in Fig. 6.25 in the HV and LV chambers. The fluctuations of axial velocities of Lagrangian discrete particles and Euler granular phase are large because bubbles pass through the HV chamber and LV chamber. However,

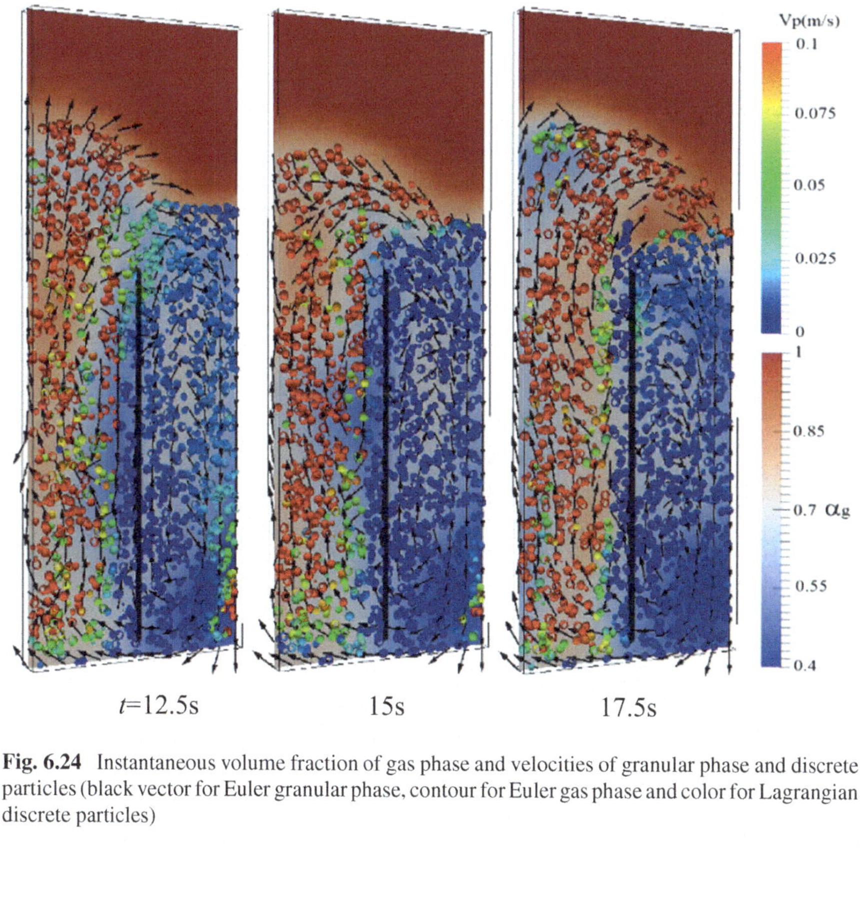

Fig. 6.24 Instantaneous volume fraction of gas phase and velocities of granular phase and discrete particles (black vector for Euler granular phase, contour for Euler gas phase and color for Lagrangian discrete particles)

they are low along *y*-direction due to the narrow space between two side-walls of the ICFB. This indicates that both Lagrangian discrete particles and Euler granular phase flow down in the LV chamber and flow up in the HV chamber. The circulation of Lagrangian discrete particles and Euler granular phase is formed within the HV chamber and the LV chamber.

Figure 6.26a shows the instantaneous volume fractions of Lagrangian discrete particles and Euler granular phase in the HV chamber and LV chamber. The volume fraction difference between the HV and LV chambers is developed because of the uneven aeration, which influences the solid flux through the slot beneath the baffle. These result in a motion of Euler granular phase and Lagrangian discrete particles from the LV chamber to the HV chamber.

The circulating solids mass flux through the slot beneath the baffle is very important to control the distributions of volume fraction and velocity in the HV and LV chambers. Figure 6.26b shows the circulating solid mass flux of Lagrangian discrete particles and Euler granular phase with times. The fluctuation of solid mass flux

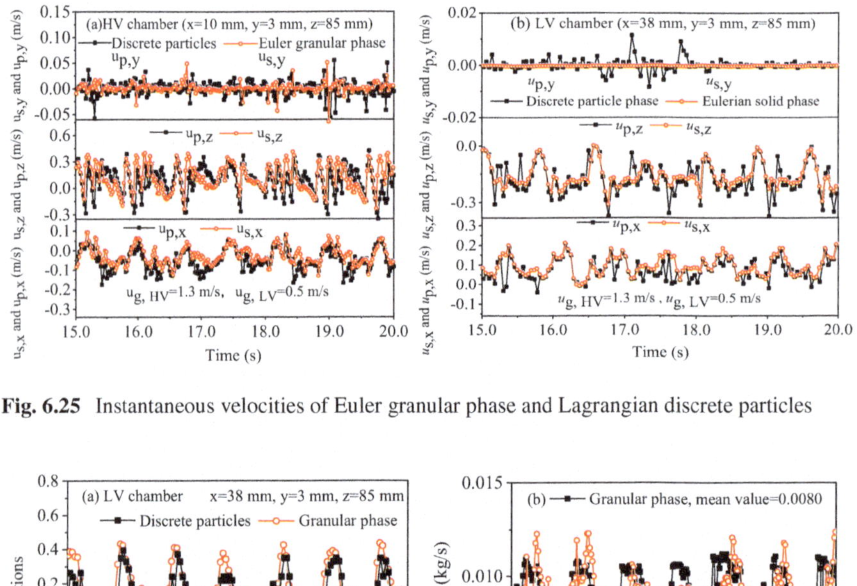

Fig. 6.25 Instantaneous velocities of Euler granular phase and Lagrangian discrete particles

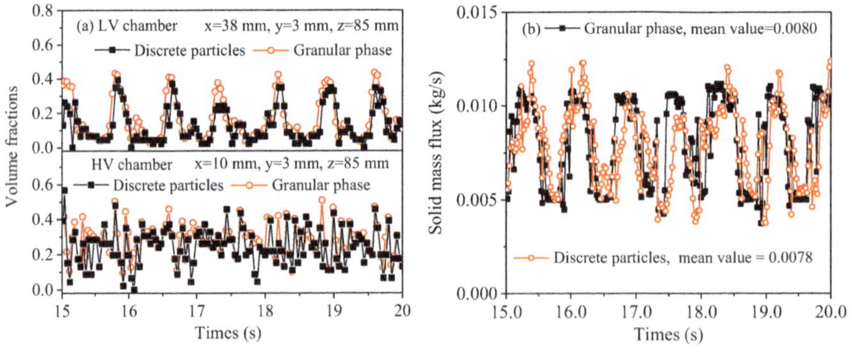

Fig. 6.26 Instantaneous volume fractions and circulating solid mass fluxes

of Lagrangian discrete particles and Euler granular phase is caused by the motion and coalescence of bubbles formed in the HV chamber and LV chamber. The calculated time-averaged solid mass fluxes are 0.0080 kg/s of Euler granular phase and 0.0078 kg/s of Lagrangian discrete particles, indicating both circulation solid mass fluxes are similar using CFD-KTGF-DEM approach in the ICFB.

6.2.3 Secondary Gas Injection Fluidized Bed

The lateral dispersion of gas and solid phases exerts great, sometimes even crucial, influence on the performance of fluidized bed reactors. Incomplete mixing of secondary air streams into the gas-solids suspension rising from the bottom region of a CFB combustor may lead to hydrocarbon emissions from sub-stoichiometric zones as well as nitrogen oxide gaseous emissions from over-stoichiometric zones. A secondary (sometimes even a tertiary) air stream is used in coal combustion, where the coal practices of staged combustion are crucial for reducing nitrogen oxide emissions. Secondary air is injected into the combustor to create staged combustion leading to lower nitrogen oxide emissions of fluidized bed reactor. Optimal radial dispersion of the gas and solid phases may be crucial in achieving high conversion and/or high selectivity in gas-conversion processes and combustions. An adequate understanding of the gas and particles mixing process may shed light on the evolution of volatiles from fuel particles or the optimal distribution of feed points of a reactant. It is therefore essential for successful modelling of reactor behavior of secondary gas jet fluidized beds.

Figure 6.27 shows the instantaneous volume fractions of granular phase and velocities of gas and discrete particles in a secondary gas injection fluidized bed. The bed thickness and width are 0.02 m and 0.145 m. The primary gas flux enters from the bottom distributor and the secondary gas are ejected from the left and right jet ports. The particle diameter and density are 3.0 mm and 2505 kg/m^3. The Young's moduli and Poisson ratio are 7×10^{10} Pa and 0.33, respectively. The inlet gas phase velocity is 3.6 m/s from the bottom. The width of the secondary air injector is 10.0 mm at a height of 15 mm from the bed bottom. The secondary air velocity is 7.2 m/s from the secondary air injector. Gas phase enters from the inlet and secondary injection. The velocity vectors show that gas enters at the bottom, and bubbles are formed through the bed, and they break at the bed surface. Finally, gas is discharged from the exit at the bed top. Near the secondary gas injection, both granular phase and discrete particles tend toward the bed center with high gas volume fractions. Larger gas velocities cause greater momentum of the rising suspension, which in turn leads to larger sizes of bubbles and higher radial dispersion of bed materials. These indicate that the laterally injected gas stream can substantially affect the performance of fluidized bed reactor.

The total gas flux consists of the inlet air flux part from the bed bottom and the secondary air part from the injection. The former is called as the primary air. Secondary air is injected at some distance above the bed bottom. Simulations show

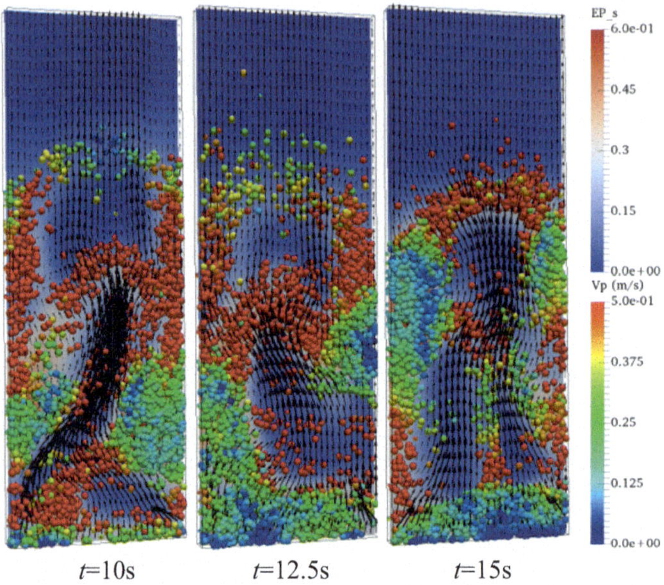

Fig. 6.27 Instantaneous volume fractions and velocities of gas and granular phases and discrete particles (black vector for gas phase, color vector for discrete particles, contour for granular phase)

that the gas phase volume fractions are low near the bed bottom because the bed material is only fluidized by primary air. The gas phase volume fractions become larger above the secondary air port due to the total gas flux. These result in a dense zone near the bed bottom and dilute zone at the upper zone of the bed. The gas dispersion in the upper zone is enhanced by the secondary air jets, with the extent of this improvement depending on bed volume fraction of particles. A larger solids volume fraction causes greater momentum of the rising suspension, which in turn leads to faster decay of the radial gas dispersion. Therefore, the secondary air penetration has a greater effect upon fluidization state in the bed.

Figure 6.28 shows the instantaneous granular phase and discrete particle velocity components along the x and z directions within two cells at $z = 10$ cm. Oscillations of both granular phase velocity u_{ep} and discrete particle velocity u_{dp} are observed due to the interactions of collision and hydrodynamic interactions. The granular phase velocities are obtained from Euler granular phase momentum equation, and the discrete particle velocities are obtained from the statistical analyses according to particle number in the cell. The former relates to granular coefficient of restitution that is found from the discrete particle restitution coefficient in the cell. The latter depends on gas turbulent kinetic energy and granular phase kinetic energy that are predicted from gas phase and granular phase turbulence models. The simulated velocity field calculations are the same for both the granular phase and the discrete particles in the cell. These confirm that the granular phase and discrete particles have the same volume fractions in the cell. Therefore, these distributions share the

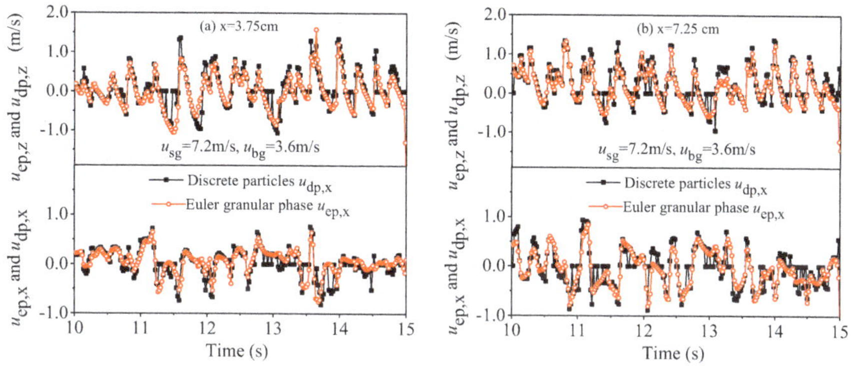

Fig. 6.28 Instantaneous granular phase and discrete particles velocity components at two cells

same configuration of granular phase and discrete particles in the cell. The simulated result enables us to understand detailed flow patterns and hydrodynamics of gas phase, granular phase and discrete particles in the gas injection fluidized bed.

Figure 6.29a shows the instantaneous granular restitution coefficients within two cells at $z = 5$ cm. Large oscillations of granular restitution coefficients are observed due to hydrodynamic interactions and interaction of collisions in the cells. The low granular restitution coefficients are found due to high gas velocities from the secondary air injection. The granular coefficient of restitution indicates the proportion of granular phase kinetic energy dissipated due to impact. Lower values correspond to practically inelastic collisions, in which most of the granular phase kinetic energy is lost. On the other hand, higher coefficients of restitution indicate almost all the kinetic energy is conserved. However, a large variation of restitution coefficient means the granular kinetic energy loss depends on impact velocity and volume fraction of discrete particles. The collision velocity of discrete particles affects granular phase coefficient of restitution. As a consequence, disregarding these factors can lead to unreliable predictions of the behavior of gas–solid fluidized beds. Thus, to correctly simulate the hydrodynamic characteristics of gas–solid flow systems, it is necessary to determine the influence of different parameters on granular phase coefficient of restitution.

The distribution of time-averaged granular phase coefficient of restitution is shown in Fig. 6.29b as a function of granular phase volume fractions. It can be seen that the granular phase coefficient of restitution decreases with increasing granular volume fractions. In general, the granular phase coefficient of restitution decreases with the increase of impact velocity for the majority of particles. The strong scatter of granular phase coefficient of restitution is observed due to the effect of impact velocity that is defined as the ratio of post- to pre-collisional relative velocity at the direction connecting the mass-centroids of them. The mean value of granular phase coefficient of restitution is 0.959. The granular volume fraction has a great influence on restitution coefficient due to the collision behavior of discrete particles. The granular restitution coefficient depends on various parameters, such as the impact velocity,

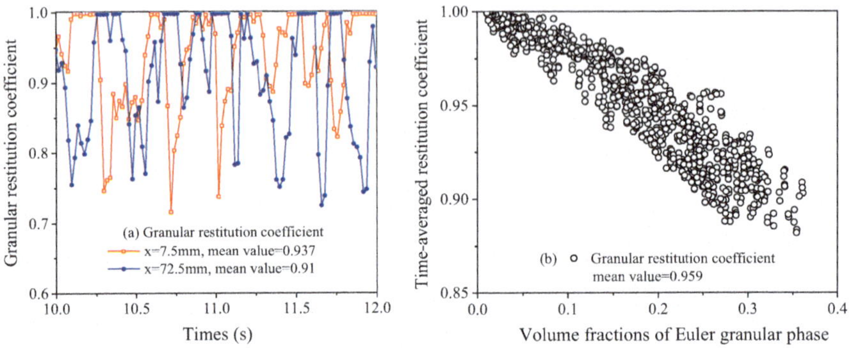

Fig. 6.29 Instantaneous and time-averaged granular restitution coefficients

volume fraction and material properties. Thus, a constant granular restitution coefficient does not accurately represent a real collisional interaction in gas–solid fluidized bed. A conventional TFM-KTGF simulation uses a constant restitution coefficient that may be modify iteratively until the obtained result matches with experimental data. However, a granular restitution coefficient from the collisions of discrete particles can be accurately predicted collisional energy dissipation in fluid–solid fluidized beds and multiphase flows.

In conclusion, the CFD-KTGF-DEM approach is proposed with KTGF and DEM models to simulate gas–solid fluidized beds. The restitution coefficient of Euler granular phase of KTGF is determined by means of DEM. The turbulence effect on motion of discrete particles is solved from Euler gas phase and Euler granular phase turbulence models. The drag force and viscous stresses on the discrete particle residing in the computational cell are determined from the momentum transfer between Euler gas phase and Euler granular phase. Needless to say, CFD-KTGF-DEM frees from empirical constants, and provides comprehensive hydrodynamics of gas and granular phases and discrete particles in a wide variety of fluid–solid fluidized bed reactors.

6.3 Supercritical Fluid Fluidized Beds

Theoretically speaking, fluids with temperature and pressure conditions below their critical point are known as subcritical fluids. While fluids are called as supercritical fluid as their pressure and temperature are higher than critical point. Subcritical fluid processing, therefore, deals with system pressure and temperature below its critical temperature and pressure in subcritical fluidized bed (FB). On the other hand, supercritical fluid processing works at higher temperature and pressure than critical point. The drastic and fast variation of properties under temperatures near the pseudo-critical temperature greatly affect the multiphase flow and heat transfer processes. Hence, supercritical fluid fluidization is a kind of new and promising processes.

Supercritical fluidized bed fluidizes solid particles with supercritical fluids. The supercritical fluid is the fluid phase, and the solid particles including biomass and coal are used in specific applications. Normally, supercritical fluid fluidized bed can be divided according the different supercritical fluids, including supercritical water, supercritical carbon dioxide, and supercritical methanol and ethanol fluids. The supercritical water fluidized bed has been used for gasifying biomass and coal for hydrogen production. Supercritical water gasification of coal and biomass discards the traditional combustion and oxidation reaction to release carbon dioxide and other pollutants and replaces coal and biomass with a gasification reduction reaction in supercritical water to finally convert into a hydrogen-rich gas product with no net carbon dioxide emissions and no pollutant emissions. The supercritical carbon dioxide is treated as a solvent and fluidization medium of inert substrate or pharmaceutical particle in the fluidized bed. Supercritical carbon dioxide fluidized bed can help to improve the material performance or control the release processes of the active substance. Thus, the design and operation of fluidized beds are greatly dependent on the understanding of flow characteristics of particles-supercritical fluid in fluidized bed reactors.

6.3.1 Supercritical Water Fluidized Bed

Supercritical water (SCW) fluidized bed is the process in which water having a pressure of over 22 MPa and a temperature over 374 °C (i.e., supercritical conditions). In SCW gasification technology, the coal and biomass water slurry is fed into the reactor, and they will be gasified almost completely, and the produced gas usually contains mainly H_2 and CO_2 with a very small fraction of CH_4 and CO, which are separated easily from the water phase by cooling to ambient temperature. The polarity of supercritical water is very weak, and it can dissolve almost all gases and organic matter. Supercritical water has a high diffusion coefficient and almost zero surface tension, which can easily enter the microporous structure of the reactant coal and biomass particles. Therefore, the supercritical-water fluidized bed is one of the most promising reactors for coal and biomass gasification.

The hydrodynamics of fluid and particles are predicted in the ambient water (AW), subcritical water (SbW) and supercritical water (SCW) fluidized beds [51]. The bed height is 1.8 m with the bed width of 0.15 m. The diameter and density of the bed materials are 1.0 mm and 1300 kg/m^3. The minimum fluidization velocity u_{mf} of bed materials is calculated using empirical correlation at different fluid pressures and temperatures [50]. The parameters used during the computational simulations are summarized Table 6.1. The density ratio is in the range from 1.3 to 2.0 for AW and SbW FBs, and is close to 10.0 for SCW FB. The conservation equations and constitutive correlations of low-Stokes-number kinetic theory of granular flow (KTGF) are summarized in Table 6.2 [5, 51].

Table 6.1 Fluid minimum fluidization velocity and threshold values in water-particle FBs

Cases	T(K)	p(MPa)	u_{mf}(m/s)	ρ_{sf}	Fr number		HD number		N_{tr} number	
					Value	State	Value	State	Value	State
1. Water	298	0.1	0.0019	1.303	0.0004	PF	0.215	PF	33.5	PF
2. SbW	600	22.0	0.0155	1.910	0.0245	PF	3.414	TR	760.9	TR
3. SCW	673	23.0	0.0524	9.705	0.2808	TR	13.62	AF	4191.0	TR

The instantaneous volume fractions of fluid and solid phases are displayed in Fig. 6.30 in the AW, SbW and SCW FBs with the same inlet velocity ratio u_f/u_{mf} of 5.0 [51]. In the AW FB, uniform distributions of volume fractions are found along bed height, indicating a particulate fluidization state. Large variations of volume fractions are observed near the bed surface where the solid volume fractions trend to zero. Furthermore, it is found that the solid volume fractions are slightly larger in the center region than they are near the walls.

For flow of bed materials in the SbW FB, the low and high volume fraction bands occur alternatively, and is known as wavy-like flow or traveling waves [1, 3, 14, 22], representing the ripples flow upward from the inlet bottom. The bed materials create stratification of the fluid-particles mixture with alternating dilute layer with low solid volume fractions and dense layer with high solid volume fractions. These travelling waves tilt to the right or to the left due to hydrodynamic interactions and collisions of particles. The breakup of the ripples results in the formation of transient aggregations of particles. Thus, the pattern of fluid-particles mixture is changed as the present of particles pockets, which is known as churn-like flow at the upper part of the bed. The occurrence of churn-like flow with high solid volume fractions represents particles aggregative fluidization. The heterogeneous flow occurs from wavy-like flow to churn-like flow along bed height. The propagation of traveling wave fronts across the variations of solid volume fractions is a key characteristic near the bottom of the SbW FBs. Obviously, the formation of the wavy-like flow near the bottom and churn-like flow at the bed upper part affects of particles circulation within the bed, reflecting a most distinct characteristic of fluidization state in SbW FBs.

For flow of fluid-particles mixture in SCW, instantaneous high local solid volume fractions are observed with particles agglomeration (also known as wispy-annular-like flow near the walls), representing a high solid volume fractions near the walls and a low solid volume fraction in the center core. Individual particles aggregative flow with high solid volume fractions appears, forming an annulus near the walls. The local high volume fraction of particles (wispy) entrain into the core of the bed illustrates that particles aggregative fluidization states exists in SCW FB. The main difference between the SbW FB and the SCW FB is the particles aggregative flow near the bottom of the bed. These results clearly reflect a difference in volume fractions between the SCW FB and the SbW FB along axial and lateral directions.

The instantaneous motion of particles is illustrated in Fig. 6.31 using CFD-DEM approach at the ambient water and sub- and super-critical water fluidized beds. For all cases, the inlet fluid velocities are the same with $u_f/u_{mf} = 5.0$. The particle

Table 6.2 Conservation equations and constitutive relations of low-Stokes-number KTGF

(1) Continuity equations (i=f for fluid phase and i=s for solid phase)

$$\frac{\partial}{\partial t}(\alpha_i \rho_i)+\nabla\cdot(\alpha_i \rho_i \boldsymbol{u}_i)=0 \tag{T1}$$

(2) Momentum conservation equations

$$\frac{\partial}{\partial t}(\alpha_f \rho_f \boldsymbol{u}_f)+\nabla\cdot(\alpha_f \rho_f \boldsymbol{u}_f \boldsymbol{u}_f)=-\alpha_f \nabla p_f+\alpha_f \rho_f \boldsymbol{g}+\nabla\cdot\boldsymbol{\tau}_f-\beta(\boldsymbol{u}_f-\boldsymbol{v}_s) \tag{T2}$$

$$\frac{\partial}{\partial t}(\varepsilon_s \rho_s \boldsymbol{u}_s)+\nabla\cdot(\varepsilon_s \rho_s \boldsymbol{u}_s \boldsymbol{u}_s)=-\varepsilon_s \nabla p_f+\varepsilon_s \rho_s \boldsymbol{g}+\nabla\cdot\boldsymbol{\tau}_s+\beta(\boldsymbol{u}_f-\boldsymbol{u}_s) \tag{T3}$$

(3) Turbulence modeling of fluid phase

$$\frac{\partial}{\partial t}(\alpha_f \rho_f k)+\nabla\cdot(\alpha_f \rho_f \boldsymbol{u}_f k)=\nabla\cdot\left(\alpha_f \frac{\mu_t}{\sigma_k}\nabla\cdot k\right)+\alpha_f \boldsymbol{\tau}_f \nabla\cdot\boldsymbol{u}_f-\alpha_f \rho_f \varepsilon \tag{T4}$$

$$\frac{\partial}{\partial t}(\alpha_f \rho_f \varepsilon)+\nabla\cdot(\alpha_f \rho_f \varepsilon \boldsymbol{u}_f)=\nabla\cdot\left(\alpha_f \frac{\mu_t}{\sigma_\varepsilon}\nabla\cdot\varepsilon\right)+\alpha_f \frac{\varepsilon}{k}(C_{1\varepsilon}\boldsymbol{\tau}_f \nabla\cdot\boldsymbol{u}_f-\rho_f C_{2\varepsilon}\varepsilon) \tag{T5}$$

(4) Granular temperature equation

$$\frac{3}{2}\left[\frac{\partial}{\partial t}(\alpha_s \rho_s \theta)+\nabla\cdot(\alpha_s \rho_s \theta \boldsymbol{u}_s)\right]=\boldsymbol{\tau}_s:\nabla\boldsymbol{u}_s+\nabla\cdot(k_s \nabla\theta)-\gamma_s \tag{T6}$$

(5) Granular pressure

$$p_s=\rho_s \alpha_s[1+2(1+e)\alpha_s g_0]\theta \tag{T7}$$

(6) Granular shear viscosity

$$\mu_s=\frac{\rho_s \alpha_s \theta \tau_p}{2+\nu\tau_p \varphi(e)}+\frac{4}{5}(1+e)\alpha_s g_0 \frac{\rho_s \alpha_s \theta \tau_p}{2+\nu\tau_p \varphi(e)}+\frac{4}{5}(1+e)\rho_s {\alpha_s}^2 g_0 d_s \sqrt{\frac{\theta}{\pi}} + \left(1+\frac{4}{5}(1+e)\alpha_s g_0\right)\frac{2\rho_s \alpha_s}{2+\nu\tau_p \varphi(e)}\cdot\frac{\mu_t}{1+St}\cdot\frac{S_{fik}}{S_{sik}} \tag{T8}$$

(7) Granular bulk viscosity

$$\mu_b=\frac{4}{3}(1+e)\rho_s {\alpha_s}^2 g_0 d_s \sqrt{\frac{\theta}{\pi}}+\frac{2}{3}\left(1+\frac{4}{5}(1+e)\alpha_s g_0\right)\frac{2\rho_s \alpha_s}{2+\nu\tau_p \varphi(e)}\frac{\mu_t}{1+St}\left(\frac{S_{fik}}{S_{sik}}-\frac{S_{fii}}{S_{sii}}\right) \tag{T9}$$

(8) Granular thermal conductivity coefficient

$$\kappa_s=\rho_s \alpha_s\left[\frac{5\theta}{3+\nu\tau_p \phi(e)}\left(\frac{1}{2}+\frac{3}{5}\alpha_s g_0(1+e)\right)\tau_p+\alpha_s g_0(1+e)2d_s\sqrt{\frac{\theta}{\pi}}\right] -\frac{5\rho_s \alpha_s \theta_s}{3+\nu\tau_p \phi(e)}\left(\frac{1}{2}+\frac{3}{5}\alpha_s g_0(1+e)\right)\frac{\tau_{Lp}}{1+St}\frac{2\mu_t}{\theta}S_{fik} \tag{T10}$$

(9) Granular fluctuation energy dissipation

(continued)

Table 6.2 (continued)

$$\gamma_s = 3\left(1-e^2\right)\alpha_s^2\rho_s g_0\theta\left(\frac{4}{d_s}\sqrt{\frac{\theta}{\pi}} - \nabla\cdot\boldsymbol{u}_s\right) \tag{T11}$$

(10) Stokes number and frequency of collisions

$$St = \frac{\tau_p}{\tau_{Lp}}, \quad \nu = \frac{96}{5d_s}\alpha_s\sqrt{\frac{\theta}{\pi}} \tag{T12}$$

(11) Relaxation time and Lagrangian integral time

$$\tau_p = \frac{\rho_s d_s^2\left(1-\frac{\alpha_s}{\alpha_{sm}}\right)^{2.5\alpha_{sm}}}{18\mu_f\left[1+0.1\left(\frac{\rho_f d_s\left|u_f-u_s\right|}{\mu_f}\right)^{0.75}\right]}, \quad \tau_{Lp} = \frac{3C_\mu k}{2\varepsilon_k\sqrt{1+C_\beta\frac{3\left|u_f-u_s\right|^2}{2k}}} \tag{T13}$$

(12) Functions for restitution coefficient

$$\varphi(e) = \frac{1+e}{2}\left[1+\frac{5}{6}(1-e)\right], \quad \phi(e) = \frac{1+e}{2}\left[1+\frac{5}{4}(1-e)\right] \tag{T14}$$

(13) Strain rates of fluid and solid phases

$$S_{fik} = \frac{1}{2}\left(\frac{\partial u_{fi}}{\partial x_k} + \frac{\partial u_{fk}}{\partial x_i}\right), \quad S_{sik} = \frac{1}{2}\left(\frac{\partial u_{si}}{\partial x_k} + \frac{\partial u_{sk}}{\partial x_i}\right) \tag{T15}$$

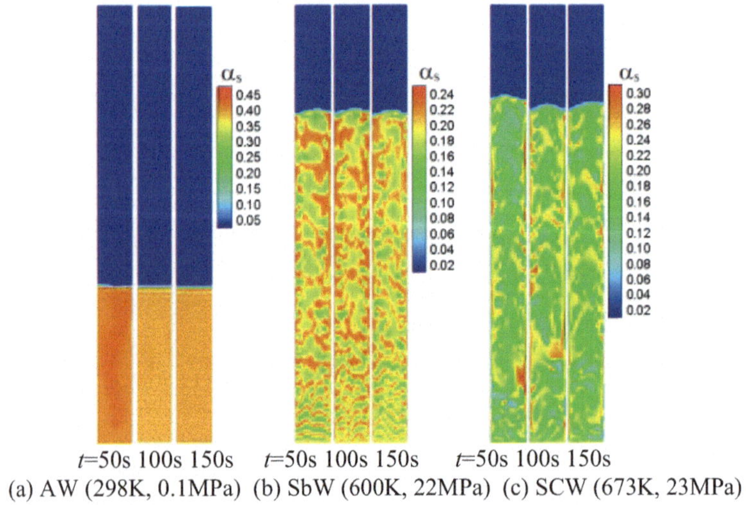

Fig. 6.30 Instantaneous solid volume fractions in AW, SbW and SCW FBs at $u_f/u_{mf} = 5.0$

diameter and density are 0.65 mm and 1450 kg/m^3. The fluid density is 998.24, 659.41 and 196.1 kg/m^3 for AW, SbW and SCW fluids, respectively. The particle number is 60,480. The normal spring constant is 1000 N/m. For the ambient water FB with the density ratio of 1.5, the distributions of particles are uniform, representing a particulate fluidization (PF). For the subcritical water FB with the density ratio of 2.2, the bands of the high large particle numbers are observed near the bed bottom. The bands of the low particles number are expressed flows of subcritical water voids. These indicate that the dense bands and dilute bands are appeared alternatively, representing that a wave-like flow appears near the bed bottom. As the wave-like flow travels upward, it is broken into particles aggregations due to the hydrodynamic interactions between the fluid phase and particles and interactions of collisions of particles. These particles aggregations have large particle numbers, representing a churn-like flow with particles aggregations at the upper part of the bed. CFD-DEM simulations indicate that the coexistence of the wave-like flow at the bed bottom and the churn-like flow at the upper part of the bed are occurred along bed in the subcritical water FB.

For flow of supercritical water and particles mixture with the density ratio of 7.4, it can be seen that the small particles numbers are appeared neat the bottom, and pass through bed center regime with the increase of their sizes. Finally, they are broken down at the bed surface. Furthermore, the large particles numbers are occurred near the walls. The large-scale turbulence with particles aggregations are characterized by an aggregative fluidization in supercritical water FB.

The particulate fluidization (PF) and aggregative fluidization (AF) of fluid-particles systems are recognized in early experimental studies [27, 47, 52]. The thresholds are proposed to distinguish PF from AF in fluidized beds, and are expressed using Froude number [52].

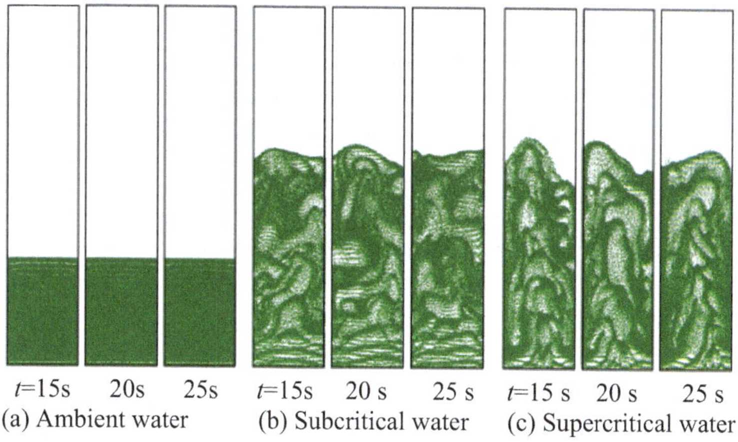

Fig. 6.31 Motion of particles in ambient and sub- and super-critical water FBs

$$Fr = \frac{u_{mf}^2}{gd_p} = \begin{cases} < 0.13 & \text{particulate} \\ > 1.30 & \text{aggregative} \end{cases} \tag{6.6}$$

where u_{mf} is the minimum fluidization velocity. d_p is the particle diameter. The fluid-particles mixture behaves particulately if the Fr value is less than the PF threshold of 0.13, and aggregatively, otherwise.

An empirical correlation of the dimensionless number HD is suggested to distinguish PF from AF according to fluid and particle densities of ρ_s and ρ_f and fluid viscosity μ_f [27].

$$HD = \frac{71.3\mu_f^2}{gd_p^3\rho_f^2}\left[\left(1 + \frac{gd_p^3\rho_f(\rho_s - \rho_f)}{54\mu_f^2}\right)^{1/2} - 1\right]^2 \left[\frac{\rho_s - \alpha_{mf}(\rho_s - \rho_f)}{(\rho_s - \rho_f)(1 - \alpha_{mf})}\right]$$
$$= \begin{cases} < 1 & \text{particulate} \\ > 10 & \text{aggregative} \end{cases} \tag{6.7}$$

According to fluid and particles properties, an empirical threshold N_{tr} is proposed to identify PF and AF [47].

$$N_{tr} = \sqrt{gd_p^3}\left(\frac{\rho_s - \rho_f}{\mu_f}\right)\begin{cases} < 50 & \text{particulate} \\ > 5000 & \text{aggregative} \end{cases} \tag{6.8}$$

The values of three dimensionless numbers are listed in Table 6.1. All correlations show that the dimensionless numbers for the first case of AW FB are lower than the PF threshold, indicating a PF state. From the three dimensionless numbers, it can be seen that a transitional regime (TR) exists between the PF threshold and the AF threshold. The HD dimensional number shows that the SCW FB is at AF state, whereas the Fr and N_{tr} dimensionless numbers show that the SCW FB is at TR state. However, the values of the dimensionless number N_{tr} are closer to the AF thresholds compared to the Fr dimensionless number. These results confirm that AW FB fluidizes particulately, while SCW FB fluidizes aggregatively. For the second case of SbW FB, it is at PF state using Fr dimensionless numbers. On the other hand, the computed results show that both calculated results using HD and N_{tr} dimensionless numbers indicate that fluid-particles mixture is at TR state with wavy-like flow near the inlet bottom and churn-like flow at the upper part from PF state to AF state in the SbW FB. These results further indicate that intermediate-state fluidization exist, utilizing a subcritical water FBs.

The instantaneous solid volume fractions are displayed in Fig. 6.32 along bed height. The uniform profile of volume fractions is found along bed height in the AW. The mean value and standard deviation of solid volume fractions are 0.405 and 0.0053. Two sections are defined according to the variation of solid volume fractions in the SbW FB. The section I characterizes a period solid volume fraction variation near the bottom. On the contrary, the solid volume fraction fluctuations are loss of periodicity in the section II at the upper part of the bed. The mean solid volume

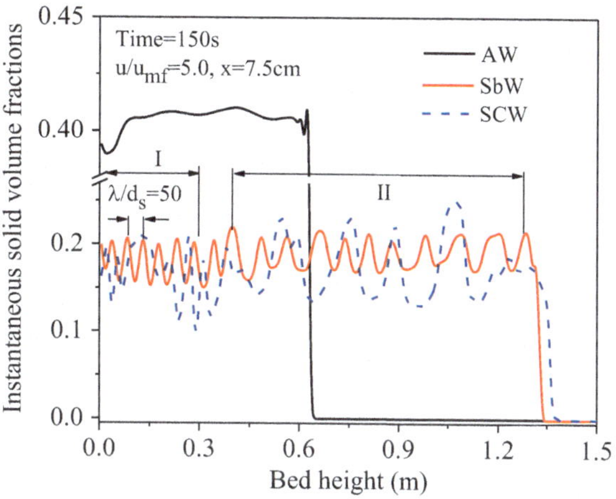

Fig. 6.32 Axial profiles of instantaneous solid volume fractions in AW, SbW and SCW FBs

fraction and its standard deviation are 0.179 and 0.023 in the section I. The mean length is identified between two peaks of volume fractions, representing a wavelength of wave-like flow in the section I. The space–time plot of SbW illustrates that the travelling waves are formed with a single hump having a wavelength λ of 50.0 mm (i.e. scaled wavelength $\lambda/d_p = 50$), reflecting that waves are evolved near the bed bottom at the section I in the SbW FB. For flow of churn-like flow, the mean value and standard deviation of solid volume fraction are 0.187 and 0.019 in the section II. For flows of particles in the SCW FB, the mean value and standard deviation of solid volume fraction are 0.172 and 0.0344, indicating that the mean solid volume fraction decrease, and the standard deviation increases. The large amplitudes of fluctuations in the solid volume fractions are due to the occurrence of particles agglomeration within the bed of SCW FB.

6.3.2 Supercritical Carbon Dioxide Fluidized Bed

Carbon dioxide (CO_2) is one of the fluids most commonly used in industrial applications for the extraction of active compounds, including tocopherols and phytosterols due to its low critical pressure and moderate temperature of 7.38 and 304.13 K. High pressure CO_2 fluid has also attracted wide usage as a working fluid for energy process engineering, including refrigeration systems [17, 40] and closed power cycles [2, 11], thanks to its thermal stability at wide ranges of pressure and temperature. In the Rankine cycles for power generation, CO_2 fluid is pumped from a sub-cooled liquid state, therefore requiring low power for it to be compressed, compared to gas compression. The supercritical CO_2 working fluid provides high temperature operability of power generation using fluidized bed (FB) reactors. Solid particles with supercritical carbon dioxide fluid are an effective storage method because the stable and low-cost particles can work at higher temperature conditions and can store energy in hot particles at relatively low storage containment cost, leading to better efficiency, compared to conventional power cycles, using fluidized bed technology. The supercritical CO_2 fluid is traditionally used as an alternative food-safety sterilization procedure/treatment due to its non-thermal process capable of inactivating microorganisms. The supercritical CO_2 fluid, with its liquid-like density, very low viscosity, and no surface tension and capillary force, could be vital for its role in shale gas fracturing. It allows for higher penetration and faster clean-up due to higher mobility, leaves the proppant inside the fractures upon vaporization and does not lead to clay swelling phenomenon, making it appropriate for shale gas recovery. The supercritical CO_2 is used as the heat transfer fluid in concentrated solar power. The thermal energy storage exchanger is needed to operate continuously using supercritical CO_2 as fluidizing agent in fluidized bed reactor.

The hydrodynamics of carbon dioxide fluid-particles mixture are simulated in ambient, high pressure and supercritical carbon dioxide FBs [51]. A two-dimensional model is used, and the geometrical parameters are 2.0 m in height and 0.2 m in width. Particle diameter and density are 0.001 m and 1600 kg/m^3. The calculated minimum fluidization velocities are listed in Table 6.3. The bottom of the bed is defined as a velocity inlet boundary, and the top is defined as an outlet with fixed pressure.

The fluidization state of carbon dioxide-particles mixture is displayed in Fig. 6.33 at three different carbon dioxide temperatures and pressures. For case I of carbon dioxide temperature and pressure of 298 K and 0.1 MPa at ambient condition, the density ratio ρ_s/ρ_f is 898.8, representing a high Stokes number condition. Small-sized bubbles are found at the bottom, while larger ones are found at the upper part, representing a large-scale aggregative fluidization (AF) state.

For case II at carbon dioxide pressure and temperature 7.0 MPa and 295 K (subcritical condition) with ρ_s/ρ_f of 2.03, showing a low Stokes number condition. The layers having high volume fractions of fluid and particles are found near the bottom. These bands with high solid volume fraction are formed alternatively, and separated by high liquid volume fraction bands, representing a wave-like flow. The waves break down as they traveled upwards. Finally, they vanish, and particle aggregates having high solid volume fractions are formed. These particle aggregates are appeared, representing chunk-like flow at the upper part. Evidently, two different regimes are occurred with wave-like and chunk-like flows along bed height.

With increase in carbon dioxide temperature and pressure to 315 K and 8.0 MPa at supercritical condition for case III with ρ_s/ρ_f of 6.12, particle strands or aggregates having high volume fractions of particles are observed in the bed, representing small

Table 6.3 Minimum fluidization velocities and dimensionless numbers in carbon dioxide FBs

Case	*T* (K)	*p* (MPa)	u_{mf} (cm/s)	*Fr* number		*HD* number		N_{tr} number	
				Value	State	Value	State	Value	State
I	298	0.1	37.98	14.7	AF	1137.2	AF	10,606.8	AF
II	295	7.0	1.759	0.032	PF	4.60	TR	1174.1	TR
III	315	8.0	4.257	0.185	TR	10.01	AF	6104.9	AF

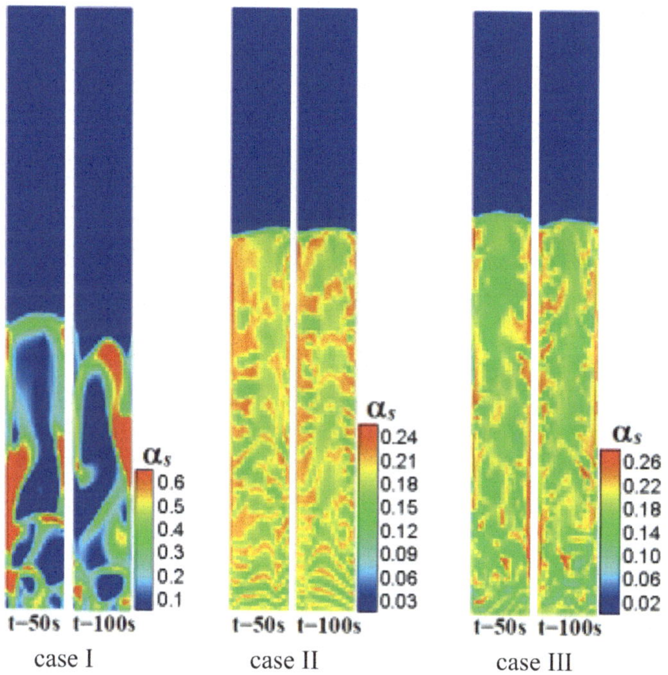

Fig. 6.33 Fluidization of carbon dioxide fluid and particles at different pressures and temperatures

scale aggregative fluidization state. More particle strands exist, forming an annulus with high solid volume fractions and a center core with low solid volume fractions along lateral direction. Compared to bubbles in case I at the ambient temperature and pressure condition, particle aggregates are substituted at supercritical carbon dioxide fluid-particles FB. These simulations indicate that the fluidization process is varied with changing carbon dioxide pressure and temperature in the bed.

The snapshots of instantaneous solid volume fractions applying CFD-DEM approach are illustrated in Fig. 6.34 at ambient, subcritical and supercritical carbon dioxide fluidized beds. The dots with green color represent transient position of discrete particles at $t = 15$ s and 30 s. These reveal that fluidization process is fully developed in the carbon dioxide fluidized bed. For all cases, the inlet velocity of carbon dioxide fluid is $u_f = 5u_{mf}$, and the particle diameter and density are 1.0 mm and 1800 kg/m^3. The fluid density is 1.8161, 810.19 and 147.03 kg/m^3 at ambient, subcritical and supercritical conditions. The total particle number is 41,650. The normal spring constant is 1000 N/m, and the spring normal/tangential ratio is 2/7.

For the ambient carbon dioxide fluid fluidized bed with the density ratio of 991.1, bubbles exist with low particles numbers. We see that bubbles are appeared in the bed when CFD-DEM is used, representing an aggregative fluidization state. Large size bubbles are observed at the upper part due to high inlet carbon dioxide velocity, representing a bubble-like flow.

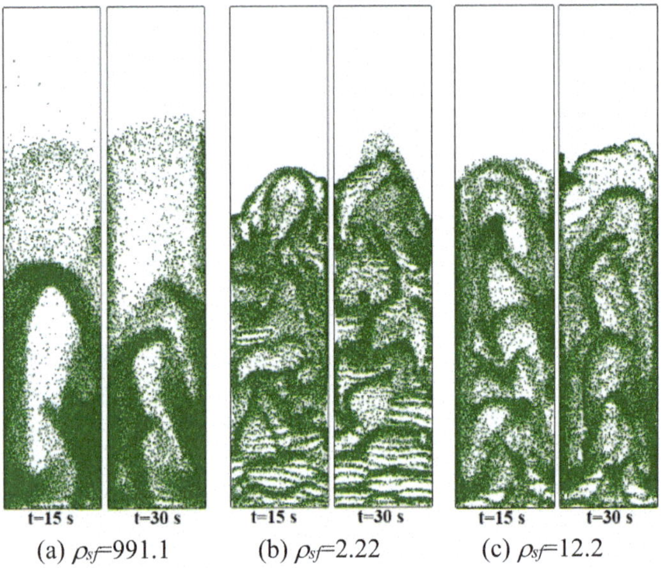

Fig. 6.34 Transient distributions of discrete particles at $t = 15$ and 30 s

For the subcritical carbon dioxide fluid condition with the density ratio of 2.22. The strip region with low particles numbers occurs alternately near the bottom from simulations using CFD-DEM. These bands and strips represent flow of waves near the bottom. As the bands are travelled upward, they are broken, and particle aggregates with high particle number are occurred, revealing a chunk-like flow at the upper part. These indicate that different flow behavior of particles exists with wave-like flow at the bottom and chunk-like flow at the upper part along bed height. The coexistence of wave-like and chunk-like flows characterizes a complex fluidization state in the subcritical carbon dioxide fluid FBs.

For the supercritical carbon dioxide fluid condition with the density ratio of 12.2, bubbles with the low local particles numbers are observed. These small bubble-like flows travel upward from the bottom to the upper part along height. Particle aggregate-like flows with local large particles numbers are also noticed, illustrating an aggregative fluidization in the supercritical carbon dioxide fluid FBs.

Table 6.3 summarizes the values of dimensionless numbers for three cases. The transitional regime (TR) is clearly shown between the PF state and the AF state from three dimensionless numbers. For case I at the ambient temperature and pressure condition, the bed materials fluidize aggregatively with large scale bubbles in the bed, representing an AF state from three dimensionless numbers of *Fr*, *HD* and N_{tr}. For fluidization process of subcritical carbon dioxide fluid in case II, both the dimensionless numbers of N_{tr} and *HD* indicate that the fluidization process is behaved by TR state. However, it is a PF state using dimensionless number *Fr*. For case III, both the dimensionless numbers of N_{tr} and *HD* indicate that the supercritical carbon dioxide-particles mixture fluidize aggregatively with small scale particle strands. On the other hand, the TR state is found using *Fr*. These indicate that fluidization process of carbon dioxide-particles mixture changes from AF, TR to AF states with the change of fluid temperature and pressure from ambient fluid to subcritical and supercritical fluid. Keeping in mind that the difference of AF state is clear between atmospheric carbon dioxide fluid and supercritical carbon dioxide fluid. The former is represented by large scale bubbles in the bed, while the latter is characterized by small particle strands in the bed. The simulated results are consistent with predictions using the thresholds of N_{tr} and *HD*. These indicate that both computational simulations and dimensionless numbers reveal distinct characteristics of carbon dioxide-particles mixture in the bed.

Figure 6.35 displays the instantaneous solid volume fractions at three carbon dioxide temperatures and pressures. The difference is clearly seen along bed height. At $H = 0.1$ m from the inlet, the volume fraction fluctuations are observed for case I at the ambient temperature and pressure. The mean solid volume fraction and standard deviation are 0.236 and 0.172. The low solid volume fractions represent flow of bubbles. From simulations of instantaneous volume fractions, the bubble can be identified if the fluid volume fraction is larger than 0.85 [13, 33]. Bubble occurs at time t_s, and vanishes at time t_e. The time interval of the existence of the bubble is $\Delta t = t_e - t_s$. Large bubbles have high time interval. Contrastingly, small bubbles have short existence time interval. Hence, the fluidization characteristics exhibit large and small bubbles flow.

For case II at subcritical carbon dioxide pressure and temperature, periodic-like oscillations are observed from the variation of volume fractions. The standard deviation of solid volume fractions is 0.023 with the mean solid volume fraction of 0.182. The large volume fractions of particles represent appearance of particle bands, and the small volume fractions of particles represent existence of liquid voids, indicating that the occurrence of particle bands and liquid voids are alternated. With increasing

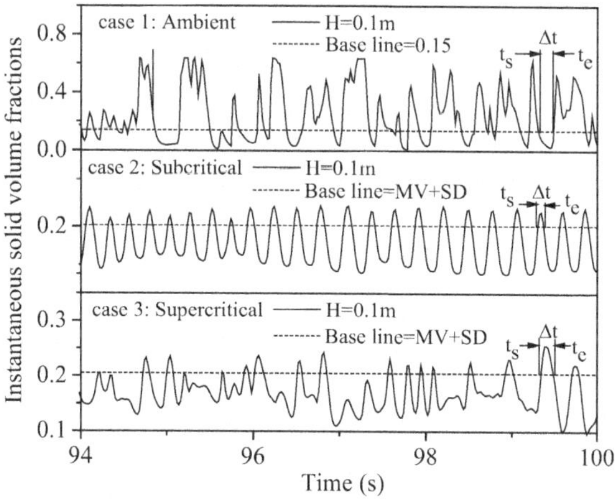

Fig. 6.35 Instantaneous solid volume fractions at three CO_2 pressures and temperatures

carbon dioxide pressure and temperature for case III at supercritical condition, large fluctuations of volume fraction are observed with the standard deviation of 0.0344 at the mean solid volume fraction of 0.171, representing the appearance of particle strands.

The transitional state from AF state to PF state was studied using the theory of linear stability in fluidized beds [3, 31]. The expansion characteristics of fluid-particles fluidized beds is described by the empirical Richardson-Zaki relation [46].

$$U = u_t \alpha_f^n \tag{6.9}$$

where n is Richardson-Zaki parameter [46], and u_t is the particle terminal velocity. Wallis [49] suggested that the instability arises whenever the continuity wave velocity u_k exceeds the dynamic wave velocity u_d through the bed materials of fluidized beds. The one-dimensional, linearized stability approach has been applied to the analytical study of transition in fluidized beds [19]. The kinematic wave velocity u_k is the limiting value as the amplitude of the imposed fluid flux approaches the infinitesimal limit [20].

$$u_k = (1 - \alpha_f)\frac{dU}{d\alpha_f} \tag{6.10}$$

where α_f is the fluid volume fraction. On evaluating the derivative $dU/d\alpha_f$ from Eq. (6.9), the kinematic wave velocity becomes:

$$u_k = n u_t (1 - \alpha_f)\alpha_f^{n-1} \tag{6.11}$$

The sonic (dynamic-wave) velocity is given at the adiabatic condition:

$$u_d = \sqrt{\frac{\partial p}{\partial \rho}} = \sqrt{-\frac{1}{\rho_s}\frac{\partial p}{\partial \alpha_f}} \tag{6.12}$$

The net force acting on fluidized particles in equilibrium is simply the sum of drag and effective weight, and is expressed

$$u_d = \sqrt{-\frac{1}{\rho_s}\frac{\partial (Nf)}{\partial \alpha_f}} \tag{6.13}$$

where N is the number of particles. f is the net force experienced by each of these particles as a result of the local fluid volume fraction change.

$$Nf = \frac{2(1 - \alpha_f)d_p}{3}(\rho_s - \rho_f)g\left[\left(\frac{U}{u_t}\right)^{\frac{4.8}{n}}\alpha_f^{-3.8} - \alpha_f\right] \tag{6.14}$$

Substituting this expression in Eq. (6.12) and using Eq. (6.9), we obtain the final relation for the dynamic wave velocity [20]:

$$u_d = \sqrt{3.2 g d_p (1 - \alpha_f)(\rho_f - \rho_s)\rho_s^{-1}} \tag{6.15}$$

The homogeneous fluidization occurs when the velocity of the perturbation wave is greater than that of the kinematic wave u_k, i.e. $u_d > u_k$. Otherwise, the unstable, heterogeneous fluidization exists. Hence, the stability criterion S relates to dynamic wave velocity u_d and kinematic wave velocity u_k of fluid-particles suspension [20, 49].

$$S = \frac{u_d - u_k}{u_k} = \begin{cases} + : & \text{Homogeneous} \\ 0 : & \text{Stability limit} \\ - : & \text{Bubbling} \end{cases} \tag{6.16}$$

Figure 6.36 shows the profiles of dynamic wave velocities and kinematic wave velocities at the different carbon dioxide temperatures and pressures. For both case I and III at the ambient and supercritical carbon dioxide fluid conditions, both dynamic wave and kinematic wave velocities reduce with increasing fluid volume fractions. The dynamic wave velocity is smaller than kinematic wave velocity. On the contrary, for case II at subcritical carbon dioxide fluid condition, the stability limit exists at two fluid volume fractions α_{fi} ($i = 1$ and 2) where the dynamic wave velocity equals to kinematic wave velocity. The critical volume fraction α_{fi} can be found [20]:

$$\frac{\alpha_{fi}^{1-n}}{(1 - \alpha_{fi})^{0.5}} = \frac{n}{1.79}\left(\frac{u_t^2}{g d_p}\right)^{0.5}\left(\frac{\rho_s}{\rho_s - \rho_f}\right)^{0.5} \tag{6.17}$$

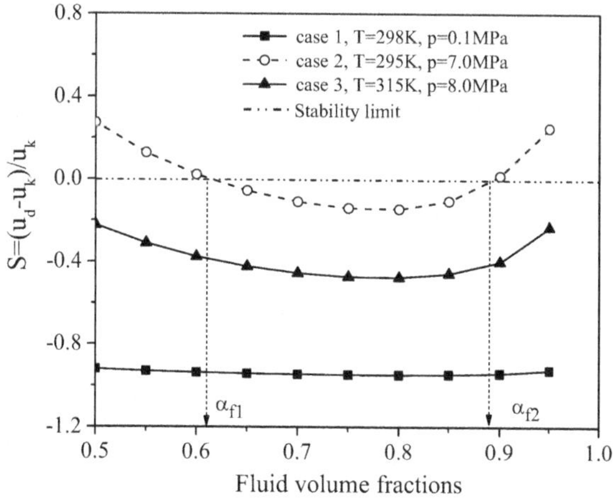

Fig. 6.36 Dynamic and kinematic wave velocities at three carbon dioxide pressures and temperatures

The critical volume fractions of α_{f1} and α_{f2} are 0.614 and 0.894, respectively. For case II, Eq. (6.16) predicts that the carbon dioxide-particles mixture is in the initial region of PF state ($u_d > u_k$), passing through the first critical fluid volume fraction α_{f1}, enter into the second region of AF state ($u_k > u_d$). Finally, the carbon dioxide-particles mixture extends to the last region of PF state ($u_d > u_k$) as the fluid volume fraction exceeds the second critical volume fraction α_{f2}. These indicate that homogeneous fluidization states exist at $\alpha_f < \alpha_{f1}$ and $\alpha_f > \alpha_{f2}$. On the other hand, bubbling fluidized bed exists at $\alpha_{f1} < \alpha_f < \alpha_{f2}$, and the stability S is negative. Both small-scale and large scale particles strands appear in the bed.

From Fig. 6.36 for case II, the largest and smallest solid volume fractions are about 0.22 and 0.13 at $H = 0.1$ m. Both them are close to the second critical volume fraction α_{f2} comparing to the first critical volume fraction α_{f1}. These indicate that the dense-dilute structures are formed as the fluid volume fractions transit from $\alpha_f < \alpha_{f2}$ to $\alpha_f > \alpha_{f2}$ through the second critical fluid volume fraction. Hence, the stability S and computational simulations suggest that both wave-like flow with particle bands and particle aggregates flow are coexisted along bed height. These reveal the fluidization state of carbon dioxide-particles mixture is changed from the PF state to the AF state through the TR state with changing pressure and temperature in the fluidized bed reactors.

6.3.3 *Supercritical Methanol Fluid Fluidized Bed*

Methanol fluid is useful in several processes that take place advantageously at elevated temperatures. Supercritical methanol has been used in biomass processing to produce biodiesel fuels in supercritical extraction, supercritical chromatography and synthetic chemistry for transesterification, dissolution and conversion of cellulose. The thermochemical processes of coal and biomass materials with supercritical methanol fluidized bed (ScM FB) are also applied because of environmentally friendly aspects in fluidized bed reactors. ScM fluid exhibits outstanding transport properties coupled with highly tuneable solvent properties, and documented an increase in reaction rate at fluid temperatures and pressures above the critical point ($T_{cr} = 239$ °C, $p_{cr} = 8.09$ MPa) without the use of auxiliary chemicals in reactors. Due to the unique properties of supercritical methanol, it has attracted much attention and widely used in the green chemical processes. The supercritical methanol satisfies the following roles in the biomass liquefaction; dissolving the decomposed products, stabilizing the reaction intermediates inhibiting the generation of tar and char, and removing the oxygen to improve the energy content of bio-oil. Also, it can donate hydrogen acting as hydroxyalkylating or hydrogenating reagents for the triple bond in the reaction. Because of these factors, supercritical methanol fluid can be a good candidate for the upgrading of coal and biomass materials, too. The thermochemical processes of solid raw materials with ScM FB reactors has several challenges in its research and development. The key operating parameters are pressure, temperature and superficial velocity of reactors. In fact, the high pressure and temperature reaction conditions effect on fluidization state of FB reactors, resulting in complex flow behavior of fluids and particles suspension. Hence, the conventional ScM FB reactor design requires knowledge of fluidization quality of bed materials.

The computational simulations of hydrodynamics of bed materials are performed in the ScM fluidized bed [51]. The bed diameter and height are 0.25 m and 2.0 m, respectively. The particle diameter and density are 0.001 m and 1400 kg/m^3. The initial bed height and solid volume fraction are 0.4 m and 0.6, respectively. One inlet–outlet bed mode is used at the bottom and the top of the bed. The methanol fluid is fed with the fluid volume fraction of unity with a constant temperature at the bottom of the bed. The values of minimum fluidization velocity are listed in Table 6.4 using Wen and Yu empirical correlation at different fluid pressures and temperatures [50].

The instantaneous solid volume fractions are illustrated in Fig. 6.37 at different fluid pressures and temperatures with the superficial velocity of $u_f/u_{mf} = 5.0$. Different fluidization behavior is clearly observed for three different cases. For case 1 with methanol fluid pressure and temperature of 0.1 MPa and 293 K (i.e. atmosphere pressure and temperature), the uniform solid volume fraction distributions are observed in a smooth expansion of the bed. Simulations show that the bed height increases from the initial bed height at first, and then expands uniformly with decreasing solid volume fractions. The cross flow instabilities are damped, and heterogeneity, or large-scale void of fluid are not observed, reflecting a steady state regime of particulate fluidization.

Increasing methanol fluid pressure and temperature for case 2 at 8.5 MPa and 503 K, bands with high solid volume fractions are observed near the bottom. Low solid volume fractions appear between two bands, indicating that a solid volume fraction wave (i.e. void wave) occurs near the bottom of the bed. When the bed materials move upwards, particles collide and aggregate due to wave instability. The dense particles packets with high solid volume fractions occur at the upper part of the bed. These particle strands induce high solid volume fractions, representing a churn-like flow near the walls and strand-like flow at the bed center. Apparently, the wavy-like flow, strand-like flow at the bed center and churn-like flow are found along the bed height, unlike in the homogeneous fluidization with low pressure and temperature as in case 1. Different four regimes were identified in the water–glass beads fluidized bed [14]. The regimes included wavy flow, wavy flow with transverse structure, fine-scale turbulent flow and bubbling states with increasing fluidizing

Table 6.4 Minimum fluidization velocities and dimensionless numbers in methanol-particle FBs

Case	T (K)	p (MPa)	u_{mf} (cm/s)	Fr number		HD number		N_{tr} number	
				Value	State	Value	State	Value	State
1	293	0.1	0.555	0.003	PF	3.808	TR	102.7	TR
2	503	8.5	2.369	0.057	PF	5.762	TR	1453.8	TR
3	523	8.5	5.674	0.329	TR	15.871	AF	5793.4	AF

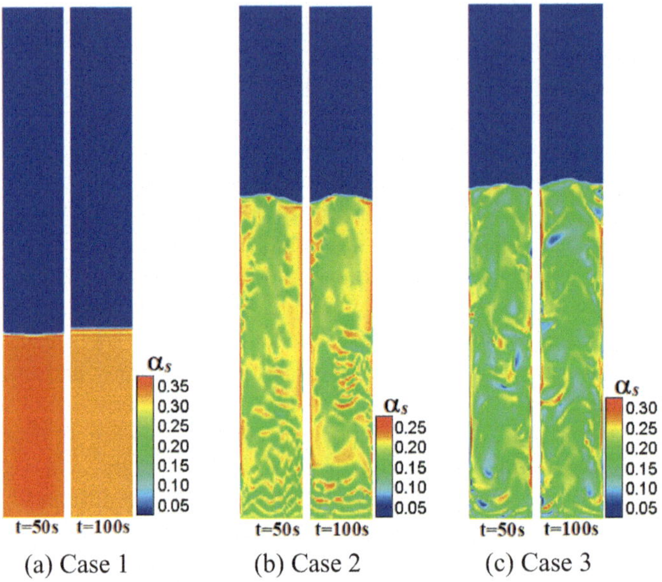

Fig. 6.37 Instantaneous solid volume fractions at different pressures and temperatures at $u_f/u_{mf} = 5.0$

velocity in the bed. Hence, the existence of the wavy-like flow near the bottom, and strand-like flow at the upper part is a new regime at the subcritical methanol fluid fluidized bed.

For case 3 at the supercritical pressure and temperature, dense particles packets with high solid volume fractions occur within the bed. These particle strands with high solid volume fractions are enclosed by low solid volume fraction fluid-particles suspension, representing a fine-scale turbulent flow of fluid-particles mixtures. Furthermore, the local void of fluid with high fluid volume fractions are observed within the bed. These local high fluid volume fractions represent a flow of fluid bubble, reflecting a bubbling state at high superficial fluid velocity. The number of fluid bubble is few, and the particles fine-scale turbulent flow is dominant within the bed.

The instantaneous solid volume fractions and discrete particles motions are illustrated in Fig. 6.38 using CFD-DEM approach at three cases of methanol fluid FBs at $u_f/u_{mf} = 5.0$. The particle diameter and density are 1.0 mm and 1500 kg/m^3, respectively. The methanol fluid densities are 791.15, 563.28 and 98.65 kg/m^3, respectively. For the ambient methanol fluid with the density ratio of 1.89, particles are fluidized as methanol fluid enters from the bottom. Uniform distributions of particle number density from CFD-DEM simulations are observed, indicating that the bed materials expand uniformly. The discrete particle number density drops sharply at the bed surface with a distinct bed height.

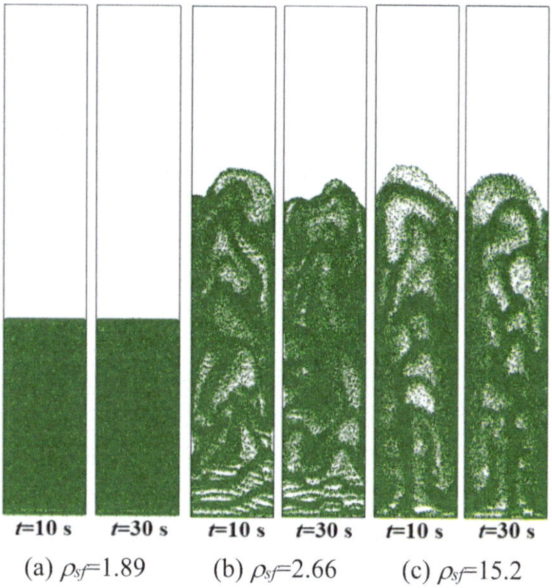

Fig. 6.38 Instantaneous solid volume fractions and particles motion

For the subcritical methanol fluid with the density ratio of 2.66, the bands with large discrete particle density numbers represent wave-like flow at the bottom. The strips with high local particle number densities are also observed, representing a churn-like flow at the upper. Strips are produced by disruptions of waves due to hydrodynamic interactions between methanol fluid and particles and collisional interactions of particles. The development of wave-like flow is the first stage in the growth of non-uniform structures of churn-like flow. CFD-DEM simulations indicate the coexistence of wave-like flow and churn-like flow along height in the subcritical methanol fluid FB.

For the supercritical methanol fluid with the density ratio of 15.2, CFD-DEM simulations indicate that the particle aggregates with high local particle number densities are observed, representing aggregate-like flow. Fluid voids with low particle number densities are also occurred in the bed. These indicate that CFD-DEM suggest that particle aggregates and fluid voids are appeared in the supercritical methanol fluid FB.

The values of dimensionless numbers of Fr, HD and N_{tr} are listed in Table 6.4. For case 1 at methanol fluid atmospheric pressure and temperature, the bed materials fluidized particulately using the threshold of Fr, and the fluidization state is at TR using the thresholds of HD and N_{tr}. For flow of subcritical methanol fluid in case 2, both calculated results using HD and N_{tr} indicate that fluid-particles mixtures are at TR state. It is also found that the ScM fluid and particles mixtures fluidize aggregatively for case 3 using HD and N_{tr}. These results indicate that methanol fluid-particles mixture fluidization state changes from PF to AF from atmospheric fluid to supercritical fluid.

From the kinematic wave velocity u_k and dynamic wave velocity u_d, the fluidization state is identified according to properties of fluid and particles. The kinematic wave velocities are calculated according to Archimedes number Ar and Richardson-Zaki parameter n.

$$u_k = \frac{\mu_f}{\rho_f d_p} n(1 - \alpha_{mf})\alpha_{mf}^{n-1}\left[-3.809 + \left(3.809^2 + 1.832Ar^{0.5}\right)^{0.5}\right]^2 \tag{6.18}$$

$$Ar = \frac{g d_p^3 \rho_f (\rho_s - \rho_f)}{\mu_f^2} \tag{6.19}$$

$$n = \frac{4.8 + 0.1032Ar^{0.57}}{1 + 0.043Ar^{0.57}} \tag{6.20}$$

The threshold indicates that the bed materials are homogeneous when the dynamic wave velocity is larger than the kinematic wave velocity, otherwise bubble occurs. When the two wave velocities are equal, the limit of stability is reached. The distributions of stability function are displayed in Fig. 6.39 at three different fluid pressures and temperatures. For case 1 at methanol fluid atmospheric pressure and temperature, the dynamic wave velocity is always in excess of the kinematic wave velocity, representing a particulate fluidization. Contrarily, for case 3 at supercritical methanol fluid pressure and temperature, the kinematic wave velocity is larger than dynamic wave velocity, representing an aggregative fluidization. On the other hand, for flow of subcritical methanol fluid in case 2, the stability limit (where the dynamic and kinematic wave velocities intersect) occurs at the fluid volume fractions of $\alpha_{f,1}$ and $\alpha_{f,2}$. The bed materials are in the initial region of homogeneous fluidization ($u_d > u_k$), extending to the first critical fluid volume fraction $\alpha_{f,1}$, follows by the second region of bubbling behavior ($u_k > u_d$). Finally, the bed materials are in the last region of homogeneous fluidization when the fluid volume fraction is larger than the second critical fluid volume fraction $\alpha_{f,2}$. These indicate that the transition contributes to hydrodynamic interactions between the fluid phase and solid phase and collisional interactions of particles in the bed.

Taken as a whole, the numerical simulations provide overwhelming support for the essential fluidization state from PF at the ambient methanol fluid FB to AF at the supercritical methanol fluid FB. The transition from PF to AF is accompanied by a marked change in fluidization state. The transition is characterized by the coexistence of the wavy-like flow near the bottom and the strand-like flow at the upper part at the subcritical methanol fluid FB. The stability criterion illustrates to differentiate from PF to AF in methanol fluid-particles FBs. The threshold provides reasonable estimates of particle strands in subcritical and supercritical methanol fluid fluidized beds.

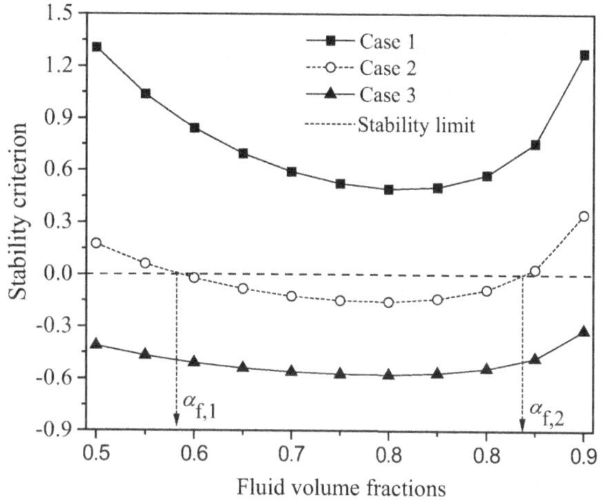

Fig. 6.39 Stability at three fluid pressures and temperatures

6.3.4 *Supercritical Ethanol Fluid Fluidized Bed*

Ethanol is a very important liquid biofuel which can be used both separately and as a component of fuel compositions. Furthermore, ethanol is used as an active medium for carrying out various chemical processes. The critical temperature T_c and pressure P_c of ethanol fluid are about $T_c = 240.9$ °C and $P_c = 6.1$ MPa, indicating that the supercritical ethanol fluidized bed reactors could be operated low pressure condition comparing to supercritical water and methanol fluidized bed reactors. Supercritical ethanol can be used as an active medium for the various chemical processes, including transesterification of triglycerides. The main advantage is lipophilization of glycerol liberated directly during the process. The treatment of coals, oil shales and biomass with supercritical ethanol solvents may yield liquid fuels of better environmental impact after possible desulfurization and denitrification. The economic evaluation of an ethanol process indicates the ethanol production using thermo-chemical processes could be possible using different oil shales and biomass materials in supercritical ethanol fluidized bed reactor.

The bed pressures and volume fractions are predicted in a supercritical ethanol fluid-particles fluidized beds [51]. The width and height of the supercritical ethanol fluid fluidized bed are 0.18 m and 2.1 m. Supercritical ethanol fluid enters from the bed distributor, and exits from the bed top. The diameter of particles is 1.0 mm with the density of 1450 kg/m^3. Initially, the bed height is about 0.4 m from the bed distributor.

The instantaneous solid volume fractions are illustrated in Fig. 6.40 at three different cases of the ambient ethanol fluid (AE), subcritical ethanol (SbE) and supercritical ethanol (SCE) FBs. The bed temperature and pressure of operations are summarized in Table 6.5. The minimum fluidized velocity u_{mf} of bed materials is calculated using Wen and Yu correlation [50]. The inlet velocities u_f for three different cases are to be $u_f/u_{mf} = 5.0$. For case 1 of AE fluid FB at $T = 293$ K and p

= 0.1 MPa, the bed materials gradually fluidize with low solid volume fractions at $t = 50$ s, and tend to fully fluidize with uniform volume fractions in the bed at $t = 100$ s. It is observed that particles expand uniformly within the bed, characterizing a particulate fluidization (PF) state.

For flow of case 2 of SbE fluid FB at the temperature $T = 473$ K and pressure $p = 8.0$ MPa, bands are observed with an alternatively low and high solid volume fractions near the bed bottom. The wavy-like flow occurs alternatively with low and high solid volume fraction layers, representing traveling waves [12, 16, 20, 26, 42]. These travelling waves tilt right or left directions as they travel toward the bed top. The rippled waves are broken due to collisional interactions of particles and

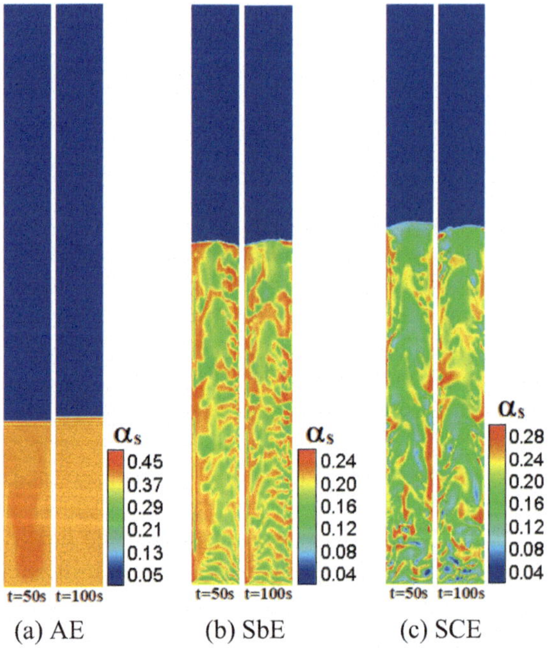

Fig. 6.40 Instantaneous solid volume fractions in ambient, sub- and super-critical ethanol fluid FBs

Table 6.5 Minimum fluidization velocity and dimensionless numbers

Case	Ethanol fluid physical parameters					Dimensionless numbers					
	T(K)	p(MPa)	ρ_f(kg/m^3)	μ_f(Pa·s)	u_{mf}(m/s)	Fr	State	HD	State	N_{tr}	State
Case1	293	0.1	789.80	1.199×10^{-3}	0.0032	0.0010	PF	2.58	TR	54.52	TR
Case2	473	8.0	582.90	1.055×10^{-4}	0.0191	0.0370	PF	4.07	TR	813.82	TR
Case3	553	8.0	133.22	2.152×10^{-5}	0.0577	0.3392	TR	15.14	AF	6059.89	AF

hydrodynamic interactions, and the transient aggregation of particles and fluid voids are formed at the upper part. The occurrence of particle pockets with large solid volume fractions exhibit a churn-like flow, representing an aggregative fluidization (AF) at the upper part. These results indicate that the heterogeneous flow of SbE fluid is characterized by the coexistence of wavy-like flow and churn-like flow from the bottom to the top of the bed. From experiments in ambient water fluidized bed [12, 14], the different regimes were observed, and they include the wavy flow, fine scale turbulent fluidized bed and bubbling fluidization state. They demonstrated that the development of particle pockets originates from instability of traveling waves. Evidently, the occurrence of the wavy-like flow and churn-like flow reflect a unique characteristic of fluidization state in SbE fluidized beds.

For flow of case 3 of SCE fluid FB at $T = 553$ K and $p = 8.0$ MPa, the high instantaneous local volume fractions are observed with fluid voids. These voids are generated at the bed bottom, and grow as they pass through the bed. Particles accumulate with large solid volume fractions near the walls, and pile up with low solid volume fraction at the center core, representing a wispy-annular-like flow in the bed. The wispy particles from local high solid volume fractions are entrained from the wall regime into the core with fluid voids, representing aggregative fluidization (AF) states in SCE fluid FB. These illustrate that the heterogeneous flow of SCE fluid is characterized by a wispy-annular-like flow with particles aggregation and fluid voids in the bed.

The instantaneous solid volume fractions are illustrated in Fig. 6.41a in AE, SbE and SCE fluid fluidized beds. For case 1 of AE fluid fluidized bed, the local instantaneous solid volume fractions are nearly constant at two monitor points. The mean value (MV) and standard deviation (SD) of solid volume fractions are similar at the height of 0.1 and 0.4 m, representing a PF state with a uniform volume fraction distribution in AE fluidized bed.

With increasing ethanol fluid pressure and temperature in case 2, periodic solid volume fractions are occurred between the maximum and minimum solid volume fractions at the height $H = 0.1$ m. The large and small solid volume fraction bands

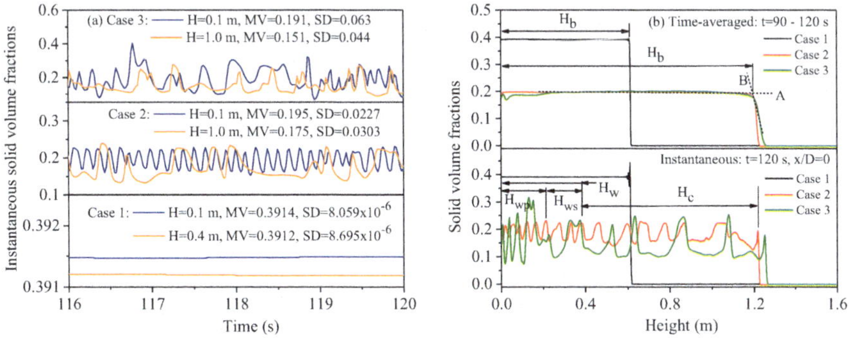

Fig. 6.41 Profiles of solid volume fractions at AE, SbE and SCE fluid fluidized beds

represent the occurrence of wave-like flow at the bottom. However, the periodicity of waves of solid volume fractions disappears, and large volume fraction fluctuations are appeared. High solid volume fractions are behaved by aggregation of particles, and low solid volume fractions act as ethanol fluid voids emerged, indicating a churn-like flow at $H = 1.0$ m in the SbE fluidized bed.

The calculated dimensionless numbers of *Fr*, *HD* and N_{tr} are summarized in Table 6.5 in AE, SbE and SCE fluidized beds. For case 1 of AE fluidized bed, the TR state is predicated from both dimensionless numbers of *HD* and N_{tr}. On the other hand, the PF state is forecasted from dimensionless number *Fr*, and is similar to computational simulations. For fluidization state in case 2 of SbE fluidized bed, the TR state is predicted using both dimensionless numbers of *HD* and N_{tr} in comparison with a PT state from dimensionless number *Fr*. From computational simulation, the TR state is characterized by a wave-like flow near the bed bottom and churn-like flow at the upper part of the bed. For case 3 of SCE fluid bed, the AF state is predicted from both dimensionless numbers of N_{tr} and *HD*, whereas the TR state is identified from dimensionless number *Fr*. From computational simulations, the AF state indicate the occurrence of particles aggregates and fluid voids in the bed.

From the straight lines A and B of solid volume fractions illustrated in Fig. 6.41(b), the bed height H_b is identified. The bed height is low in AE fluidized bed compared to the value in SbE and SCE fluidized beds. For case 2 in SbE fluidized bed, the bed height H_b includes a height H_w of the wave-like flow and height H_c of the churn-like flow. The height of wave-like flow is divided into two parts of H_{wp} and H_{ws}. The waves are originated from the first segment near the inlet at the bed bottom. The wave amplitudes grow gradually with the maximum and minimum solid volume fractions, representing a primary stage of wave-like flow. The wave amplitude reaches constant as they travel upward, and the wave-like flow reaches a stable stage, representing flow of saturated waves. The amplitudes of wave-like flow become larger, and eventually saturate. These indicate that the wave-like flow is transited from primary waves at the height H_{wp} to the saturated waves at the height H_{ws}. The wavelength is computed in terms of the distance between two adjacent major dense or dilute bands. The values of wavelength λ are 0.0459 and 0.0603 m, and the dimensionless wavelength λ/d_p is 45.9 and 60.3 at the primary and saturated wave stages. These indicate that the wavelength is 30.0% longer in the saturated waves than that in the primary waves in the wave-like flow of SbE fluidized bed. The frequency of waves is identified according to the number of peaks per second. The computed wave frequency is 6.92 Hz in wave-like flow of the SbE fluidized bed.

The power spectrum of the volume fraction fluctuations is computed from the simulated local instantaneous volume fractions of particles. Results are illustrated in Fig. 6.42 in SbE and SCE fluidized beds. The sampling frequency of the computational simulations is 50 Hz. Generally, the power spectrum densities are low at low frequency, and they decrease at high frequencies. The peak of power spectrum density is observed, highlighting a dominant frequency f_z. For case 2 at $H = 0.1$ m, the value of dominant frequency f_z is 6.86 Hz for wave-like flow, and it agrees with wave frequency f_w computed from the number of peaks. Furthermore, it is about 2.46 Hz

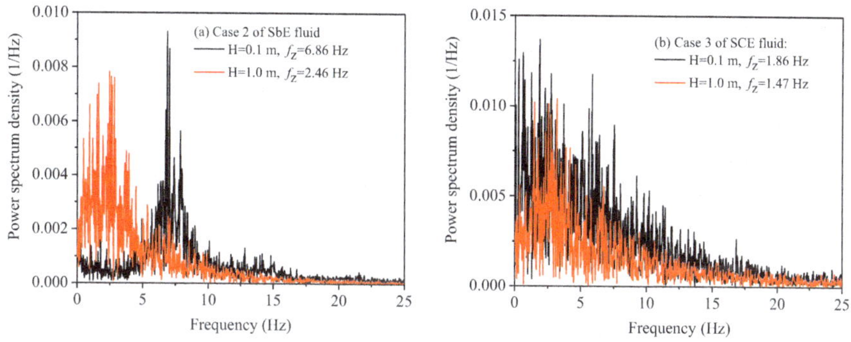

Fig. 6.42 Power spectrum density of solid volume fraction fluctuations in SbE and SCE fluid FBs

for churn-like flow at $H = 1.0$ m in SbE fluidized bed. For another thing, the dominant frequency for flow of supercritical ethanol fluid-particle mixtures is 1.86 Hz and 1.47 Hz at $H = 0.1$ m and 1.0 m, respectively. From experimental measurements, Du et al. [13] computed power spectrum densities between bubbling bed and turbulent regime. They suggested that the dominant frequencies are about 3.0 Hz and 2.0 Hz in bubbling fluidized bed and the transition regime. For the turbulent regime, the dominant frequency decrease to 1.5 Hz due to the turbulent movement of fluid voids in the fluidized bed.

6.4 Non-Newtonian Fluid and Cuttings Transport of Well Drilling

High-efficiency drilling is essential for the exploitation of oil and gas fields. Cuttings transport is one of the biggest challenges during drilling wells with horizontal and highly inclined sections. Inadequate hole cleaning can result in many drilling problems, including cuttings accumulation, premature bit wear, stuck pipe and high torque. In the process of drilling, the drilling fluid is pumped into the well through the drill pipe and across the drill bit, and then fluid with rock cuttings are lifted up to the ground through the annulus between the drill string and wellbore wall. In horizontal or highly inclined sections of the well, the cuttings are settled down in the lower wall of the wellbore due to gravity, and as a result build up a fixed cuttings bed. Pressure drop through the annulus is one of the most significant factors controlling borehole cleaning in well drilling due to its effect on shear stress acting on the cuttings bed. In addition to effects of wellbore structure characteristics (well deviation angle and drill pipe eccentricity.) and cuttings characteristics (cuttings size and volume fractions.), the transportation of cuttings in the annulus is strongly affected by drilling fluid rheological properties.

Drilling fluids usually exhibit non-Newtonian rheological behavior, namely the relationship between shear stress τ and shear rate γ is non-linear. The two parameters

Bingham plastic model that can be simply expressed

$$\tau = \tau_p + \mu_p \gamma \tag{6.21}$$

where τ_p and μ_p are the yield point and plastic viscosity. The Bingham plastic model tends to estimate the yield stresses. The power law model is a two-parameter model for a wide variety of shear-thinning ($n < 1$) or shear-thickening ($n > 1$) fluid, and is expressed as [23]

$$\tau = K\gamma^n \tag{6.22}$$

where K and n represent the consistency coefficient and flow behavior index. They are used to describe the rheological behavior of drilling fluids because the predictions coincide with rheological behavior fairly [7]. The three parameters model is expressed as below.

$$\tau = \tau_0 + K_{HB}\gamma^n \tag{6.23}$$

where τ_0, K_{HB} and n are the yield stress, consistency coefficient and flow behavior index, respectively. It is also referred to the yield power law or modified power law model.

6.4.1 Drag Model for Power-Law Fluid-Particle Mixture

In the non-Newtonian fluid-particle flow, the motion of particles alters the shear rate and thus the apparent viscosity of the fluid, which in turn influences the drag between the two phases. Furthermore, at high solids concentration the presence of neighboring particles leads to larger velocity gradients, which further highlights the effect of rheological parameters on the drag force. The pressure loss of non-Newtonian fluids flowing through the porous media can be described by Blake-Kozeny-Carman equation [4]. The complex cross-section of the interstitial void space is characterized in terms of the hydraulic diameter D_e [23]. For characterizing the tortuous nature of the pore channel, the ratio of the actual length of the channel to its linear distance can be defined as the tortuosity factor K_1 [28]. Then, the flow space in the bed of particles is made equivalent to a bundle of tortuous conduits with complex cross-sections and lengths described by specific parameters. For this purpose, it is customary to introduce the dimensionless friction factor f_d to characterize the flow resistance

$$f_d = \frac{d_p}{\rho_l u_0^2} \frac{\Delta P}{L} \frac{\alpha_l^3}{\alpha_s} \tag{6.24}$$

where L is the linear distance of the fluid flowing through the bed. ρ_l and α_l are the fluid density and fluid volume fraction. ΔP is the pressure drop caused by the fluid flowing through the particle pores with the fluid volume fraction α_l and particles volume fraction α_s. u_0 is the superficial velocity of the fluid. In transitional and turbulent flow regime, the inertial effect begins to manifest, and together with the viscous effect dominate the flow resistance. For this purpose, the dimensionless friction factor f_d using Ergun equation characterizes the flow resistance at high solid volume fractions [4, 18].

$$f_d = \frac{150}{Re_p} + 1.75 \tag{6.25}$$

where Re_p is the pore Reynolds number considering the rheological properties of non-Newtonian fluids.

$$Re_p = \frac{\rho_l u_0 d_p}{\eta_{eff}\alpha_s} \tag{6.26}$$

where μ_{eff} is the effective viscosity of fluid. The shear stress of power-law fluid can be expressed at the pore wall of the particle bed:

$$\tau_w = K'(\gamma_w)^{n'} \tag{6.27}$$

where K' and n' are the apparent consistency coefficient and the apparent flow behavior index at the particle wall surface. For power-law fluids, they can be expressed as:

$$K' = K\left(\frac{3n+1}{4n}\right)^n \tag{6.28}$$

$$n' = n \tag{6.29}$$

where K and n are the generally normal rheological parameters. Noting that the coefficient $((3n + 1)/4n)^n$ is added to correct the fluid consistency coefficient at the pore wall created by particles.

The effective viscosity of power law fluids in the pores is given by [8, 43]:

$$\eta_{eff} = K'\left(\frac{8u_0}{D_e\alpha_l}\right)^{n'-1} \tag{6.30}$$

where D_e is the equivalent diameter of the pore created by particles and it is defined as:

$$D_e = \frac{2d_p}{3}\left(\frac{\alpha_l}{1-\alpha_l}\right) \tag{6.31}$$

From Eq. (6.30), the effect viscosity of power law fluids is given by

$$\eta_{eff} = K'\left(\frac{12u_0\alpha_s}{\alpha_l^2 d_p}\right)^{n'-1} \tag{6.32}$$

Substituting Eqs. (6.31) and (6.32) into Eq. (6.26), the Reynolds number with the corrected apparent rheological parameters at the wall surface of the pore created by particles can be obtained as:

$$Re_p' = \frac{(12)^{1-n}\rho_l u_0^{2-n} d_p^n \alpha_l^{2n-2}}{K\left(\frac{3n+1}{4n}\right)^n \alpha_s^n} \tag{6.33}$$

Substituting formula (6.33) into Eq. (6.25), the Ergun type equation of the pressure drop of the power-law fluid flowing through the particles system is obtained:

$$\frac{\Delta P}{L} = 150\frac{K\left[\frac{(3n+1)}{4n}\right]^n \alpha_s^{n+1}|\boldsymbol{u}_l - \boldsymbol{v}_s|^n}{(12)^{1-n} d_p^{n+1}\alpha_l^{n+1}} + 1.75\frac{\rho_l|\boldsymbol{u}_l - \boldsymbol{v}_s|^2\alpha_s}{d_p\alpha_l} \tag{6.34}$$

where u_l and v_s are the fluid velocity and solid velocity. The first term on the right-hand side of the above formula represents the viscous drag loss due to the friction between the power-law fluid and the particle surface, and the second term represents the inertial loss due to the turbulent vortex and the expansion and contraction of the pore channel created by particles.

In the momentum conservation equations of the liquid–solid flow, the pressure drop and the relative velocity satisfy the following relationship:

$$\alpha_l\frac{\Delta P}{L} = \beta|\boldsymbol{u}_l - \boldsymbol{v}_s| \tag{6.35}$$

Substituting the above equation into Eq. (6.31), the drag coefficient of the power-law fluid-particle flow under high particle concentration conditions is obtained as [6]:

$$\beta_{PL1} = 150\frac{K\left[\left(\frac{3}{4} + \frac{1}{4n}\right)\right]^n \alpha_s^{n+1}|\boldsymbol{u}_l - \boldsymbol{v}_s|^{n-1}}{(12)^{1-n} d_p^{n+1}\alpha_l^n} + 1.75\frac{\rho_l\alpha_s|\boldsymbol{u}_l - \boldsymbol{v}_s|}{d_p} \tag{6.36}$$

This gives a relation between the drag coefficient of power-law fluids and the rheological parameters under high particle concentration conditions.

For Reynolds number Re_p of the power law fluid flowing through the pore channels created by particles, [37] obtained a slightly different wall shear rate model:

$$\gamma_w = 6K_0K_1\left(\frac{1-\alpha_l}{\alpha_l^2}\right)\left(\frac{u_0}{d_p}\right) \tag{6.37}$$

where K_0 is the cross-sectional shape factor of the particle pore channel. K_0 can be defined as 2.0 for the flow in the circular tube and 3.0 for the flow in the slit. K_1 is the tortuosity factor of the pore channel, which is defined as $K_1 = L_e/L$, that is the ratio of the average flow distance L_e in the tortuous channel to the linear distance L of the bed.

Adopting the cross-sectional shape factor K_0 of 2.5 and using the tortuosity factor K_1 of $2^{0.5}$ to correct the flow distance and the local velocity, the apparent viscosity of the power-law fluid in the pore channel created by particles [37]:

$$\eta_{eff} = K'\left(15\sqrt{2}\left(\frac{1-\alpha_l}{\alpha_l^2}\right)\left(\frac{u_0}{d_p}\right)\right)^{n'-1} \tag{6.38}$$

Combining Eqs. (6.26) and (6.38), the Reynolds number for the power law fluid flowing through the pore channel can be obtained as:

$$Re_p' = \frac{\left(15\sqrt{2}\right)^{1-n} \rho_l u_0^{2-n} d_p^n \alpha_l^{2n-2}}{K\left(\frac{3n+1}{4n}\right)^n \alpha_s^n} \tag{6.39}$$

Substituting the above formula into Eq. (6.25), the power law fluid-particle drag coefficient under high particle concentration is obtained as [6]:

$$\beta_{PL2} = 150\frac{K\left[\frac{(3n+1)}{4n}\right]^n \alpha_s^{n+1}|\boldsymbol{u}_l - \boldsymbol{v}_s|^{n-1}}{\left(15\sqrt{2}\right)^{1-n} d_p^{n+1}\alpha_l^n} + 1.75\frac{\rho_l \alpha_s |\boldsymbol{u}_l - \boldsymbol{v}_s|}{d_p} \tag{6.40}$$

This reveals a relation between the drag coefficient of power-law fluid and the consistency factor K and power-law index n.

With decreasing solid volume fractions, the drag force between the non-Newtonian fluid and particles can be calculated by the sum of the resistance of the fluid, F_d, acting on a single particle. The pressure drop of the fluid passing through the particle system can be expressed as:

$$\Delta P = NF_d \tag{6.41}$$

where F_d is the force on a single particle exerted by the fluid, and N is the number of particles within the designated flow space. They can be calculated as follows:

$$N = \frac{6\alpha_s L}{\pi d_p^3} \tag{6.42}$$

$$F_d = C_d \rho_l \frac{\pi}{4} d_p^2 \frac{u_0^2}{2} \tag{6.43}$$

Combining with Eq. (6.41) and (6.43) using the function $f(\alpha_l)$ considering the effect of particle aggregation, the pressure drop is expressed:

$$\frac{\Delta P}{L} = \frac{3C_d \rho_l \alpha_s |\boldsymbol{u}_l - \boldsymbol{v}_s|^2}{4d_p} \alpha_l^{-2.65} \tag{6.44}$$

Then the expression of the drag coefficient for the power-law fluid and particle mixture in Wen-Yu form:

$$\beta_{WY} = \frac{3C_d \rho_l \alpha_l \alpha_s |\boldsymbol{u}_l - \boldsymbol{v}_s|}{4d_p} \alpha_l^{-2.65} \tag{6.45}$$

For the non-Newtonian fluid-particle flow, a treatment method is to still use the Newtonian C_d-Re correlation using Reynolds number Re_n. For the power-law fluid, Re_n can be expressed as:

$$\mathrm{Re}_n = \frac{\rho_l |\boldsymbol{u}_l - \boldsymbol{v}_s|^{2-n} \alpha_l^{2-n} d_p^n}{K} \tag{6.46}$$

The drag coefficient C_d is expressed as the function of Re_n and flow behavior index n at low particle concentrations:

$$C_d = 3^{\frac{3n-3}{2}} \left[\frac{33n^5 - 63n^4 - 11n^3 + 97n^2 + 16n}{4n^2(n+1)(n+2)(2n+1)} \right] \frac{24}{Re_n} \quad Re_n < 1.0 \tag{6.47}$$

$$C_d = \left[\frac{33n^5 - 63n^4 - 11n^3 + 97n^2 + 16n}{4n^2(n+1)(n+2)(2n+1)} \right] \frac{3^{(1.5n-1.5)} 24}{Re_n} + \frac{10.5n - 3.5}{Re_n^{(0.32n+0.13)}} \quad 1 \le Re_n < 1000 \tag{6.48}$$

From Eqs. (6.36) and (6.45), the drag model of power-law fluid and particles mixture expresses as follows

$$\beta = (1 - \phi)\beta_{PL1} + \phi \beta_{WY} \tag{6.49}$$

and using Eqs. (6.40) and (6.45) the fluid-particles momentum exchange coefficient is:

$$\beta = (1 - \phi)\beta_{PL2} + \phi \beta_{WY} \tag{6.50}$$

where the blending function, ϕ, is [29]

$$\phi = \{\arctan[150 \times 1.75(0.2 - \alpha_s)]/\pi\} + 0.5 \tag{6.51}$$

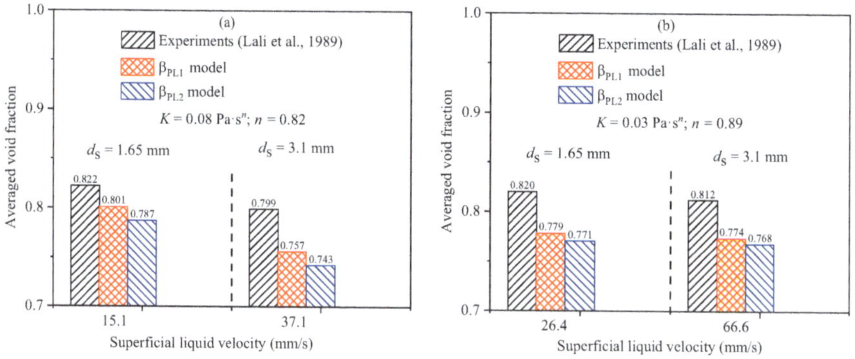

Fig. 6.43 Comparison of simulated and experimental bed void fractions in fluidized bed

These drag models consider effects of rheological properties of power-law fluids on the interaction between fluid and solid phases, which includes the correction of rheological properties at the wall of the pore channel at high particle concentrations and adopts the single particle drag coefficient correlation considering the influence of the flow behavior index at low particle concentrations. In particular, as the flow behavior index n is equal to 1, the fluid exhibits Newton friction characteristics, and the proposed model can be degraded to Huilin-Gidaspow drag model.

To verify the drag models, both the drag model of Eq. (6.49) and the drag model of Eq. (6.50) are used to simulate the characteristics of bed expansion fluidized by the power-law fluid, and compared with experimental data in fluidized bed [38]. Figure 6.43 shows the comparison of simulated and experimental bed void fractions. The bed void fractions calculated by Eq. (6.49) are in good agreement with experimental measured values. The Eq. (6.50) drag model under-predict fluid volume fractions in the bed. It is a more reasonable combination to calculate the drag between power-law fluid and particles that using Eq. (6.49) in fluidized beds.

6.4.2 *Cuttings Transport in Horizontal and Inclining Wells*

Flow pattern of cuttings is shown in Fig. 6.44 along the horizontal well hole, including cuttings volume fraction, cuttings tangential velocity and axial velocity [5]. The hole diameter D_h is 203.2 mm, and the pipe diameter D_p is 101.6 mm. The eccentric distance E is 25 mm. The eccentricity defines according to eccentric distance E, and it is $e = 2E/(D_h\text{-}D_p)$. The eccentricity is 0.5. The particle diameter and density are 3.0 mm and 2550 kg/m^3. The mud density is 1437.6 kg/m^3. The consistency factor K and power-law index n are 0.9488 Pa.s^n and 0.4097, respectively. The inlet solid concentration is 4.0% at the inlet velocity of 1.016 m/s. The wellbore wall is stable without the cracks or the pore channels. The axial length of the annulus is within the length of a single drill pipe and there is no tool joint over the entire simulated length.

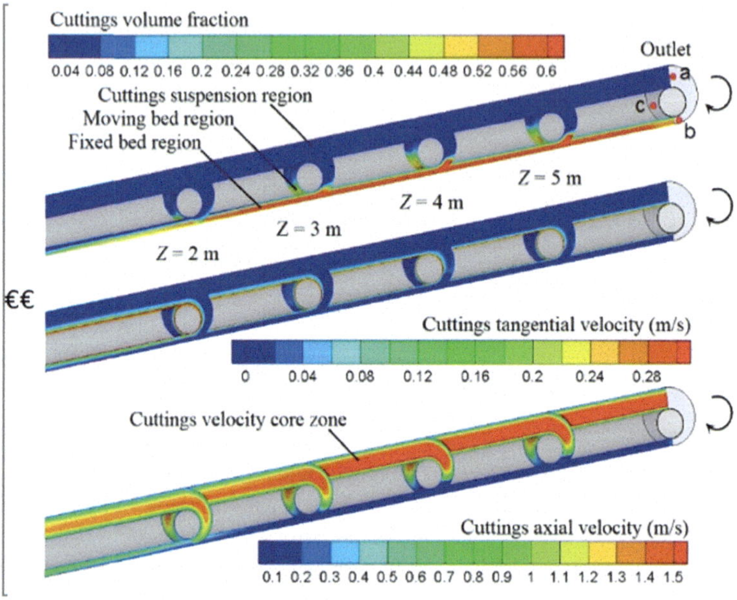

Fig. 6.44 Cuttings deposition along the annulus with drill pipe rotation ($\omega = 100$ rpm, $d_p = 3$ mm)

The inner pipe rotates around self-axis at a specified rotational speed. To simulate the effect of drill pipe rotation on cuttings transport, the sliding mesh is applied in the annulus. The annulus is divided into the inner and the outer flow zones. The outer flow zone is static, and the inner flow zone is rotating at the same rotation speed with drill pipe.

The cutting particles are continuously injected into the wellbore with the drilling fluid from the inlet. Due to the difference of density between the cuttings and the drilling fluid, the cuttings is gradually deposited to the bottom of well wall and the cuttings bed forms along the well length by gravity force. Simulated results show that the rotation of drill pipe produces spiral flow of drilling fluid and cuttings through the annulus. The cuttings tangential velocity is largest near the wall of drill pipe and it decreases gradually along the radial direction of the annulus. The profile of cuttings axial velocity indicates that the high velocity core zone locates in upper side of the annulus, which contributes to the vast majority of cuttings transport. The cuttings velocity is very low, and the fixed bed with high volume fraction of cuttings is gradually formed near the lower side of the borehole wall due to the resistance caused by cuttings accumulation.

Both velocity and concentration of cuttings present a relative uniform distribution in the region near inlet attributed to the uniform inlet boundary conditions. Due to the density difference between cuttings and drilling fluid, cuttings gradually deposited to the bottom of well wall and the cuttings bed forms with the increase of well length. The bed height and velocity magnitude of cuttings reach a stable value at an approximate distance of 3.0 m from the inlet [5]. A stable deposition and flow conditions are achieved along the total length of the annulus ultimately. Both bed heights and velocity of cuttings reach a stable state at an approximate distance of 3.0 m from the inlet. A stable deposition of cutting particles is achieved along the length of the annulus ultimately. From the distributions of volume fraction and velocity of cuttings, the flow of cuttings can be divided into three regions, namely, the cuttings suspension region, the moving bed region and the fixed bed region. The cuttings suspension region with low solid volume fraction and high velocity of drilling fluid is formed at the upper region of the annulus. The fixed bed region with high solid volume fraction and low velocity of drilling fluid is located at the bottom of the annulus. Furthermore, the visualization of core zone of cuttings volume fraction and velocity clearly depicts swaying phenomena along the direction of drill pipe rotation.

The annular velocity of drilling fluid is a key factor in the process of cuttings transport. To analyze the impact of inlet fluid velocity, the distributions of velocity and volume fraction in both horizontal well and vertical well are simulated. The drill pipe rotation produces spiral flow of drilling fluid through the annulus, while increasing the fluid inlet velocity reduces such effect. For horizontal well, the simulated results show different patterns of fluid velocity. The velocity magnitude contours of liquid phase are illustrated in Fig. 6.45 at different inlet fluid velocities in horizontal annulus. Unlike vertical wells, effect of cuttings accumulation on fluid velocity pattern was exposed. The velocity core zone locates in the larger gap of the annulus. There is almost no flow near the lower side of the borehole wall especially for lower inlet velocities due to the resistance caused by cuttings accumulation. Furthermore, the shift of velocity core zone along the direction of drill pipe rotation weakens due to the cuttings accumulation near the bottom of the wellbore. The cuttings accumulation decreases with the increase of inlet fluid velocity.

The annular velocity of drilling fluid is one of the major factors affecting wellbore cleaning efficiency, which is usually measured by the ultimate cuttings volume fraction in the annulus. The cuttings volume fraction is characterized by the volume occupied by cuttings per unit volume in the annulus. The lower the cuttings volume fraction is, the better the cuttings transport performance is. The cuttings volume fraction contours of a fully developed flow in the vertical annulus are shown in Fig. 6.46 at different inlet fluid velocities. As it is seen that the cuttings bed is rarely formed in vertical wells, which is significantly different from horizontal wells. Cuttings concentration is approximate to the feed one, and more uniform is found in vertical annulus.

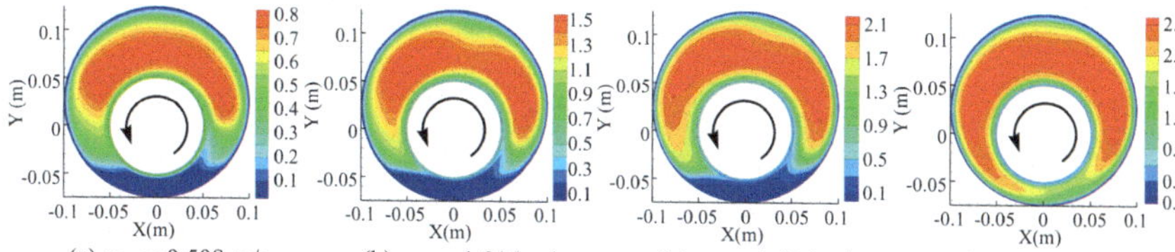

(a) $u_{f,in}$ = 0.508 m/s (b) $u_{f,in}$ = 1.016 m/s (c) $u_{f,in}$ = 1.524 m/s (d) $u_{f,in}$ = 2.032 m/s

Fig. 6.45 Profile of fluid velocity pattern in horizontal annulus

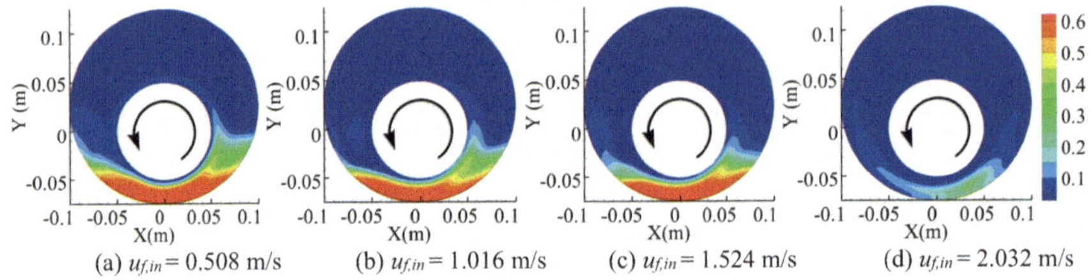

Fig. 6.46 Profile of cuttings volume fraction in horizontal annulus

Comparing to fluid velocity, the particle concentration is low, while the fluid velocity is high. The differences of fluid annular velocities between the larger gap and the narrower gap increase with a higher inlet velocity. Consequently, the vast majority of cuttings are transported out of wellbore by high-velocity fluid at the larger gap of annulus and the residence time of cuttings at the narrower gap becomes longer.

For a constant drill pipe rotation of 100 rpm, both the height of cuttings bed and the cuttings volume fraction decrease as the inlet fluid velocity increases. Furthermore, at the inlet fluid velocities of 0.508 and 1.016 m/s the drill pipe rotation results in a higher volume fraction of cuttings bed on one side, which is in consistency with the swaying of cuttings along the direction of drill pipe rotation. With the increase of the inlet fluid velocity, the drag force exerted on the cuttings increases and more cuttings move from the fixed bed. They enter into the suspended region, and then are carried out of borehole by drilling fluid. Thus, the increasing inlet fluid velocity leads to improvement of cuttings transport. Furthermore, increasing inlet fluid velocity results in weakening of the swaying phenomenon of cuttings distribution which is caused by drill pipe rotation.

The orientation of the wellbore varies considerably from horizontal to vertical sections. The inclination angle of the well is defined as the deviation angle of the well from the vertical axis. Figure 6.47 shows the distribution of cuttings volume fraction through the annulus at different lower side of the hole provides upholder and the friction force exits between cuttings and walls. Unlike the surface of cuttings bed in the horizontal annulus, in the inclined annulus ($\theta = 30$ degree and 60 degree) the distributions of cuttings bed are irregular and its surface shows a wavy shape. This is because that the motion of cuttings is in three major ways, namely, suspension, rolling and a combination of suspension and rolling in the inclined wellbore annulus.

In order to evaluate the transport efficiency of cuttings, the cuttings transport ratio (CTR) is defined according to inlet particles concentration $\alpha_{s,in}$ and averaged particles concentration in the annulus.

$$CTR = \frac{\alpha_{s,in}}{\alpha_{s,\infty}} \tag{6.52}$$

Figure 6.48 shows the cuttings transport ratio as a function of inclination angles. The cuttings transport ratio decreases, reaches minimum, and then increases with the increase of inclination angles. It is also found that the cuttings transport ratio increases

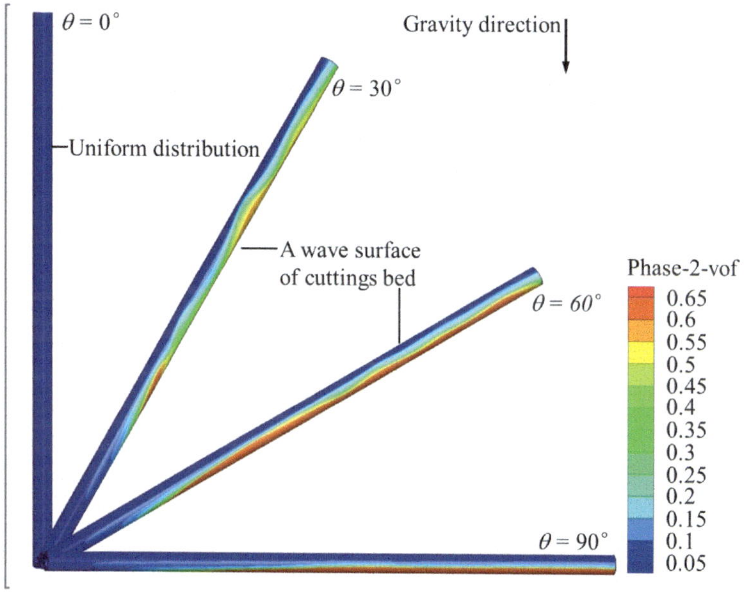

Fig. 6.47 Cuttings volume fraction through the annulus at different inclination angles

with the increase of inlet fluid velocity. For given inlet fluid velocity, the vertical annulus gives a high cuttings transport performance. There is a critical inclination angle of the well (θ_{ctr}) at which the cuttings transport ratio is lowest. When the inclination angle is smaller than θ_{ctr}, the cuttings transport ratio is decreased. The critical inclination angle for cuttings transport ratio is about 50–65 degree. The *CTR* is increased as the inclination angle is larger than θ_{ctr}. It reduces at the low inclining angle. The reason is the transition of cuttings flow from the suspension mechanism to the rolling one takes place around θ_{ctr} in the annulus. Furthermore, the lower the inlet velocity is, the smaller the critical inclination angle is, and the earlier the critical inclination angle θ_{ctr} appears.

The profile of pressure drop indicates that the pressure drop increases, reaches maximum, and then decrease with the in-crease of inclination angles through the annulus. The pressure drop is the lowest at higher inclination angles near horizontal annulus. There is a critical inclination angle (θ_{dp}), at which the pressure drop is highest through the annulus. The value of θ_{dp} is in the range from 35 to 45 degree. The pressure drop increases as the angle of inclination is less than θ_{dp}. However, it is decreased at inclinations larger than θ_{dp}. According to the value of θ_{ctr} and θ_{dp} obtained, it can be concluded that hole cleaning is difficult to be handled during drilling wells with the inclinations around 35 degree to 45 degree.

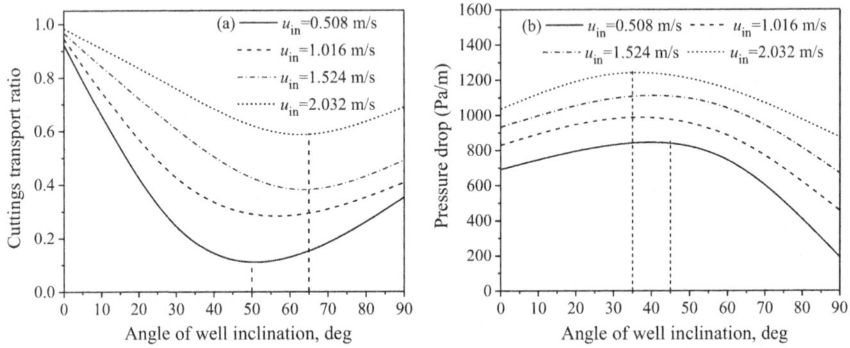

Fig. 6.48 Effect of well inclination on cuttings transport ratios and pressure drop of the annulus

6.4.3 Orbital Motion Effect of Horizontal Drill Pipes

Drill pipe rotation induces tangential velocities of drilling fluids and cuttings and it is a factor that can not be ignored during hole cleaning due to the drill pipe rotation effect. At actual drilling operations, the motion of drill pipe changes at different positions along the well. In most cases the drill pipe is not constrained with centralizers and has both rotary and orbital motion. First, the flow-induced orbital motion for a rotating cylinder is one of the reasons. Flow-induced vibrations of the rotating rigid body are commonly encountered in fields of engineering. These oscillations are often not limited to one direction, which lead to an orbital motion of the body following an elliptical path. The dominant motion of the cylinder is an orbit combined with a steady offset, and gyroscopic effects occur when the orbiting cylinder is inclined to the horizontal. The orbital motion will occur, and in general, in the same direction as that for its self-rotation when the rotation is driven only by an external torque. On the other hand, the orbital motion is induced by the rotation of the crooked drill string. The drill string cannot be treated as a rigid rod due to the high ratio of its length to diameter, and it can be regarded as an absolutely flexible cable by geometrical similarity. Mechanical phenomena typical for elastic rods and the drill string stability loss can be observed. Due to the local geometrical imperfections of the axial trajectories of bore-holes, the drill string is subjected to the contact and friction forces exerted by the borehole wall additionally. The drill pipe is bent and deformed in the axial direction, and the rotation of the crooked pipe produces orbital revolution around the axis of the wellbore. As a consequence, the cuttings transport is more affected by the orbital motion of drill pipe rather than the self-rotation.

The configuration of the computational domain is a horizontal annulus consisting of the orbiting pipe and the wellbore. As shown in Fig. 6.49a, the orbital motion of drill pipe is specified as the circular movement at a constant revolution speed Ω_d around the axis of well-bore. At the same time, the drill pipe rotates around its own axis at the rotation speed ω_d. The orbital motion of drill pipe is in the same or opposite direction as that for its self-rotation. The orbital angle α is defined as

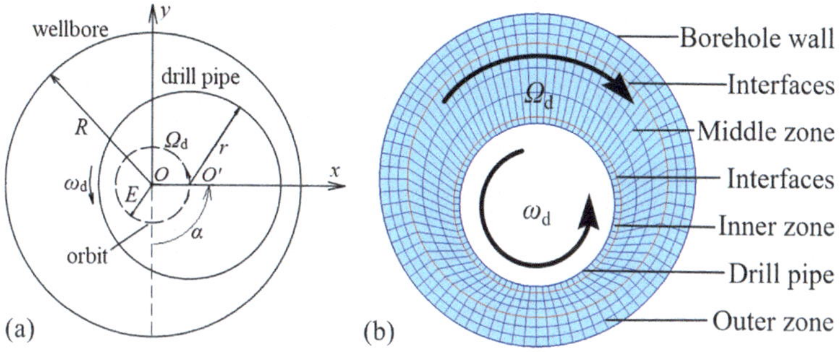

Fig. 6.49 Sketch of the orbiting drill pipe in the drilling annulus

the angle at which the drill pipe axis deviates counter-clockwise from the negative y-axis. At the initial time of the calculation, the drill pipe locates at the lowest point of its revolution orbit and $\alpha = 0$. The trajectory equations of the axis of the drill pipe can be expressed as [7]:

$$O'_x = E\cos(\alpha - 90^\circ) \tag{6.53}$$

$$O'_y = E\sin(\alpha - 90^\circ) \tag{6.54}$$

$$\alpha = \Omega_d t \times 180^\circ \tag{6.55}$$

in which O'_x and O'_y are the x-component and the y-component of the coordinates of the drill pipe axis trajectory respectively, E is the eccentric distance of the drill pipe from the axis of the wellbore and t is the calculating time. In general, the higher the rotation speed, the higher the orbital revolution speed.

For the orbital movement of the drill pipe wall along horizontal direction, the embedded sliding mesh method is applied. The computational domain is divided into the inner, middle and outer flow zones. The geometry of the annulus and the three flow zones are sketched in Fig. 6.49b. The outer flow zone and the adjacent borehole wall are stationary, the mesh on middle flow zone rotates around the axis of the wellbore to provide orbital revolution of the drill pipe, and the mesh on inner flow zone self-rotates around the axis of the drill pipe while revolving with the middle zone. The drill string is a moving wall and makes synchronous movement with its adjacent inner zone at every moment. The connections between the outer and middle zones and between the middle and inner zones are achieved with two pairs of interfaces. The annulus cross-section is meshed into quadrilateral grid cells that are radially distributed. The tangential sizes of the meshes gradually increase from the inside to the outside. The entire axial length of the annulus is swept in a cooper grid type with this cross section as the source surface to create hexahedral structured grids throughout the annulus.

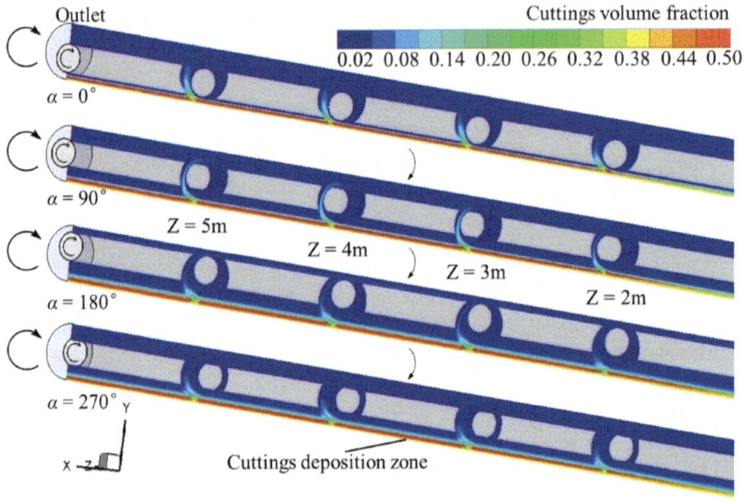

Fig. 6.50 Cuttings deposition along the horizontal annulus with an orbiting pipe at four instantaneous moments of a drill pipe orbital period

In order to show flow patterns of cuttings along the annulus in a complete orbital period, the distributions of cuttings concentration are sketched in Fig. 6.50 [5]. When the orbital angle of the drill pipe α is 0°, the drill pipe passes by the lowest point of its rotation orbit and the radial clearance between the drill pipe and the lower wall of wellbore is minimum. As the orbital angle is 180°, the drill pipe passes by the highest point of its rotation orbit and the radial clearance between the drill pipe and the lower wall of wellbore is maximum. When the orbital angle of the drill pipe is 90° and 270° respectively, the drill pipe passes by the right-most and left-most ends of the rotation orbit. The gap between the pipe and the lower wall of wellbore varies with the change of orbital angles and then affecting the local volume fractions of cuttings.

The profiles of instantaneous cuttings concentration at the axial distance of 3.0 m in the annulus with an orbiting pipe at four orbital angles are shown in Fig. 6.51. Roughly, the deposition of cuttings is formed at the lower wall of the well, and the low volume fraction of cuttings exists at the upper region. Furthermore, affected by the orbital motion of drill pipe, the core zone distribution of cuttings concentration reveals swaying phenomena. Interestingly, a high concentration region of cuttings near the right side of drill pipe is observed during the counterclockwise orbital motion of the drill pipe, especially when the orbital angle α is from 0° to 90°. This indicates that the orbital motion of drill pipe produces an entrainment effect on the cutting particles.

Figure 6.52 shows profiles of tangential velocity and streamlines of cuttings in the horizontal annulus with an orbiting pipe. As the orbital angle α is 0°. The wide gap in the upper part of the annulus is the velocity core zone where most of the cuttings are transported. Due to the resistance caused by cuttings accumulation, the

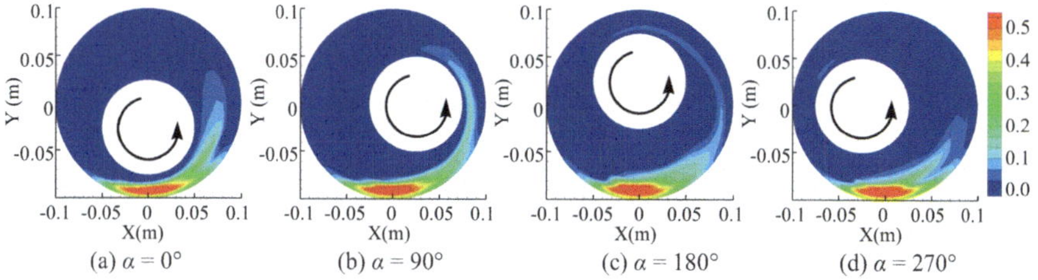

Fig. 6.51 Profiles of cutting volume fractions in the horizontal annulus with an orbiting pipe

velocities of cuttings are very small near the lower wall of the annulus. Obviously, the distribution of cuttings velocity is closely related to the orbital angle. The maximum axial velocity of cuttings is always in the wide gap of the annulus and it rotates following the orbital motion of the drill pipe, and lagging behind that. The tangential velocity is the highest at the wall of drill pipe and it is the lowest at the inner wall of the casing. The secondary flow appears with the orbital motion of drill pipe. As shown, the secondary flow exists at the inner wall of the casing. The cuttings and drilling fluids possess a large tangential velocity component and the secondary flow exists because of the stirring effect of the orbital motion. In the secondary flow regime, the direction of tangential velocity is opposite to that of the mainstream. Furthermore, it can also be observed from the four instantaneous results that the regime of high tangential velocity and secondary flow rotates with the orbital motion of drill pipe.

A comparison of cuttings concentration is shown in Fig. 6.53 in the horizontal well with a motionless pipe, a self-rotating pipe, or an orbiting pipe. In the annulus with a motionless pipe, the cuttings accumulation is extremely severe. In more than half the volume of the annulus, the concentration of cuttings exceeds 50%. Extremely dense cuttings bed is formed in lower wall of the well, which is to the disadvantage of hole cleaning. While in the annulus with a self-rotating pipe, the tangential force introduced by the rotating pipe improves hole cleaning. The volume fraction of retained cuttings is significantly reduced. Moreover, in the annulus with an orbiting pipe, the cuttings concentration is even lower. The orbiting drill pipe enhances the spiral flow of drilling fluid and periodically scrapes the cuttings bed at the lower side of the annulus, which causes more cuttings to be carried out of the wellbore.

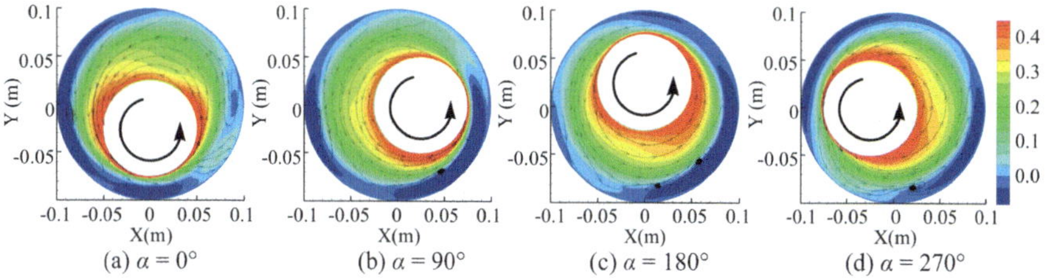

Fig. 6.52 Tangential velocity and streamlines of cuttings in the horizontal annulus with an orbiting pipe

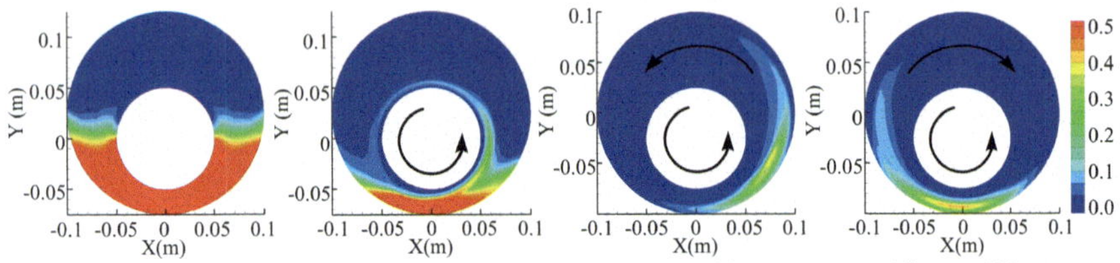

(a) $\omega_d = 0$ rpm $\Omega_d = 0$ rpm (b) $\omega_d = 100$ rpm $\Omega_d = 0$ rpm (c) $\omega_d = 100$ rpm $\Omega_d = 100$ rpm (d) $\omega_d = 100$ rpm $\Omega_d = -100$ rpm

Fig. 6.53 Cuttings concentrations with a motionless pipe, a self-rotating pipe (orbiting pipe)

The delamination phenomena of cuttings concentration are obvious, especially in the annulus with a motionless pipe or a self-rotating pipe. According to the distribution of solid concentration, three flow zones can be defined, namely the cuttings suspension zone in the upper side of the wellbore, the fixed bed zone in the lower side of the wellbore and the moving bed zone between them.

In summary, comparing with self-rotation, the orbital motion of drill pipe effects on hole cleaning significantly due to its periodical stirring and entrainment effect on cutting particles. Instead of the self-rotation, the orbital motion of drill pipe plays the dominant role in the tangential flow in the annulus. The direction of cuttings bed swaying phenomenon is the same as that of the orbital motion. The core zone of axial velocity rotates following the orbital motion, and lagging behind that. When the drill pipe makes orbital motion in the opposite direction to its self-rotation, the tangential flow is divided into two distinct eddies, which are in the opposite direction. The direction of the eddy in wide gap of the annulus is the same as the direction of the orbital motion, while the direction of the eddy in narrow gap is the opposite.

6.5 Molten Salt-Particles Fluidization

The primary uses of molten salt in energy technologies are in power production and energy storage. Molten salts remain a single-phase liquid even at very high temperatures and atmospheric pressure, which makes molten salt well-suited to advanced energy technologies, such as molten salt reactors. Fluidized beds (FB) have been widely applied in thermochemical processes of coal and biomass in order to obtain homogeneous thermal distributions in the reactors. These homogeneous thermal profiles permit better reaction performance when compared to the fixed bed thermochemical processes. The objective of the thermochemical conversion is to convert coal and biomass to combustible gases, suitable for use as fuel of combustion. Furthermore, it is very important to control the temperature of thermochemical processes. The fluidized bed reactor bed temperature keeps at near constant to avoid bed materials aggregation due to local high temperature because that leads to the reduction of conversion efficiency. Thus, using high temperature molten salt as fluidization agent, fluidized bed thermochemical processes of solid materials cover endothermic reactions at relative high temperatures because these temperatures aid in the breaking down of organic materials into methane, hydrogen and carbon monoxide gaseous species with steam. Therefore, a constant search in the engineering community for high temperature working fluid acting on fluidization agent became essential to improve the overall conversion efficiency in fluidized bed reactors.

The nitrate molten salt works well in the 240–565 °C temperature range using a 60–40%wt mixture of $NaNO_3$ and KNO_3 at ambient pressure. The concentrated solar power can be coupled with thermal energy storage and conventional fuels using molten salt. The molten salts are used as heat transfer fluid, and is heated and used in a conventional steam generator to produce electricity. The operating molten salts temperature was about 566 °C in CRTF's plant [24] and 550 °C in Archimede's plant [34]. The viscosity of commercial molten salt used in solar thermal energy processes was nearly 4.6 mPa·s at 250 °C (above its melting point) and 1.8 mPa·s at 400 °C [15], illustrating that the fluidity of molten salt is similar to water. The density of molten salt is about from 1700 to 1900 kg/m^3, and decreases with the increase of temperature. For coal particles, the density is about 2200–2400 kg/m^3. The density ratio ρ_{sf} of coal particles and molten salt fluid is in the range of 1.16–1.42 ($\rho_{sf} = \rho_s/\rho_f$, where ρ_s and ρ_f are the densities of coal particles and molten salt fluid.). Apparently, hydrodynamics of molten salt fluid and particles suspension depend on density ratio in molten salt-particles fluidized bed reactors.

6.5.1 *Molten Salt-Coal Particle Conventional Fluidized Bed*

In order to study the conventional fluidization state of molten salt-coal particles mixture, three different cases are simulated in the fluidized bed (FB). The parameters are listed in Table 6.6. The bed width and height of the molten salt-particles FB is 0.25 and 2.5 m. The coal particles diameter is 3.0 mm with the coal particles density of 2000 kg/m^3 (such as bituminous coal) and 2400 kg/m^3 (example for anthracite coal). The properties of molten salt fluid are constant in the bed, and the molten salt fluid viscosity is 0.0018 Pas with two different densities of 1700 kg/m^3 and 1900 kg/m^3. The molten salt fluid enters from the inlet of the bed bottom, and exits at the top of the bed [51]. The initial bed height and solid volume fraction are 0.3 m and 0.6. The minimum fluidization velocity is calculated using empirical correlation [50] according to Archimedes number.

The instantaneous solid volume fractions are displayed in Fig. 6.54 in molten salt fluid-particles conventional fluidized bed. For base case 1 with coal particles and molten salt fluid densities of 2000 and 1900 kg/m^3, the density ratio is 1.05. The distributions of solid volume fractions are uniform along lateral and axial directions, representing a particulate fluidization. For base case 2 with particles and molten salt fluid densities of 2400 and 1700 kg/m^3, the density ratio is 1.41. The low and high solids volume fraction bands appear alternately near the bed bottom. These particle bands are known as wave-like flow or traveling waves. Ripples are observed at the bottom of bed, and pass through the bed with alternating dilute layer with low solid

Table 6.6 Parameters used in simulations in molten salt-particles conventional fluidized bed

Cases	ρ_f	μ_f	ρ_s	u_{mf}	u_f	*Fr* number		N_{tr} number		*VH* number	
	kg/m³	Pa s	kg/m³	cm/s	cm/s	Value	State	Value	State	Value	State
Base case 1	1900	0.0018	2000	0.265	1.26	0.0002	PF	28.592	PF	2.996	TR
Base case 2	1700	0.0018	2400	1.332	6.33	0.0060	PF	200.14	TR	3.668	TR
Base case 3	1900	0.0018	2400	1.007	7.32	0.0035	PF	142.96	TR	3.461	TR

volume fractions and dense layer with high solid volume fractions. It is also found that these travelling waves tilt either to the left or to the right because of hydrodynamic interactions between the molten salt fluid and particles and the collisional interactions of particles. The travelling waves are broken down at the middle along bed height due to the instability of the interface separating the dilute layer and the dense layer. As the waves are interrupted, particles aggregations occur, representing a flow of particle pockets. Here, these formation of particle aggregates with high solid volume fractions is known as chunk-like flow, representing a large scale turbulent flow at the upper part of the bed. Experimental observation showed that the different flows of wavy with transverse structure, and large scale turbulent and bubbling fluidization states were observed with changing inlet velocities in fluid and particles fluidized beds [12, 14, 16, 22, 26]. They suggested that the occurrence of small and large turbulence originated from a wave instability. Obviously, the formation of the wave-like flow near the bottom and chunk-like flow at the upper part characterize a distinct fluidization state of molten salt fluid-particles mixtures in the conventional fluidized bed reactors.

For base case 3 with the densities of particles and molten salt fluid of 2400 and 1900 kg/m^3, the density ratio is 1.26. The uniform distributions of solid volume fractions appeared near the bed bottom. However, particles aggregations occurred at the upper part of the bed. The solid phase exhibits large and random displacements of aggregates, characterizing a large scale turbulent regime with particle aggregates at the upper part of the bed. Homogeneous fluidization near the bottom and bubbling fluidization at the upper part appear along bed height.

The instantaneous solid volume fractions are displayed in Fig. 6.55 at three base cases in the molten salt and particles conventional FB. For base case 1, the solid volume fractions are nearly constant at the bed height H_h. For base case 2, large fluctuations of solid volume fractions are observed due to the occurrence of wave-like flow near the bed bottom and chunk-like flow at the upper part of the bed. The bed divides into two regimes along bed height. One regime is denoted by a wave-like flow with the height H_w, while the other is represented by a chunk-like flow with the height H_c. The wave-like flow occurs with periodic low and high solid volume fractions within the height H_w. The wavelength is expressed between two peaks of volume fractions. The mean wavelength λ is 0.036 m, and the dimensionless

Fig. 6.54 Instantaneous solid volume fractions in molten salt-particles conventional FB

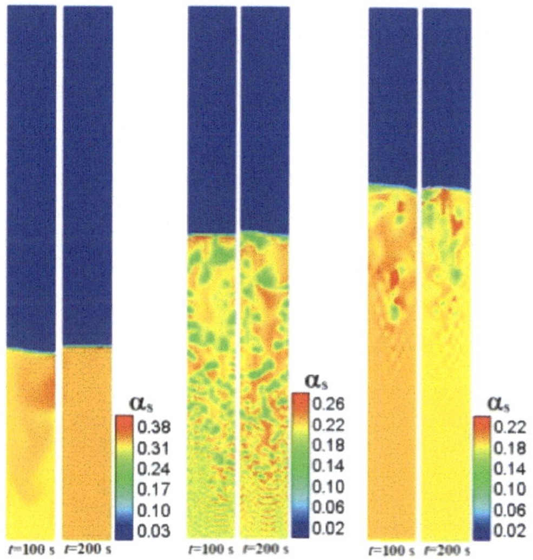

(a) base case 1 (b) base case 2 (c) base case 3

wavelength λ/d_p is 12.0. The wave-like flow is vanished, while the chunk-like flow with large size of particle aggregates occur at the height H_c. The travelling wave is broken, and the flow of molten salt fluid-particle mixtures transits from wave-like flow to chunk-like flow. For base case 3, two different regimes are observed along bed height. One represents a regime of homogeneous flow, and the solid volume fractions are almost equal near the bed bottom with the height of H_h. The other is large fluctuations of solid volume fractions, and it appears at the height of H_c due to particle aggregates, reflecting that particulate fluidization exists near the bed bottom while aggregative fluidization formed at the upper part of the bed.

The values of three dimensionless numbers of *Fr*, *HD* and N_{tr} are summarized in Table 6.6 for three base cases. For base case 1, the dimensionless numbers of *Fr* and N_{tr} are lower than the PF threshold, while the dimensionless number of *HD* is larger than the PF threshold, and these results represent a PF state in the bed. For base case 2, the value of dimensionless number *Fr* is less than the PF threshold. However, both dimensionless numbers of *HD* and N_{tr} are larger than the PF thresholds. These results suggest that the molten salt fluid and particles fluidize at TR state, and is similar to simulations with a wave-like flow near the bottom and chunk-like flow at the upper part of the bed. For base case 3, the bed materials fluidize at PF state using *Fr* dimensionless number, while both dimensionless numbers of *HD* and N_{tr} suggest that molten salt fluid-particle mixtures are at TR state which is similar to simulations with a homogeneous flow near the bed bottom and particle aggregates at the bed upper part. These results further indicate that intermediate-state fluidization exists with changing properties of bed materials in molten salt fluid conventional fluidized beds.

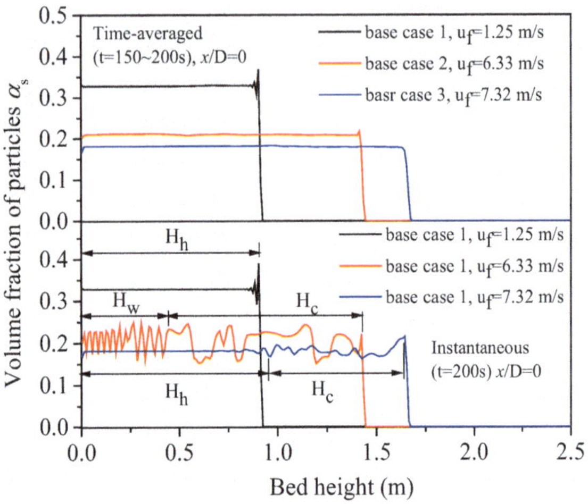

Fig. 6.55 Profiles of solid volume fractions of molten salt FB

6.5.2 Molten Salt-Biomass Particle Inverse Fluidized Bed

Biomass is regarded as a sustainable energy and promising alternative energy of fossil energy. Gasification is an effective thermochemical process converting biomass into high quality syngas. However, traditional thermochemical process needs burning partial of the biomass feedstock for energy supplement to drive the gasification process keeping a high temperature condition of the reactor. The use of concentrated solar energy as a heat source for pyrolysis and gasification of biomass is an efficient mean for production of hydrogen rich synthesis gas. Utilizing molten salts as a reaction and heat transfer media promises enhanced stability to solar transients and faster reaction rates.

Figure 6.56 shows the principle of a concentrated solar power with biomass gasification in inverse fluidized bed (IFB) reactor. Using concentrated solar power in thermochemical processes, the combustible gases can be produced from biomass particles pyrolysis using high temperature molten salt fluid heated by solar energy in the IFB reactor. Molten salt fluid acts as heat transfer fluid, similar to water or steam and thermal oils. Generally, the application of water/steam as both heat transfer fluid and working fluid simplifies the thermochemical systems and decreases the cost of electricity production. Thermal oils like Dowtherm A, Therminol 66 and Therminol VP-1 have been used as heat transfer fluids in thermal energy storage and concentrating solar power applications. The highest temperature of thermal oils as heat transfer fluid limits to around 400 °C in a thermodynamic cycle. Molten salt fluid acts as heat transfer fluid mainly due to their thermal stability at high temperatures (generally larger than 500 °C). The operating temperature range is between 300 and 565 °C in the currently available solar power tower plants. The high temperature molten salt fluid can act as heat source. Molten salt fluid absorbs the concentrated sunlight using the solar power tower. Biomass particles are heated by molten salt.

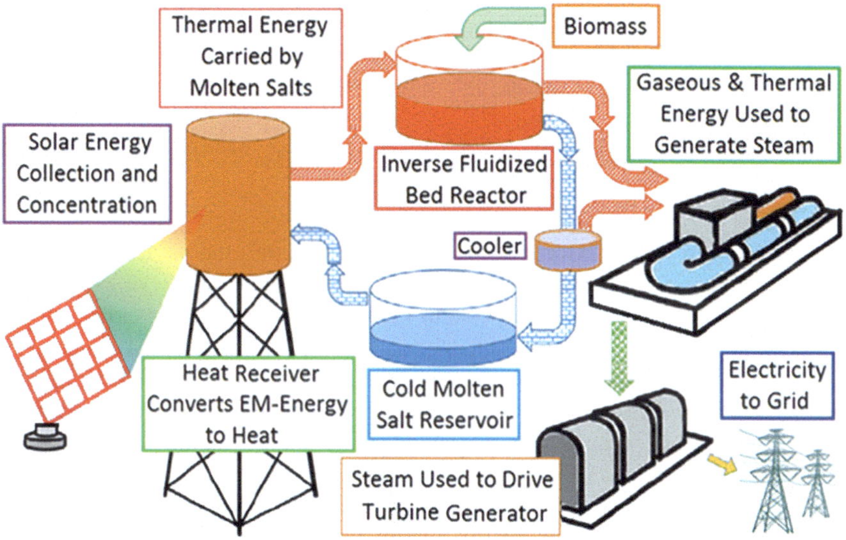

Fig. 6.56 Principle of a concentrated solar power and biomass gasification with inverse fluidized bed reactor system

Pyrolysis process occurs, and syngas is produced in the IFB reactor. Thus, the knowledge of hydrodynamics of molten salt fluid and particles is essential for reliable design of IFB gasification reactors [48].

In the molten salt-biomass particles IFB reactors, low density biomass particles become mobile, or fluidize, when the drag exerted by a molten slat fluid flowing downwards through the column overcomes the buoyancy force on the biomass particles. Inverse fluidization is the reverse of what is considered to be normal fluidization, where heavier particles are fluidized by the upwards flow of a lighter gas or liquid. The IFBR is also known as a downflow fluidized bed reactor due to the direction of movement of liquid inside the reactor.

Fluidizing light weight particles by heaver medium is advantageous in many industrial applications in biotechnology and catalytic chemical reaction engineering, where enhanced multi-phase mixing can improve heat and mass transfer performance. In these systems, light weight particles are fluidized by the countercurrent flow of liquid downwards.

The hydrodynamics of molten salt fluid and biomass particles are simulated in an IFB reactor. The height and width of the IFB reactor are 2.5 and 0.25 m. Biomass particles are taken as bed materials with the particle density range of 700 to 900 kg/m^3. The molten salt fluid is supplied from the inlet at the bed top [51]. The density ratio is about 0.41–0.53 at the molten salt fluid density of 1700 kg/m^3, indicating that the biomass particle density is about half of molten salt fluid. The parameters are summarized in Table 6.7 in computational simulations.

Table 6.7 Properties and fluid minimum fluidization velocities

Cases	ρ_f (kg/m^3)	μ_f (Pa s)	ρ_s (kg/m^3)	d_p (mm)	u_{mf} (cm/s)	u_f (cm/s)	u_f/u_{mf}
Case 1	1700	0.0018	900	3.0	1.468	8.84	6.0
Case 2	1700	0.0018	700	3.0	1.721	8.84	5.15
Case 3	1772	0.00074	700	3.0	2.265	15.29	6.75

The base cases are performed in a molten salt fluid and biomass particles IFB reactor. Table 6.7 summarizes molten salt and biomass particles properties. Figure 6.57 shows the instantaneous volume fractions of biomass particles in a molten salt fluid-biomass particles IFB reactor. For all cases, biomass particle diameter is 3.0 mm. In case 3, the solid volume fractions are distributed uniformly along height. The fluidized bed height H increases from the initial fixed bed height H_0, and the solid volume fractions are reduced from the initial solid volume fraction. This represents a particulate fluidization (PF) state.

For both case 1 and 2 with two different biomass particle densities, bands with low and high volume fractions of biomass particles are observed near the inlet regime of the bed. These bands incline from the walls toward the bed center, and are moved downward from the bed inlet. Here, it is known as a wave-like flow near the inlet regime. The waves with biomass particles bands are dissolved at the mid height of the bed, and the particles aggregations are formed. These biomass particles aggregations are known as a chunk-like flow at the upper part, simulate to particles clusters in fluidized bed risers due to hydrodynamic interactions and collisional interactions of particles.

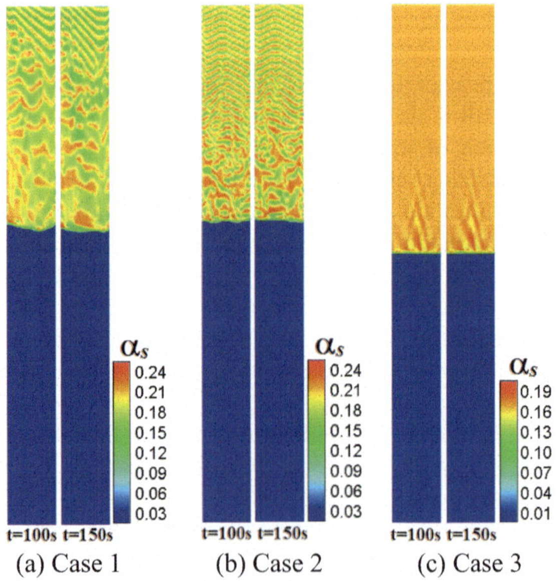

Fig. 6.57 Instantaneous solid volume fractions in a molten salt fluid-biomass particles IFB reactor

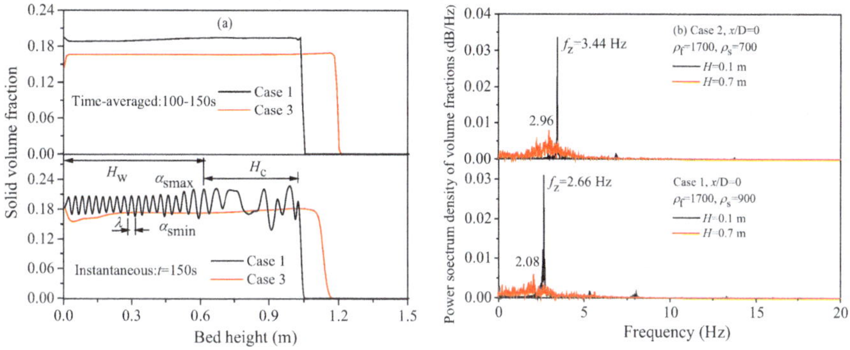

Fig. 6.58 Profiles of solid volume fractions at center axis of bed and power spectrum densities

The axial distributions of volume fractions are shown in Fig. 6.58a at two cases in the molten salt-biomass particles IFB reactor. For case 3, the solid volume fractions decrease from the inlet at first, reach minimum and increase gradually. Then, it tends nearly to constant along bed height. The biomass particles volume fractions are reduced, and then tend to zero close to the bed surface. For case 1, two regimes are observed along bed height in terms of the variation of volume fractions of biomass particles. The periodic volume fractions are appeared at the height H_w. Both high and low volume fraction layers of biomass particles represent the bands of biomass particles and molten salt fluid, displaying the wave-like flow regime. As the distance exceeds the height H_w, the periodic volume fraction is disappeared, and the random oscillations of volume fractions is occurred, representing a chunk-like flow of biomass particle aggregates with the height of H_c. These indicate that the bed can be divided into a wave flow and an aggregate flow along bed height.

The wavelength can be determined in terms of high solid volume fraction α_{smax} and low solid volume fraction α_{smin} in the height H_w. The solid volume fraction variation is $\Delta\alpha_s$. From simulated results, we find that the height H_w is 0.18 and 0.55 m, and the values of wavelength λ are about 44.17 and 113.7 mm for case 1 and 2. The dimensionless wavelength $(\lambda/d_p)_1$ is 37.9 and 14.7, respectively. These indicate that the waves travel with a wavelength λ of volume fraction perturbation.

The power spectrum density can be computed by means of the fast Fourier transform (FFT) technique in terms of instantaneous solids volume fractions. Figure 6.58b illustrates the power spectrum densities corresponding to the fluctuations of biomass particles volume fractions at two locations. Both cases show that the fluctuation magnitudes are low at the low and high frequencies. The diagram highlights a dominant frequency according to the magnitudes of power spectrum density fluctuations. They are 2.66 and 2.08 Hz for case 1, and 3.44 and 2.96 Hz for case 2 at $H = 0.1$ and 0.7 m, indicating that the dominant frequency in the wave-like flow regime is higher than in the chunk-like flow regime.

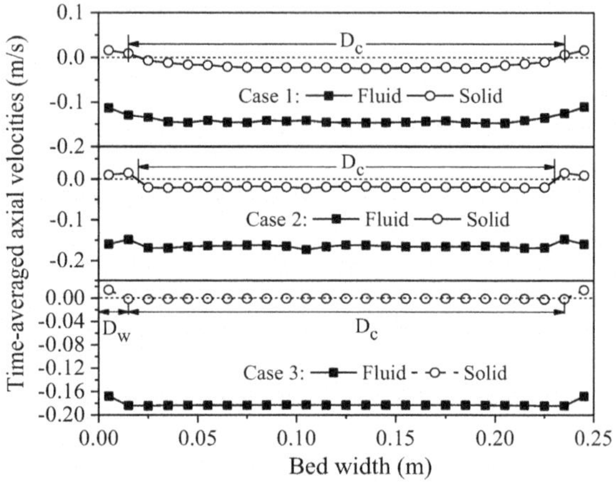

Fig. 6.59 Solid volume fractions and axial velocities of molten salt fluid and biomass particles

Figure 6.59 illustrates the profiles of biomass particles axial velocities along lateral direction. All cases indicate that both molten salt and biomass particles axial velocities are negative in the center regime at D_c, reflecting a concurrent down-flow of molten salt fluid-biomass particles mixture. However, the axial velocities are negative for molten salt fluid, and positive for biomass particles at D_W. These indicate that the solid phase moves upward, and the molten salt fluid phase passes downward through the bed at D_W, reflecting biomass particles circulation with a down-flow at D_W and up-flow at D_c in the bed.

The influence of molten salt fluid temperature on fluidization process of molten salt fluid and biomass particles is displayed in Fig. 6.60 at three molten salt temperatures. The particle density and diameter are 700 kg/m^3 and 3.0 mm, respectively. All cases show that the wave-like flow is occurred with an alternative dense and dilute layers near the inlet regime. Conversely, the chunk-like flow appears with particles aggregations as the waves are broken at the upper part. In the molten salt fluid and biomass particles IFB reactor, the bed materials are suspended partly by the down-flowing molten salt fluid due to the drag force. On the other hand, the biomass particles trend upward motion due to the buoyancy. Both drag and buoyancy forces relate to solid volume fractions, and cause instability of molten salt fluid-biomass particles suspension in the bed. From plots, it can be seen that the waves incline to the right or to the left, and these tilted waves are broken easily due to the instability of the interface between the dilute layer and the dense layer, representing a transitional state in the bed.

Figure 6.61 shows instantaneous volume fractions at three inlet molten salt fluid velocities in an IFB reactor with the biomass particle density and diameter of 700 kg/m^3 and 3.0 mm. The bed temperature is 723 K. All cases show that the waves with the bands of low and high biomass particles layers are occurred near the inlet regime, and the chunk-like flow with biomass particle aggregates is emerged at the upper part. The bed expansion heights are increased with increasing inlet molten salt fluid velocities. Furthermore, the wave-like flow regime is vanished gradually as the inlet

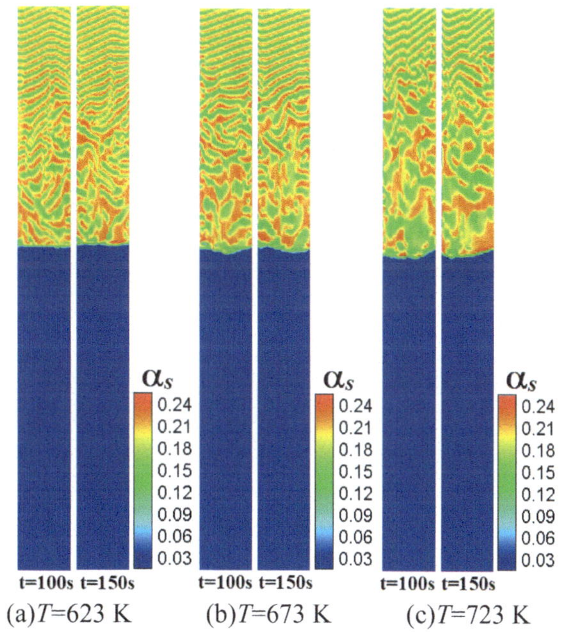

Fig. 6.60 Instantaneous solid volume fractions at three different molten salt fluid temperatures

molten salt fluid velocity increases because more energy is supplied from the bed inlet.

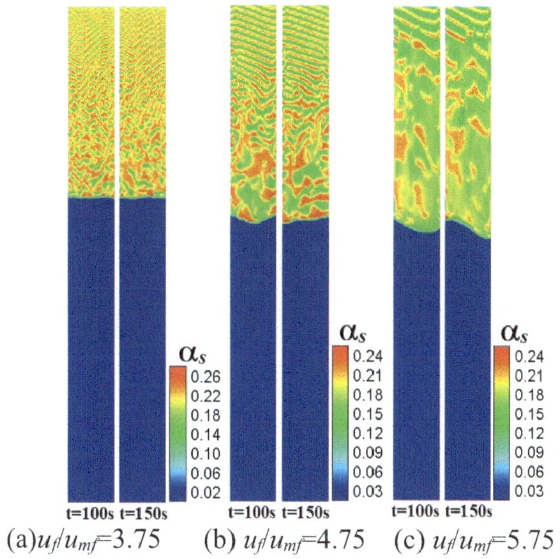

Fig. 6.61 Instantaneous solid volume fractions at different inlet molten salt fluid velocities

In summary, for flow of the molten salt-coal particles conventional FB reactor, the coexistence of wavy-like flow near the bottom and large scale turbulent regime with chunk-like flow at the upper part is found. With increasing molten salt fluid temperatures and inlet fluid velocities, the molten salt fluid and particles suspension develops from particulate fluidization state to transitional state.

For flow of the molten salt-biomass particles IFB reactor, the wave-like flow consists of an initial wave-like flow stage with small wavelength and a saturated wave-like flow stage with large wavelength. The chunk-like flow stems from the end of saturated wave-like flow stage in the molten salt-biomass particles IFB reactors.

6.6 Particles Brownout Induced by Hovering Helicopter

During hovering flight, a helicopter maintains a constant position over a selected point. When the helicopter is hovering or flying close to the ground with widely distributed discrete sand and snow, the rotor-induced flow field will cause the rolling up of the discrete particulates, resulting in a brownout phenomenon. Brownout obscures the pilot's field of vision, which can lead to a serious safety threat to helicopters flying close to the ground. The appearance of the phenomenon of brownout originates from dust particles tend to aggregate in regions of high vorticity in the hovering state. The complex lifting of sand particles is caused not only by boundary-layer shear and uneven pressure distribution, but also by bombardment effects when suspended particles impact the bed and eject new particles.

A helicopter flying through an atmosphere containing particulates may accumulate high electrostatic charges which can challenge its operational safety. In a fixed-wing aircraft a short-circuiting element to ground is very unlikely, the presence of high electrostatic potentials constitutes a safety hazard for helicopters since they are frequently operated close to the earth's surface. In most cases, the take-off and landing site of helicopter is desert covered with loose dusts. When the helicopter flies at low altitude or hovers, a large amount of sand and dust is rolled up by the swirling air stream, and the sand and dust content in the air is increased. A large dust cloud is formed and causes severe visual obscurations that affect the pilot's ability to safe the flight. The air velocity distribution, the dust and gas turbulent behavior and gas pressure are fundamental parameters to be investigated on fluid dynamics. These parameters are interrelated to each other, resulting in the flow behavior of gas and particles very complex.

The configuration of the helicopter rotor in the present study consists of two blades. To predicate hydrodynamics of air and particles induced by the rotation of the two-blade rotor from the particles layer, the idealized geometric model is shown in Fig. 6.62 with the revolving two-blade rotor arrangement. The open chamber has a 12.0 m by 12.0 m floor area and a 4.0 m high ceiling. The particles layer has a height of 0.5 m with solid volume fraction of 0.6 from the bottom of the chamber. The diameter and density of particles are 50.0 μm and 2000 kg/m^3. The computational domain segments into two parts. One is the two-blade rotor part with a central cylindrical solid element. The surrounding external zone forms the second part in the chamber. The central cylindrical solid element is modeled by a rotating reference frame with the diameter and height of 2.3 and 0.3 m. The two-bladed rotor diameter D is 2.286 m, and the motor housing diameter is 0.05D. The blade diameter is 1.11 m with the blade chord of 0.1905 m. The pitch angle is set at 8.0 degree. It is desirable that the central cylindrical solid element is as large as possible, but without intersecting any non-rotational fixed surfaces. The surface of the blade and the hub (or housing) surface of the two-blade rotor are set to rotate at the rotation rate of the rotating reference frame to reflect the real two-blade rotor operation. Special attention is paid to the meshing of the blades and the rounding at the tips. They are meshed with a mixture of unstructured and structured patterns to enable alignment of the grid with the air flow direction. The inlet is assigned with the air pressure inlet boundary condition, and the exit is assigned with the air pressure outlet boundary condition. The bottom of the particles lay is the wall boundary condition. During the rotation of the blades, the non-slip boundary condition is used for gas phase and solids phase at the walls of the blades. The velocities of air and particles are equal to the moving velocity of the blades. The surrounding external zone is modeled in a stationary frame. The flow field around the blades is numerically calculated using the multiple reference frame method to model the swirl components with the revolving movement of the two-blade rotor.

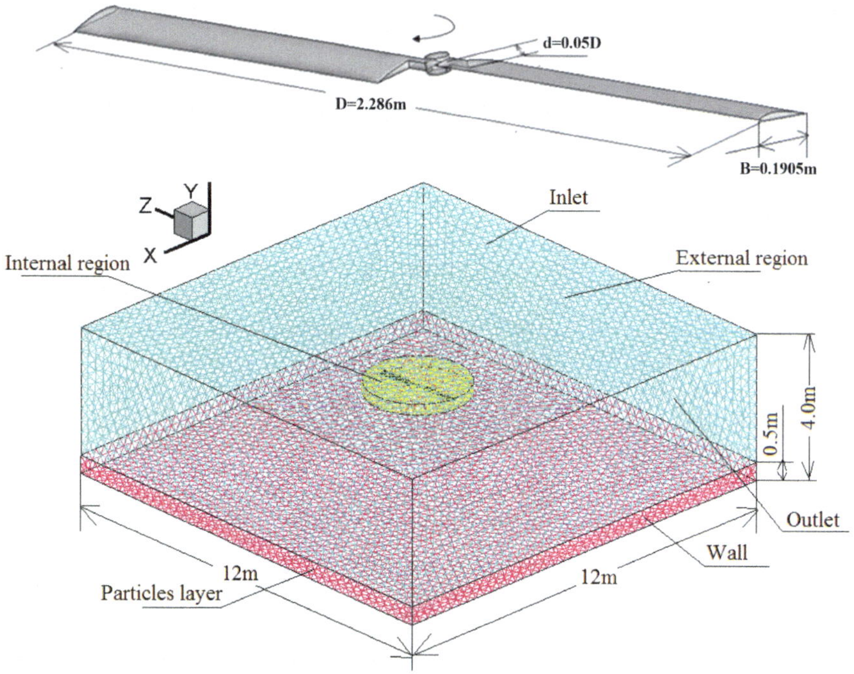

Fig. 6.62 Schematic diagram and geometrical revolving two-bladed rotor

The streamlines of air velocities are shown in Fig. 6.63 using the kinetic theory of granular flow and hybrid theories of granular flow at two hovering heights of the two-blade rotor. The rotor rotational speed is 1000 rpm. Air passes through the two-blade rotor with negative velocity, and then air plume moves downward the surface of the particles layer. When air plume impacts on the surface of the particles layer, it changes the direction of flow from the vertical direction to the horizontal direction along the surface of the particles layer. The high negative air velocity is located at the tip of the blades. Near the housing of the two-blade rotor, air velocity becomes positive, indicating that air does not be driven downward due to the blockage of the motor housing. It is also found that the air vortexes are generated at the base of the cylindrical domain, where air flows upward from the surface of the particles layer. The air vortexes formed at the left and right corners are caused by the inlet–outlet effect where the air velocity changes from the inlet to the exit. The air horizontal flow results in the motion of particles located at the surface of the particles layer through the interactions between air and particles, leading to entrainment of particles from the surface of the particles layer. These confirm that the helicopter brownout is generated by the entrainment of dust and ground debris due to the helicopter downwash and outwash during hovering, take-off and landing.

Figure 6.64 shows the instantaneous velocities contours of air and particles at the two-blade rotor hovering height of 2.5 m over the particles layer. Both velocities of air and solids phases are high beneath the rotor and low near the surface of the particles layer. The difference of velocities between the air phase and the solids phase is obvious. At the dimensionless height h/H of 0.8 (where h is the distance from the bottom of the particles layer to the plane selected to analyze the computational results.), the air velocity is high and the velocity of particles is zero at the center of the two-blade rotor. On the other hand, the air velocity is low and the velocity of

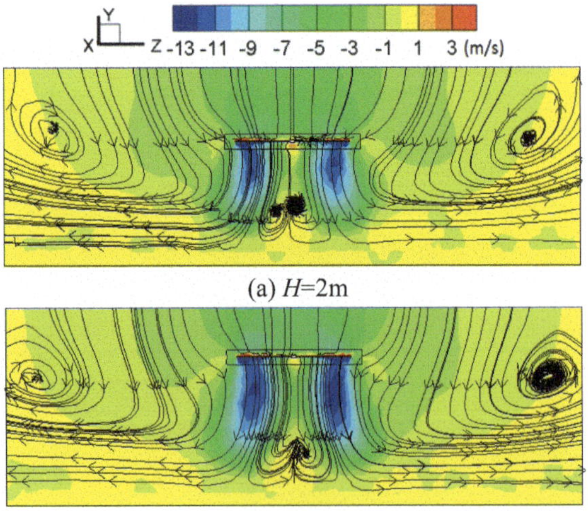

Fig. 6.63 Streamlines of air velocity at two hovering heights

(a) H=2m

(b) H=2.5m

particles is large near the exit walls, indicating that the difference of velocity between the air phase and the solids phase is caused by the two-blade rotor. From simulated volume fraction of particles, the main flows of particles exist in the cylindrical domain and near the chamber walls region. In the cylindrical domain, the air plume pushes particles downward the surface of the particles layer. It is found that the size of the downward flow of air is close to the diameter of the two-blade rotor. Moreover, the simulated air velocity distribution shows a cylindrical domain below the two-blade rotor where air is driven downwards caused by the rotation of the rotor. The cylindrical domain and perimeter region are characterized by velocities and volume fraction of air and particles induced by the two-blade rotor. Thus, the TFM model is able to replicate the main flow region induced by the two-blade rotor hovering over the particles layer.

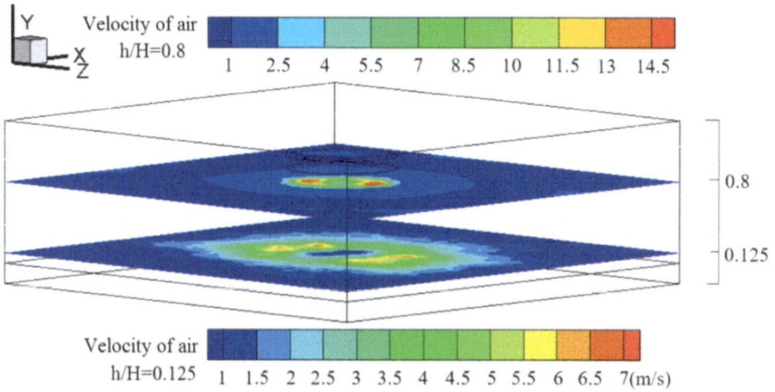

(a) Air velocity

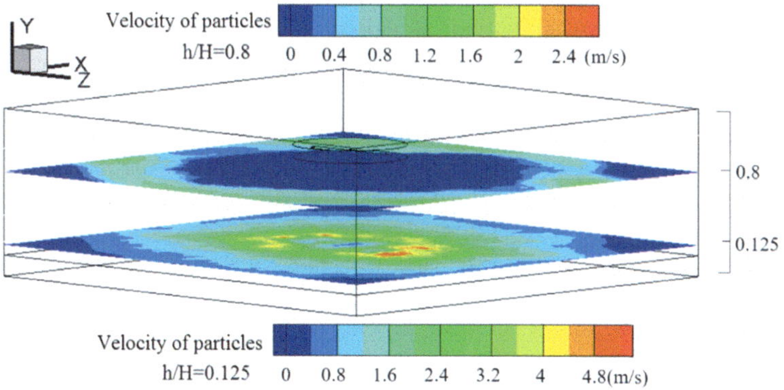

(b) Velocity of particles

Fig. 6.64 Instantaneous velocity contours of air and particles at two hovering heights

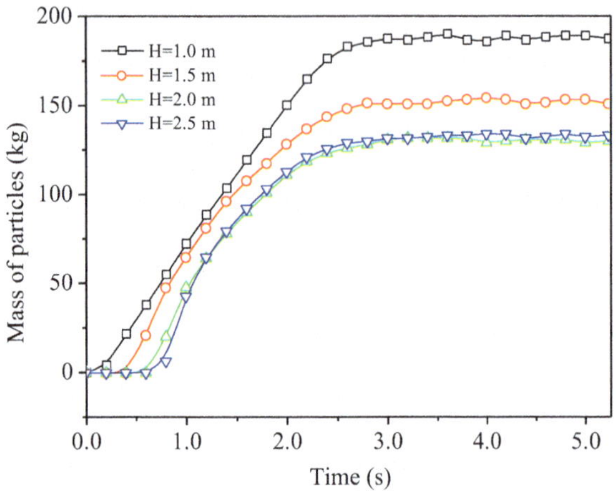

Fig. 6.65 Profile of total mass of entrained particles

The initial particles mass of the particles layer m_0 is calculated from the initial volume fraction of particles and the initial height of the particles layer. The instantaneous particles mass of the particles layer m_t is also calculated based on instantaneous volume fraction of particles at the time t at the same initial height. The reduction between two masses of particles, $(m_0 - m_t)$, gives the instantaneous entrained particles mass from the particles layer. The instantaneous mass of entrained particles is shown in Fig. 6.65 at four hovering heights. The instantaneous mass of entrained particles increases with decreasing hovering height of the two-blade rotor, indicating that more particles will be entrained from the particles layer. The spread of particles relates to the hovering height of the two-blade rotor, and the perimeter region affects the overall distribution of particles. The horizontal flow of air and particles forced by the rotor enhances the entrainment of particles.

In summary, numerical simulations of hydrodynamics of air and particles induced by the revolving rotor are conducted by means of the kinetic theory of granular flow and hybrid theories of granular flow from the particles layer. The simulated results confirm that a revolving rotor entrains air flow by producing positive air pressure beneath the blades, and drives the down-flow as a swirling air plume impinges on the particles layer. This study provides an example of such an approach to predict the air and particles flow induced by a two-blade rotor. Simulated results indicate that the high tangential velocity induced by the two-blade rotor enhances the entrainment of particles from the particles layer. The entrained particles mass increases with decreasing hovering height of the rotor over the particles layer.

In the study presented herein, we predicte distributions of particles volume fractions and velocities of an idealized geometry of the helicopter during hovering in a particles layer. The entrainment of particles induced in the rotor blades increases with the decrease of hovering height. By providing this kind of data, the developed numerical tool is capable to facilitate the design of the system of the electrostatic discharge attributed to friction with the surrounding ionized particles. Currently, this model is used for the study of hydrodynamics of gas and particles. In the future we plan to implement the model in a more advanced TFM-KTGF method for the hydrodynamics around a helicopter. This includes the implementation of a more detailed geometry in order to account for a larger variety of helicopter flight conditions.

Nomenclature

A_s	Particle upstream projection area, m^2
C_d	Drag coefficient for a single particle
d_p	Particle diameter, m
D_e	Equivalent diameter, m
D_{in}	Inside diameter, m
f	Fraction of clusters
f_c	Cluster frequency, Hz
f_{in}	Frequency of inlet gas velocity, Hz
F_c	Existence time of clusters, s
F_d	Force acting on a single particle by fluid, N
Fr	Froude number
k	Turbulence kinetic energy, m^2/s^2
K	Consistency coefficient
K_0	Cross-sectional shape factor
K_1	Tortuosity factor
M	Second order moment of fluctuation velocity, m^2/s^2
N	Flow behavior index
N_{df}	Rate of energy dissipation, W/kg
Re_p	Reynolds number in the pores
Re_n	Reynolds number
S	Stability criterion
T	Duration of clusters, s
u_{mf}	Minimum fluidization velocity, m/s
u_0	Superficial velocity, m/s
u_t	Particle settling velocity, m/s
V	Fluctuating velocity, m/s

Greek Letters

β	Drag coefficient between the two phases, kg/m^3s
γ	Shear rate, 1/s
γ_w	Shear rate at the pore wall, 1/s
α_f	Fluid volume fraction
α_l	Liquid concentration
α_s	Particle concentration

a_{sc}	Cluster solid volume fractions in the dense phase
η_{eff}	Effective viscosity, kg/m·s
θ	Granular temperature, m^2/s^2
μ_f	Fluid viscosity, kg/m·s
μ_s	Solid shear viscosity, kg/m·s
μ_t	Turbulent eddy viscosity, kg/m·s
ν	Frequency of collisions, 1/s
λ	Wave length, m
ρ_l	Liquid density, kg/m^3
ρ_s	Particle diameter, kg/m^3
τ_w	Shear stress at the pore wall, N/m^2

Abbreviations

AF	Aggregative fluidization
AW	Ambient water
CFB	Circulating fluidized bed
CFD	Computational fluid dynamics
DEM	Discrete element method
FB	Fluidized bed
HV	High velocity
HD	Dimensionless number
KTGF	Kinetic theory of granular flow
LV	Low velocity
PF	Particulate fluidization
SbE	Subcritical ethanol
SbW	Subcritical water
SCE	Supercritical ethanol
ScM	Supercritical methanol
SCW	Supercritical water
TFM	Two-fluid model
TR	Transition regime

References

1. Anderson, T. B., & Jackson, R. (1968). Fluid mechanical description of fluidized beds: Stability of the state of uniform fluidization. *Industrial and Engineering Chemistry Research Fundamentals, 7*, 12–21.
2. Angelino, G. (1968). Carbon dioxide condensation cycle for power production. *Journal of Engineering for Power, 90*, 287–295.

3. Batchelor, G. K. (1993). Secondary instability of a gas-fluidized bed. *Journal of Fluid Mechanics, 257*, 359–371.
4. Bird, R. B., Stewart, W. E., & Lightfoot, E. N. (1960). *Transport Phenomena.* Wiley.
5. Boxue, P. (2021). *Kinetic theory and numerical simulation of Non-Newtonian fluid particle flow*, Ph. D. Thesis, Harbin: Harbin Institute of Technology.
6. Boxue, P., Shuyan, W., & Huilin, L. (2022). A modified drag model for power-law fluid-particle two-phase flow used in computational fluid dynamics simulation. *Advanced Powder Technology., 32*, 1207–1218.
7. Boxue, P., Shuyan, W., Xiaoxue, J., & Huilin, L. (2019). Effect of orbital motion of drill pipe on the transport of non-Newtonian fluid-cuttings mixture in horizontal drilling annulus. *Journal of Petroleum Science and Engineering, 174*, 201–215.
8. Brea, F. M., Edwards, M. F., & Wilkinson, W. L. (1976). The flow of non-Newtonian slurries through fixed and fluidized beds. *Chemical Engineering Science, 31*, 329–336.
9. Buijtenen, M. S., Dijk, W., Deen, N. G., Kuipers, J. A. M., Leadbeater, T., & Parker, D. J. (2011). Numerical and experimental study on multiple-spout fluidized beds. *Chemical Engineering Science, 66*, 2368–2376.
10. Cherntongchai, P., Chaiwattana, S., & Leruk, R. (2018). Bed expansion characteristics in sound assisted fluidization of Geldart's group A powder. *Powder Technology, 340*, 243–252.
11. Crespi, F., Gavagnin, G., Sanchez, D., & Martinez, G. S. (2017). Supercritical carbon dioxide cycles for power generation: A review. *Applied Energy, 195*, 152–183.
12. Didwania, A. K., & Homsy, G. M. (1981). Flow regimes and flow transitions in liquid fluidized beds. *International Journal of Multiphase Flow, 7*, 563–580.
13. Du, B., Warsito, W., & Fan, L. S. (2003). Bed nonhomogeneity in turbulent gas–solid fluidization. *AIChE Journal, 49*, 1109–1126.
14. Duru, P., & Guazzelli, E. (2002). Experimental investigation of a secondary instability of liquid fluidized beds and the formation of bubbles. *Journal of Fluid Mechanics, 470*, 359–382.
15. Ejima, T., Sato, Y., Yamamura, T., Tamai, K., Hasebe, M., Bohn, M. S., & Janz, G. J. (1987). Viscosity of the eutectic dilithium carbonate-disodium carbonate-dipotassium carbonate melt. *Journal of Chemical & Engineering Data, 32*, 180–182.
16. El-Kaissy, M. M., & Homsy, G. M. (1976). Instability waves and the origin of bubbles in fluidized beds. *International Journal of Multiphase Flow, 2*, 379–395.
17. Elbel, S., & Lawrence, N. (2016). Review of recent developments in advanced ejector technology. *International Journal of Refrigeration, 62*, 1–18.
18. Ergun, S. (1952). Fluid flow through packed columns. *Chemical Engineering Progress, 48*, 89–94.
19. Foscolo, U., & Gibilaro, L. G. (1984). A fully predictive criterion for the transition between particulate and aggregate fluidization. *Chemical Engineering Science, 39*, 1667–1674.
20. Gibilaro, L. G. (2001). *Fluidization Dynamics: The formulation and applications of a predictive theory for the fluidized state.* Butterworth-Heinemann.
21. Gidaspow, D. (1994). *Multiphase flow and fluidization: Continuum and kinetic theory descriptions.* Academic Press, San Diego.
22. Glasser, B. J., Kevrekidis, I. G., & Sundaresan, S. (1996). One- and two-dimensional traveling wave solutions in gas-fluidized beds. *Journal of Fluid Mechanics, 306*, 183–221.
23. Govier, G. W., Aziz, K., & Schowalter, W. R. (1972). The flow of complex mixtures in pipes. *Journal of Applied Mechanics, 40*, 404–412.
24. Guillot, S., Faik, A., Rakhmatullin, A., Lambert, J., Veron, E., Echegut, P., Bessada, C., & Calvt, N. (2012). Corrosion effects between molten salts and thermal storage material for concentrated solar power plants. *Applied Energy, 94*, 174–181.
25. Guodong, L., Pengwei, L., Junnan, Z., Shuai, W., Yao, W., Xiaolong, Y., & Huilin, L. (2020). CFD-DEM study of the effects of direct current electric field on gas-solid fluidization. *Powder Technology, 362*, 416–427.
26. Ham, J. M., Thomas, S., Guazzelli, E., Homsy, G. M., & Anselmet, M. C. (1990). An experimental study of the stability of liquid-fluidized beds. *International Journal of Multiphase Flow, 16*, 171–185.

27. Harrion, D., & Davidson, J. F. (1961). On the nature of aggregative and particulate fluidization. *Transection of Institute of Chemical Engineering, 39*, 202–211.
28. Hemphil, T., Campos, W., & Tehrani, M. A. (1993). Yield power law model accurately predicts mud rheology. *Oil & Gas Journal, 91*, 45–50.
29. Huilin, L., & Gidaspow, D. (2003). Hydrodynamics of binary fluidization in a riser: CFD simulation using two granular temperatures. *Chemical Engineering Science, 58*, 3777–3792.
30. Huilin, L., Yunhua, Z., Zhiheng, S., Ding, J., & Jiying, J. (2006). Numerical simulations of gas–solid flow in tapered risers. *Powder Technology, 169*, 89–98.
31. Jackson, R. (1963). The mechanics of fluidized beds. I. The stability of the state of uniform fluidization. *Transactions of the Institution of Chemical Engineers, 41*, 13–21.
32. Jiang, P. J., Bi, H. T., Jean, R. H., & Fan, L. S. (1991). Baffle effects on performance of catalytic circulating fluidized bed reactor. *AIChE Journal, 37*, 1392–1400.
33. Jung, J., Gidaspow, D., & Gamwo, I. K. (2006). Bubble computation, granular temperatures, and Reynolds stresses. *Chemical Engineering Communications, 193*, 946–975.
34. Maccari, A., Bissi, D., Casubolo, G., & Guerrini, F. (2015). Archimede solar energy molten salt parabolic trough demo plant: A step ahead towards the new frontiers of CSP. *Energy Procedia, 69*, 1643–1651.
35. Karimi, F., Haghshenasfard, M., Sotudeh-Gharebagh, R., Zarghami, R., & Mostoufi, N. (2020). Multiscale characterization of nanoparticles in a magnetically assisted fluidized bed. *Particuology, 51*, 64–71.
36. Kashyap, M., Gidaspow, D., & Driscoll, M. (2008). Effect of electric field on the hydrodynamics of fluidized nanoparticles. *Powder Technology, 183*, 441–453.
37. Kemblowski, Z., & Michniewicz, M. (1979). A new look at the laminar flow of power law fluids through granular beds. *Rheologica Acta, 18*, 730–739.
38. Lali, A. M., Khare, A. S., Joshi, J. B., & Nigam, K. D. P. (1989). Behavior of solid particles in viscous non-Newtonian solutions: Settling velocity, wall effects and bed expansion in solid-liquid fluidized beds. *Powder Technology, 57*, 39–50.
39. Lin, W., Guoli, Q., Ming, T., Xuemin, L., Hassan, M., & Huilin, L. (2018). Simulated pulsed flow of gas and particles in a horizontal oppose-pulsed gas jets of bubbling fluidized bed. *Advanced Powder Technology, 29*, 3507–3519.
40. Lorentzen, G. (1994). Carbon dioxide as a refrigerant. *International Journal of Refrigeration, 17*, 292–301.
41. Lun, C. K. K., Savage, S. B., Jeffrey, D. J., & Chepurniy, N. (1984). Kinetic theories for granular flow: Inelastic particles in Couette flow and slightly inelastic particles in a general flow field. *Journal of Fluid Mechanics, 140*, 223–256.
42. Nicolas, M., Chomaz, J. M., Vallet, D., & Guazzelli, E. (1996). Experimental investigations on the nature of the first wavy instability in liquid-fluidized beds. *Physics of Fluids, 8*, 1987–1989.
43. Mishra, P., Singh, D., & Mishra, I. M. (1975). Momentum transfer to Newtonian and non-Newtonian fluids flowing through packed and fluidized beds. *Chemical Engineering Science, 30*, 397–405.
44. Qinghong, Z. (2020). Numerical simulation of fluidized bed using coupled Euler gas phase-Euler solid phase-Lagrangian discrete particles model. Ph. D. thesis, Harbin: Harbin Institute of Technology.
45. Qinghong, Z., Shuyan, W., Huilin, L., Qiujing, W., Ming, T., & Guodong, L. (2018). Impact velocity-dependent restitution coefficient using a coupled Eulerian fluid phase-Eulerian solid phase-Lagrangian discrete particles phase model in gas-monodisperse particles internally circulating fluidized bed. *International Journal of Multiphase Flow, 105*, 142–158.
46. Richardson, J. F., & Zaki, W. N. (1954). Sedimentation and fluidization. *Transactions of the Institution of Chemical Engineers, 32*, 35–48.
47. Verloop, J., & Heertjes, P. M. (1970). Shock waves as a criteria for the transition from homogeneous to heterogeneous fluidization. *Chemical Engineering Science, 25*, 825–832.
48. Vignarooban, K., Xu, X., & Arvay, A. (2015). Heat transfer fluids for concentrating solar, power systems–A review. *Applied Energy, 146*, 383–396.
49. Wallis, G. B. (1969). *One-dimensional two-phase flow*. McGrawHill.

50. Wen, C. Y., & Yu, Y. H. (1966). Mechanics of fluidization. *American Institute of Chemical Engineers Symposium Series, 62*, 100–111.
51. Wenjian, C. (2023) Numerical simulations of low density ratio particle-fluid two-phase fluidization characteristics. Ph. D. Thesis, Harbin: Harbin Institute of Technology.
52. Wilhelm, R. H., & Kwauk, M. (1948). Fluidization of solid particles. *Chemical Engineering Progress, 44*, 201–218.
53. Xiaoxue, J. (2022). *Minimum energy dissipation rate drag coefficient model and numerical simulation on the active and passive regulation of cluster in riser*. Ph. D. Thesis, Harbin: Harbin Institute of Technology.
54. Xiaoxue, J., Shuyan, W., Baoli, S., Marcellus, U. U., & Huilin, L. (2022). Passive method-based clusters regulation of two-stage fluidized bed riser using dynamic cluster structure-dependent drag model. *Chemical Engineering Journal, 431*, 134111.
55. Zhenhua, H., Xiang, L., Huilin, L., Guodong, L., Yurong, H., Shuai, W., & Pengfei, X. (2010). Numerical simulation of particle motion in a gradient magnetically assisted fluidized bed. *Powder Technology, 203*, 555–564.

The manufacturer's authorised representative in the EU is Springer
Nature Customer Service Centre GmbH, Europaplatz 3, 69115 Heidelberg,
Germany. If you have any concerns regarding our products, please
contact ProductSafety@springernature.com

Printed and bound by CPI Group (UK) Ltd, Croydon, CR0 4YY
01/10/2026
02225939-0003